Editors:

R. W. Johnson, Canberra (Australia)
G. A. Mahood, Stanford (USA)
R. Scarpa, L'Aquila (Italy)

Jonathan H. Fink (Ed.)

Lava Flows and Domes

Emplacement Mechanisms and Hazard Implications

With 99 Figures

Springer-Verlag Berlin Heidelberg New York
London Paris Tokyo Hong Kong

Dr. JONATHAN H. FINK
Department of Geology
Arizona State University
Tempe, AZ 85287
USA

ISBN-13:978-3-642-74381-8   e-ISBN-13:978-3-642-74379-5
DOI: 10.1007/978-3-642-74379-5

Library of Congress Cataloging-in-Publication Data. Lava flows and domes: emplacement mechanisms and hazard implications / Jonathan Fink (ed.). P. cm. − (IAVCEI proceedings in volcanology; 2) Collection of papers based on a symposium held in 1987 at the International Union of Geology and Geodesy Congress in Vancouver,B.C.ISBN-13:978-3-642-74381-8 (U.S.)1.Domes(Geology)−Congresses.2.Lava−Congresses.I. Fink, Jonathan H.   II. Series. QE611.L38   1989   551.2′1−dc20   89-21833

© Springer-Verlag Berlin Heidelberg 1990
Softcover reprint of the hardcover 1st edition 1990

Typesetting: K+V Fotosatz GmbH, Beerfelden

2132/3145-543210 − Printed on acid-free paper

Preface

This collection of papers is based on a symposium held in 1987 at the International Union of Geology and Geodesy Congress in Vancouver, British Columbia. The Symposium was planned as a follow-up to a session at the 1984 Geological Society of America Annual Meeting in Reno, Nevada, which dealt with the emplacement of silicic lava domes. In both cases, emphasis was placed on the physical and mechanical rather than chemical aspects of lava flow. The IUGG Symposium consisted of two lecture sessions, a poster session, and two discussion periods, and had 22 participants. The contributions to this volume are all based on papers presented in the various parts of the Symposium.

The motivation for studying lava flow mechanics is both practical and scientific. Scientists and government agencies seek to more effectively predict the hazards associated with active lavas. Recovering mineral resources found in lava flows and domes also requires an understanding of their emplacement. From a more theoretical standpoint, petrologists view lava studies as a way to directly observe the rheologic consequences of mixing crystals, bubbles, and solid blocks of country rock with silicate liquids. This information can then be used to constrain processes occurring in the concealed conduits, dikes, and chambers that feed flows and domes on the surface.

In recent years, knowledge of lava flow mechanics has been greatly advanced by studies of long-lived eruptive activity at several relatively accessible volcanoes: Mount St. Helens, Kilauea, and Mauna Loa volcanoes in the United States, Mount Etna in Italy, and Arenal volcano in Costa Rica, among others. These eruptions have allowed large numbers of volcanologists with a wide range of backgrounds to observe and interpret phenomena that had previously been seen by only relatively small groups of scientists. In addition, long-term monitoring programs by the United States Geological Survey (USGS) at Mount St. Helens and Kilauea (along with relatively liberal helicopter budgets) have provided longitudinal data sets with greater spatial and temporal accuracy than those obtained previously. These data sets include high resolution digital topographic information which can be electronically manipulated to give fresh insights into magma budgets and the kinetics of dome and flow emplacement. Finally, better observations have prompted a growing interaction between field-based volcanologists and laboratory-oriented physicists and engineers, leading to steady advances in the theoretical modeling of flow processes.

The first five papers of this volume deal with the emplacement of silicic lava, with a strong emphasis on the well-documented growth of the dacite dome at Mount St. Helens from late 1980 to 1986. The remaining four papers focus on modeling flows of a more mafic character, and are concerned with phenomena ranging in scale from the microscopic texture of aa at Mount Etna to the way that individual flows combine to build large-scale volcanic edifices like Arenal or Sakurajima. Each paper in the collection presents some combination of field observations and theoretical modeling. The papers are arranged with those that are primarily observational followed by those with a more theoretical orientation.

In the first paper concerning the Mount St. Helens dacite dome, Swanson and Holcomb use precise digital topographic data to document regular increases in the height, diameter, and volume of the dome. They demonstrate that despite the episodic and seemingly random nature of many aspects of lava extrusion, long-term patterns are remarkably predictable. They conclude that changes in the overall shape of the dome are most likely controlled by the tensile strength of the brecciated carapace and the viscosity of the interior.

Anderson and Fink focus their study on determining what the surface texture of the Mount St. Helens dome can tell about its emplacement conditions. Using digital topography and hydrogen isotopic data, they show that the distribution of smooth and scoriaceous textures on a given lobe depends primarily on the magnitude of the underlying slope and, secondly, on the magmatic water content. For many of the lobes, water content increases steadily toward the vent, causing the lava to vesiculate when a critical concentration of 0.3 – 0.4 wt. % is attained. The chapter suggests that this increasing water content means that the potential for endogenic explosive hazards may be enhanced during the course of an extrusion.

Iverson uses the data presented by Swanson and Holcomb as the basis for a mechanical model that relates the static shape of a dome to the strength and thickness of its brittle crust and to the viscosity and over-pressure of magma in its ductile interior. This model differs from previous ones by incorporating rheologic contrasts between the crust and interior of the dome. Iverson determines conditions necessary for rupture of the carapace and exogenous growth, and he shows that during much of the history of the St. Helens dome, its emplacement has been consistent with predictions of the model. Finally, he calculates that the potential for excessive build-up of pressure and subsequent explosive failure are greatest for domes with a spherical shape.

Denlinger builds upon Iverson's model and focuses on the mechanism by which the carapace of the Mount St. Helens dome ruptures and new lava is extruded. Drawing on the concept of subcritical crack growth, he shows how the acceleration of deformation leading up to an extrusive episode can be explained either by a build-up of pressure in the interior or by slow growth of microfractures which weaken the brittle shell. Precise deformation data associated with emplacement of the May 1982 lobe are best explained by the crack growth mechanism. Denlinger goes on to demonstrate how this model can be used to predict the time necessary to critically overpressurize a dome.

Blake's study of dome growth incorporates observations from many different field areas with a series of laboratory simulations using kaolin slurries. He extends an earlier theoretical model for the spread of a Newtonian fluid (Huppert, 1982, *Jour. Fluid Mech.*, 121:43–58) to allow for the documented viscoplastic or Bingham rheologic behavior of magma. This model provides a closer correlation between the evolution of dome shapes and rheologic properties than the earlier Newtonian models. Blake goes on to relate the potential for explosive hazards from the collapse of Pelean-type domes to their shape, which he in turn relates to their yield strength.

In the first of four papers dealing with the emplacement of basalts and andesites, Kilburn discusses the factors controlling the formation of surface textures on basalt flows at Mount Etna. He combines field and SEM data to characterize several distinct types of pahoehoe and aa textures. He further identifies two longitudinal morphological trends associated with increasing lava crystallinity: from pahoehoe to aa along single flows, and from "threaded" to "arrowhead" type pahoehoe on near-vent surfaces down a flow field. He explains how recognition of such transitions can help field geologists estimate the extent of magma undercooling and the distribution of stresses within lava flow fields.

Fink and Zimbelman take advantage of an extensive set of field data collected by the USGS and calculate rheologic properties along the lengths of four basalt flows erupted at Kilauea Volcano in 1983. All four flows show exponential increases in calculated viscosity and yield strength with respect to either distance from vent or time since onset of eruption. Fink and Zimbelman further demonstrate that the longitudinal rate of increase for both parameters is proportional to the eruption temperature and to the major oxide composition of the lavas. They suggest that these longitudinal gradients in rheologic properties may prove useful tools in the remote identification of flow composition.

In the following paper, Ishihara, Iguchi, and Kamo combine models relating the composition, cooling rates, and rheology of lava flows to their thicknesses and develop a numerical code that predicts the paths taken and distances traveled by flows with a range of compositions. They apply their model to three historic Japanese eruptions for which digital topographic data and geologic maps are available and find a good correlation between predicted and observed flow outlines and thicknesses. Discrepancies are related to limitations imposed by the resolution of the original digital topographic data sets. This approach represents a powerful tool for predicting the hazards associated with flows of more mafic compositions.

In the final contribution, Borgia and Linneman use detailed measurements from Arenal Volcano to develop a model for the ways that individual lava flows may build up a volcanic edifice. As in Iverson's model for the St. Helens dome, they find that growth of a cooled crust is a critical factor determining the morphology and the distribution of stresses within individual as well as composite flows. Borgia and Linneman ultimately arrive at a set of equations for the evolution of lava fields and the growth of volcanic structure which successfully explains the distribution, extent, and slope of the flow fields that make up the entire Arenal edifice.

These nine papers should provide the reader with a host of approaches to the modeling of effusive volcanic processes. A few conclusions emerge from a comparison of these studies. First, monitoring programs that continue throughout the course of a long-lived eruption can lead to discoveries of general principles that would not be evident from less complete records. Several of the models presented here could not have been developed without the availability of relatively simple types of field data, systematically gathered. Future automation of these data recovery capabilities holds great potential. Repetitive collection of high resolution air photos that can be transformed into digital topographic data sets and nascent plans to launch an orbiting volcanological observatory are two promising moves in this direction. A second point is that testing the validity of models developed at a single field locality should ideally involve data from many different volcanoes. The failure of a model from one location to explain phenomena at another may lead to identification of additional important physical processes. The use of data from many sets of silicic domes by Blake, and application of Ishihara et al.'s model to flows on three different volcanoes illustrate the value of this approach. Finally, in volcanology as in other areas of geology, collaborations between observers trained to recognize subtle field relationships and theoreticians with backgrounds in applied mathematics, fluid mechanics, and material science are leading to the discovery of new, quantitatively rigorous relationships. Small field conferences that allow theoreticians and volcanologists to argue on the outcrop about such issues as boundary conditions and rheologic constraints should yield significant rewards in understanding.

JONATHAN H. FINK

Contents

On the Mechanisms of Lava Flow Emplacement and Volcano
Growth: Arenal, Costa Rica

List of Contributors

ANDERSON, STEVEN W., Department of Geology, Arizona State University, Tempe, AZ 85287, USA

BLAKE, STEPHEN, Dept. of Geology, University of Auckland, Private Bag, Auckland, New Zealand and Dept. of Earth Sciences, The Open University, Walton Hall, Milton Keynes, MK7 6AA, UK (Present Address)

BORGIA, ANDREA, Department of Mineral Sciences, N.M.N.H., Smithsonian Institution, Washington, DC 20560, USA and Jet Propulsion Laboratory, 4800 Oak Grove Ave., Pasadena, CA 91109, USA (Present Address)

DENLINGER, ROGER P., U.S. Geological Survey, School of Oceanography, University of Washington, Seattle, WA 98195, USA

FINK, JONATHAN H., Department of Geology, Arizona State University, Tempe, AZ 85287, USA

HOLCOMB, ROBIN T., U.S. Geological Survey, Cascades Volcano Observatory, 5400 MacArthur Blvd., Vancouver, WA 98661, USA and School of Oceanography, University of Washington, Seattle, WA 98195, USA (Present Address)

IGUCHI, MASATO, Sakurajima Volcanological Observatory, Disaster Prevention Research Institute, Kyoto University, Sakurajima, Kagoshima 891-14, Japan

ISHIHARA, KAZUHIRO, Sakurajima Volcanological Observatory, Disaster Prevention Research Institute, Kyoto University, Sakurajima, Kagoshima 891-14, Japan

IVERSON, RICHARD M., U.S. Geological Survey, Cascades Volcano Observatory, 5400 MacArthur Blvd., Vancouver, WA 98661, USA

KAMO, KOSUKE, Sakurajima Volcanological Observatory, Disaster Prevention Research Institute, Kyoto University, Sakurajima, Kagoshima 891-14, Japan

KILBURN, CHRISTOPHER, Dipartimento di Geofisica e Vulcanologia, Università di Napoli, Largo San Marcellino 10, 80138 Napoli, and Osservatorio Vesuviano, Centro Sorveglianza, Via Manzoni 249, 80123 Napoli, Italy

LINNEMAN, SCOTT R., Department of Geology and Geophysics, University of Wyoming, Laramie, WY 82071, USA

SWANSON, DONALD A., U.S. Geological Survey, Cascades Volcano Observatory, 5400 MacArthur Blvd., Vancouver, WA 98661, USA

ZIMBELMAN, JIM, Center for Earth and Planetary Studies, National Air and Space Museum, Smithsonian Institution, Washington, DC 20560, USA

Silicic Lava Domes

Regularities in Growth of the Mount St. Helens Dacite Dome, 1980–1986

D. A. SWANSON and R. T. HOLCOMB

Abstract

The dacite dome at Mount St. Helens grew episodically between October 18, 1980, and October 22, 1986, chiefly by extrusion of thick flows but also by endogenous growth resulting from intrusion into its molten core. Typical growth episodes lasted several days and produced volumes of $1.2-4.5 \times 10^6 \, \text{m}^3$, but growth was continuous from February 1983 to February 1984. By the end of October 1986, the volume of the dome and its talus apron was about $74.1 \times 10^6 \, \text{m}^3$, and the volume of all erupted material (including tephra and debris removed from the dome by explosions and rockfalls) was about $77.1 \times 10^6 \, \text{m}^3$.

Despite episodic activity, certain aspects of the 1980–1986 dome growth were quite regular. The long-term growth rate was approximately linear during three distinct periods: $1.8 \times 10^6 \, \text{m}^3/\text{mo}$ between October 18, 1980 and the end of 1981, $1.3 \times 10^6 \, \text{m}^3/\text{mo}$ between March 1982 and March 1984, and $0.62 \times 10^6 \, \text{m}^3/\text{mo}$ thereafter. The change from one period to the next coincides with distinct changes in style of eruption, magma composition, or associated seismicity. Long-term magma supply was approximately volume-predictable during each growth period and for certain episodes was also time-predictable. The height of the dome increased according to the equations $h = 43.44 (\ln t) - 83.79$ and $h = 23.22 t^{0.32}$, where h is height in meters and t is time in days since October 18, 1980. The average diameter (including talus apron) increased according to the power law $d = 176.16 t^{0.22}$, where d is diameter in meters. The ratio h/d ranged from 0.227 to 0.292 (mean of 0.266) except for the initial period of growth, when the dome was flatter ($h/d = 0.142$) possibly owing to a weak, relatively thin crust and the lack of a significant mantle of talus. The h/d ratios fall within the field defined for Japanese domes by I. Moriya and are less than the empirical upper limit of 0.32. The general equation $C = V/hd^2$, where V is known and h and d are calculated from the above equations, yields a value for the shape factor C of 0.2583 (s. d. = 0.0282) before the year of continuous growth and 0.3341 (s. d. = 0.0196) thereafter. The shape is probably controlled by the net effective viscosity and tensile strength of the hot core, cool outer shell, and flanking talus. Modeling by Iverson (this Vol.) and Denlinger (this Vol.) suggests that the outer shell is the most important of these factors.

1 Introduction

A dacite dome began to form in the crater of Mount St. Helens on October 18, 1980, 5 months after the catastrophic events of May 18. Two earlier domes had formed in late June and early August but were explosively destroyed in late July and mid-October respectively. Helicopter observations made near the vent 40 minutes after the last explosion on October 18 revealed a new dome just beginning to form. Estimates based on photographs taken then indicate that its diameter was about 25 m and its height about 10 m. The visual observations suggested that the dome was spreading laterally from the lip of the feeding conduit, so that the diameter of the conduit at the surface was probably somewhat less than 25 m. This estimate is much less than values of 100–110 m assumed by Scandone and Malone (1985) and 105–135 m calculated by Carey and Sigurdsson (1985) for the width of the conduit on May 18. If their figures are correct for May 18, then either only the tapered end of a wider magma column was observed on October 18 or the upper part of the conduit had narrowed, presumably through filling by earlier dome lava and fallback from explosions. Geodetic measurements of ground deformation before and during dome growth in 1981–86 define radial displacements away from essentially a point source at the base of the dome. In May 1985, Fremont and Malone (1987) used high-precision techniques for locating earthquakes to define a source volume beneath the dome only 30 m in diameter. The geodetic and seismic evidence thus implies that the upper part of the feeder conduit in 1981–86 was less than 50 m and probably no more than 20–30 m wide, only 2–3% of the width of the dome in October 1986.

The dome grew in a complex series of extrusions preceded, accompanied, and at times supplanted by periods of endogenous growth (Swanson et al. 1987). The extrusions produced short (200–400 m), thick (20–40 m) flows, which we term lobes, that piled atop one another and generally did not reach the crater floor before crumbling into talus. The lobes were erupted in an overlapping, seemingly haphazard, manner that eventually built the composite dome. Most of the lobes were fed from the summit region of the dome, but a few issued from eccentric vents high on the flanks. Seventeen episodes of dome growth occurred between October 18, 1980, and October 22, 1986, inclusive. Fourteen episodes produced one lobe each, and three produced two lobes each (December 1980, March–April 1982, and February 1983–February 1984), when the dome ruptured at two different locations.

Endogenous growth began slowly 1–3 weeks before each extrusion. The rate of endogenous growth, determined by geodetic measurements of displacement of the surface of the dome, accelerated almost exponentially to the time of extrusion. The slow, pre-extrusive rise of magma up the conduit and into the dome caused radial cracking and thrust faulting of the crater floor and expansion of the dome itself (Chadwick et al. 1983, 1988; Dzurisin et al. 1983); such deformation was useful in predicting the start of each extrusion (Swanson et al. 1983). Endogenous growth generally affected only a relatively small sector of the dome, typically half or less. Commonly the oldest exposed part of the dome was the site of greatest endogenous growth, possibly because cooling and alteration had decreased the tensile strength of the crust, but many exceptions occurred. Some periods of endogenous growth caused severe fracturing, faulting, and distension of the dome. In May 1985 (Swanson 1985) and May and October 1986, sector grabens tens of meters deep and hundreds of meters long resulted from endogenous growth, and outward-directed radial displacements of as much as 70 m were measured. Endogenous growth was essentially continuous for one full year (February 1983 to February 1984) and became increasingly important during later episodes of growth as the volume of the dome and consequently its holding capacity enlarged. Overall, endogenous growth probably accounts for 30–40% of the volume of the dome.

Talus occurs as extensive aprons mantling the flanks of the dome and in irregular patches high on the dome. The talus accumulations comprise one of the most conspicuous features of the dome. Most of the talus formed from hot rockfalls during extrusion and rapid endogenous growth; only a minor amount was generated by cold rockfalls during periods of quiet. Hot talus blocks developed radial prismatic jointing during cooling. Renewed movement (slumping, rockfalls) broke the fragile, jointed blocks into several joint-bounded pieces and further contributed to the talus accumulation.

The dome slowly subsided and spread outward between episodes of growth, apparently as its hot, relatively ductile core yielded under gravitational stress. Typical maximum rates of spreading and subsidence during quiet periods were 2–5 mm/day.

Several small explosions excavated pits and wedge-shaped sectors from the dome. The total volume of rock removed from the dome by these explosions was small, probably less than $2.5 \times 10^6 \, m^3$, but the surface morphology of the dome was significantly modified until later extrusion and endogenous growth filled or disrupted the depressions.

The first lobe was confined to a 300-m-wide shallow depression in the crater floor. Later lobes filled the depression and spread onto the surrounding floor, burying older lobes in the process. The crater floor was nearly flat except north of the depression, where the first 120 m sloped northward about 12°, the next 140 m (an area often called "the rampart") about 19°, and the next 200 m about 11° (slopes determined from map 1, Table 3). These slopes may have influenced the shape of the dome to some degree, but we do not discuss them further.

By October 31, 1986, the dome stood about 267 m above its vent and about 350 m above its northern base, which rests on the northward-sloping crater floor 550–600 m from the vent. At that time the slightly elliptical dome had an east-west diameter of 860 m, a north-south diameter of 1060 m, and a volume of about $74.1 \times 10^6 \, m^3$, including the volume of talus mantling its lower flanks.

The dacitic chemical composition of successive lobes showed little net change during growth of the 1980–86 dome (Table 1); on average, however, lobes extruded after 1981 may be slightly more silicic than earlier ones (Fig. 1). The combined content of plagioclase, orthopyroxene, hornblende, Fe-Ti oxides, and minor clinopyroxene is 40–45 vol% and may have increased very slightly with time (Cashman and Taggart 1983; KV Cashman, personal communication, 1988). The relatively small changes in SiO_2 and crystallinity imply that the effective viscosity (liquid plus crystals) and yield strength of magma entering the dome probably remained almost constant during growth of the dome, although possible gas loss with time could have increased the effective viscosity slightly. We calculated an effective viscosity of 10^{10-11} poise for several lobes on the basis of flow rates down the flanks of the dome (Chadwick et al. 1988).

Table 1. Average chemical composition of Mount St. Helens dome (69 samples, October 1980 – October 1986)[a]

Oxide	Percent
SiO_2	63.0 (0.4)[b]
Al_2O_3	17.9 (0.2)
FeO^t	4.68 (0.15)
MgO	2.25 (0.13)
CaO	5.42 (0.13)
Na_2O	4.51 (0.08)
K_2O	1.29 (0.04)
TiO_2	0.69 (0.02)
P_2O_5	0.15 (0.015)
MnO	0.08 (0.005)

[a] XRF analyses by J.E. Taggart (U.S. Geological Survey).
[b] Standard deviation of the mean of the 69 analyses.

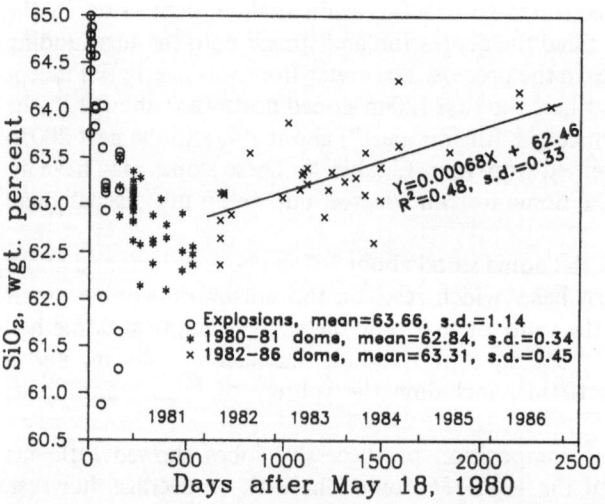

Fig. 1. SiO_2 contents for 69 samples of dacite from the 1980–86 eruption of Mount St. Helens. All analyses done in U.S. Geological Survey Denver laboratory using XRF procedures (courtesy of JE Taggart). Samples divided into three groups: products of 1980 magmatic explosions (n = 27), 1980–81 dome-building episodes (n = 37), and 1982–86 dome-building episodes (n = 32). Best-fit line calculated for 1982–86 samples only. Scatter toward low SiO_2 values may reflect random physical contamination from gabbroic inclusions and related megacrysts, which have comprised 3–4 vol% of the dome throughout its growth (Heliker 1984, 1985)

The growth of the dome was complex (Swanson et al. 1987), and it would be difficult to reconstruct the detailed history of the dome from its present characteristics. Taking account of this complexity will ultimately lead to much greater understanding of the processes involved in its formation.

In this chapter, however, we disregard short-term complexities and search for regularities in the growth process. To do this, we must in general consider longer time periods than those involved in one or two growth events, and we must be content with averaging out observations that may be important to the detailed history of growth. Ultimately we seek commonalities that bear not only on the growth of the Mount St. Helens dome but perhaps on the growth of other dacite domes.

We recognized several regularities in the growth pattern that are easily overlooked if only the details are considered. We outline in this chapter some of these regularities in order to enable other workers to formulate realistic mechanistic and rheologic models of the observed behavior. However, we do not want to leave readers with an oversimplified view of how the dome grew, and so we point out that the narrative in Swanson et al. (1987) stresses some of the realistic complexities of dome growth that may dampen exaggerated enthusiasm engendered by consideration of the regularities alone.

2 Methods

Growth of the dome was documented by frequent geodetic measurements and by a sequence of large-scale topographic maps (mostly 1:2000 with 2-m contour interval; a few 1:4000 with 5-m contour interval) photogrammetrically prepared from vertical aerial photographs taken at key times during the eruption (Tables 2 and 3). We routinely estimated the elevation of the top of the dome by theodolite measurements from one or two points whose elevation and approximate distance from the dome are known. Such elevations are probably good to no better than several meters, owing to the relatively poorly known distance (probably within 50 m, sometimes within 5 m) and the perspective. Less often, the precise elevation was determined photogrammetrically during preparation of each topographic map; such elevations are good to well within 1 m. We calculated dome heights relative to an assumed datum of 1914 m above sea level for the base of the October 1980 lobe, the oldest part of the dome.

Diameters of the dome were measured from the topographic maps. Such measurements require judgments as to how the edge of the dome is defined, because talus fans commonly extend far beyond the unfragmented bulk of the dome. Such judgments lead to errors discussed in a following section.

Volumes were calculated by differencing digitized maps using computer routines at Arizona State University, courtesy of Jon Fink and Mike Malin. These volumes substantially improve earlier estimates in Swanson et al. (1987). The volumes have not been adjusted for fractures and porosity, owing to

Table 2. Volume and linear dimensions of dome

Start of eruptive episode	Duration (days)	Erupted volume[a] (10^6 m³)	Dome volume[b] (10^6 m³)	Dome dimensions[c] (l/w/h in m)
10/18/80	1	1.2[d]	1.2[d]	290/260/39
12/27/80	7	1.6	2.8	440/320/93
02/05/81	3	3.6	6.4	520/464/120
04/10/81	3	4.1	10.5	666/470/129
06/18/81	2	4.1	14.6	674/520/161
09/06/81	5	3.9	18.5	680/590/161
10/30/81	4	3.6	22.1	680/600/184
03/19/82	24	3.4[e]	24.5	760/600/195.1
05/14/82	5	2.7	27.2	800/636/198
08/18/82	6	4.6	31.8	820/692/204.1
02/07/83	368	22.4[f]	53.2	860/780/227.3
03/29/84	5	2.0[g]	54.2	860/790/227.1
09/10/84	3	3.7	57.9	960/800/225.6
05/24/85	17	4.3	62.2	1000/800/222.9
05/08/86	5	5.8	68.0	1040/800/255.9
10/21/86	4	6.1	74.1	1060/860/267.2
Total		77.1	74.1	

[a] Total volume erupted, including tephra and rockfall debris.
[b] Net volume of dome shown at end of each extrusive episode, determined by summing differences between successive digitized maps (see text) except where noted.
[c] Dimensions at end of each extrusive episode, including talus. l, length; w, width; h, height above base of October 1980 lobe at assumed elevation of 1914 m. All measurements from topographic maps.
[d] Volume of October 1980 lobe estimated from dimensions
[e] Includes 0.8×10^6 m³ (dense rock equivalent) for pumice erupted on March 19 and deposited beyond dome; also includes 0.2×10^6 m³ for rock-avalanche of April 4.
[f] Includes 1×10^6 m³ for notch in dome blown out and then refilled.
[g] Includes estimated volume of 1.1×10^6 m³ for March 1984 lobe and 0.9×10^6 m³ for June 1984 lobe; difference between this figure and the net 10^6 m³ increase in dome volume results from 10^6 m³ (estimated) crater formed by explosions in May.

unknowns in both quantities. Each map required many hours of digitization, a tedious, potentially error-prone procedure. The estimated maximum error in the calculated volume difference between two successive maps is 10%, about 2% of which may be related to the computational procedure (M. Malin and J. Fink, personal communication, 1988). These errors are presumably random and hence do not accumulate from one map interval to the next.

3 Volumetric Rate of Growth

The calculated volume of the dome at the end of 1986 is about 74.1×10^6 m³ (Table 2), but we estimate that 3×10^6 m³ was removed from the dome and deposited beyond the area of the digitized map by explosions, rockfalls and as-

Table 3. Topographic maps used in this chapter

Number and date of map	Eruptive episodes depicted by map
1. October 23, 1980	October 1980
2. January 15, 1981	December 1980 – January 1981
3. March 9, 1981	February 1981
4. May 8, 1981	April 1981
5. July 31, 1981	June 1981
6. October 15, 1981	September 1981
7. November 4, 1981	October – November 1981
8. May 13, 1982	March – April 1982
9. June 23, 1982	May 1982
10. September 23, 1982	August 1982
14[a]. January 31, 1984	Continuous activity from February 7, 1983 to date of map
15. March 14, 1984	Continuous activity from January 31, 1984 to about February 10, 1984
16. August 29, 1984	March 1984, explosions of May 1984, and June 1984
17. September 13, 1984	September 1984
18. July 11, 1985	May – June 1985
19. July 18, 1986	May 1986
20. November 12, 1986	October 1986

[a] Maps 11 – 13 not digitized.

sociated surges and pyroclastic flows (Mellors et al. 1988), and tephra fallout. We use the total volume of erupted material, estimated as about $77.1 \times 10^6 \, m^3$, in subsequent discussions except where noted.

The dome grew in three stages distinguished by different rates of magma supply (Fig. 2A). The first stage (A), from October 18, 1980, to the end of 1981, was characterized by a constant long-term rate of supply, averaged over periods of several weeks, fit by the equation $V = 0.058t - 0.13$, where V is volume in millions of cubic meters and t is time in days since October 18, 1980. The coefficient of determination (R^2) for this equation is 0.988, the standard deviation (s. d.) is $0.89 \times 10^6 \, m^3$, and the monthly supply is about $1.8 \times 10^6 \, m^3$.

The second stage of growth (B), from the end of 1981 to mid-February 1984, followed the linear equation $V = 0.042t + 3.75$, with $R^2 = 0.997$, s. d. = 0.83, and monthly supply of about $1.3 \times 10^6 \, m^3$. Half of this period, from February 1983 to February 1984, was a time of continuous dome growth.

The third stage of growth (C), from mid-February 1984 to October 1986, followed the equation $V = 0.020t + 31.98$, with $R^2 = 0.982$, s. d. = 1.07, and monthly supply of about $0.62 \times 10^6 \, m^3$. This period was one in which the recurrence interval between episodes of growth generally increased.

In the three stages of growth, batches of magma were supplied to the dome in a predictable manner. The volume added to the dome (including tephra and rockfalls) during a particular growth episode is a linear function of time elapsed since the previous episode. Such volume predictability contrasts with time predictability, in which the time of the next eruption can be predicted

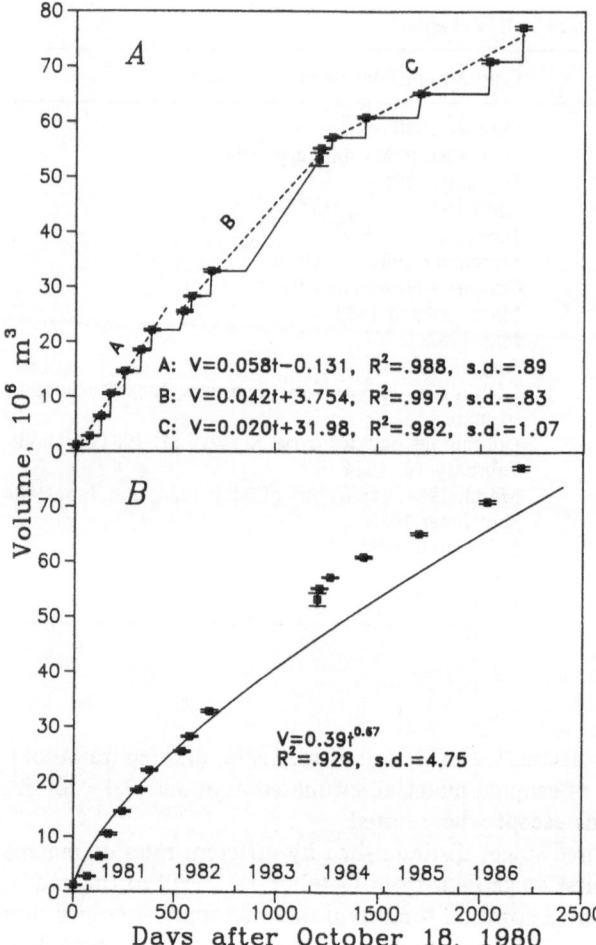

Fig. 2. Cumulative volume of dacite erupted at Mount St. Helens vs time, October 18, 1980 to October 25, 1986. Standard error *bars* shown. See text for discussion of methods and errors. *A.* Note volume-predictable nature of growth during three distinct periods, labeled *A, B,* and *C. Dashed lines* denote linear best fits to calculated volumes. *B.* Best-fit power curve and equation; fit is poor by comparison to linear fits in *A.* *V,* volume in $10^6\,\text{m}^3$; *t,* time in days after October 18, 1980

A: $V=0.058t-0.131$, $R^2=.988$, s.d.$=.89$
B: $V=0.042t+3.754$, $R^2=.997$, s.d.$=.83$
C: $V=0.020t+31.98$, $R^2=.982$, s.d.$=1.07$

$V=0.39t^{0.67}$
$R^2=.928$, s.d.$=4.75$

knowing the volume of the previous eruption (Bacon 1982; Kuntz et al. 1986). In 1981 the activity was nearly periodic, so that the activity was both volume- and time-predictable. During the year of essentially continuous growth (February 1983 to February 1984), the dome apparently enlarged at the same average rate as it did during the preceding year of episodic activity.

Casual inspection of the volume-time plot suggests that the supply rate for the entire history of the dome was regularly declining in some simple manner. We attempted to describe the volume-time data with one simple equation. The closest approximation is a power-law function (Fig. 2B), but the s.d. of the fit, $4.75\times10^6\,\text{m}^3$, is large. We made no attempt to fit the data to a higher-order polynomial.

4 Change in Height

The height of the dome increased rather systematically, except for relatively brief periods of faster or slower growth (Fig. 3). The height-time relation for 104 theodolite measurements is rather well expressed by two best-fit functions:

Logarithmic fit, $h = 48.50\,(\ln t) - 117.33$, $R^2 = 0.95$, s.d. $= 11.24$

Power fit, $h = 20.48\,t^{0.34}$, $R^2 = 0.88$, s.d. $= 16.31$

where h is height in meters, t is time in days after October 18, 1980, and s.d. is standard deviation of the residuals (Fig. 3A). We made no attempt to improve these or other fits with higher order polynomials. The heights of significant spines are shown in Fig. 3A but excluded from the best-fit calculations.

Figure 3B shows heights taken from 20 topographic maps between October 23, 1980, and November 12, 1986. The following best-fit curves were calculated:

Logarithmic fit, $h = 39.07\,(\ln t) - 56.43$, $R^2 = 0.94$, s.d. $= 14.50$

Power fit, $h = 25.51\,t^{0.31}$, $R^2 = 0.98$, s.d. $= 13.03$

The map heights are more reliable than the theodolite data, because they avoid the problems of perspective and poorly known sighting distances. However, they comprise a much more limited data set, 20 vs 104 observations. We arbitrarily weighted the map heights three times stronger than the theodolite heights and combined the two into a single data set (Fig. 3C), whose best-fit equations are:

Logarithmic fit, $h = 43.44\,(\ln t) - 83.79$, $R^2 = 0.93$, s.d. $= 13.50$

Power fit, $h = 23.22\,t^{0.32}$, $R^2 = 0.92$, s.d. $= 14.66$

We prefer these equations, because they use all of the data but acknowledge the superiority of the map heights. However, there is little basis to choose one among all of the equations. The fits of the observations to the curves are all rather good, and the numerical coefficients and exponents are similar for each type of function. We conclude that the height increased according to either a logarithmic or a power law similar to those presented here.

The height increased consistently with the volume (Fig. 4). The best overall fit is provided by an exponential expression; it has a higher standard deviation than the power fit, but only because of a relatively poor match at the high end of the curve. The height increased faster than can be accounted for by simple balloon-like expansion, because it varies more closely with the square root than with the cube root of the volume (Fig. 4). Evidently some process in addition to constant-shape volumetric expansion influenced the rate of upward growth.

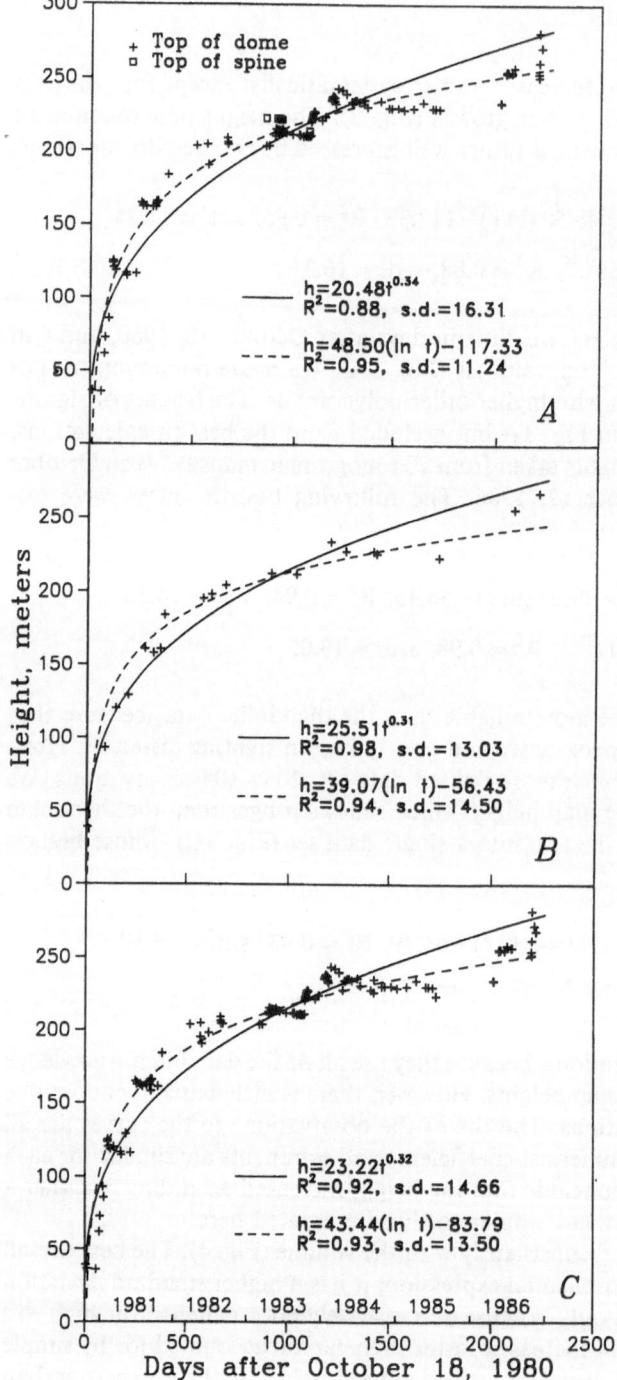

Fig. 3A–C. Change of
height with time for
Mount St. Helens
dome. Smooth curves
are best fits defined by
indicated equations,
excluding spines.
A Theodolite data;
B map data; *C* com-
bined theodolite and
map data. *h*, Height in
meters. Estimated er-
rors in height measure-
ments are ±4 m in *A*,
±1 m in *B*, and ±2 m
in combined data set of
C

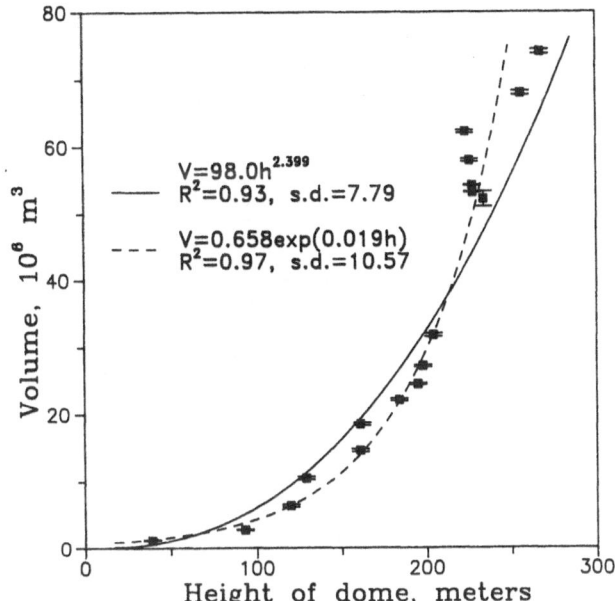

Fig. 4. Variation of dome volume with height for Mount St. Helens dome. Curves show best-fit power and exponential functions.
h, map height in meters. Standard error bars shown for volume; estimated error for map heights is ±1 m

$V = 98.0h^{2.399}$
$R^2 = 0.93$, s.d. = 7.79

$V = 0.658 \exp(0.019h)$
$R^2 = 0.97$, s.d. = 10.57

Volume, 10^6 m^3

Height of dome, meters

5 Change in Diameter

Two values for the north-south and east-west diameters were measured from each available map; one, D_d, neglects the talus apron resting on the crater floor, and the other, D_t, includes the apron. D_d and D_t are similar in the east-west direction but are different in the north-south direction, because the contact between talus and nonfragmented rock is hard to define on the north flank of the dome, where talus is more extensive than elsewhere. D_t is the more reliably estimated value and is used in subsequent discussion unless otherwise noted. Rheologic modeling of the dome may, however, need to distinguish the talus from the bulk of the edifice, because of their different physical properties (Huppert et al. 1982). Estimating even D_t is subject to large uncertainty, owing to the difficulty in defining the outer limit of the talus apron. Trial-and-error measurements suggest that a deviation of ±7% from the nominal north-south diameter is expectable. The error for the east-west diameter, however, is smaller, probably about ±4%, because of lesser development of talus.

Increase in the north-south diameter (Fig. 5A) is expressed by:

Logarithmic fit, $d = 133.55\,(\ln t) - 70.18$, $R^2 = 0.91$, s.d. = 59.94

Power fit, $d = 177.58\,t^{0.23}$, ($R^2 = 0.97$, s.d. = 40.84)

where d ($= D_t$) is diameter in meters.

Increase in the east-west diameter (Fig. 5B) is expressed by:

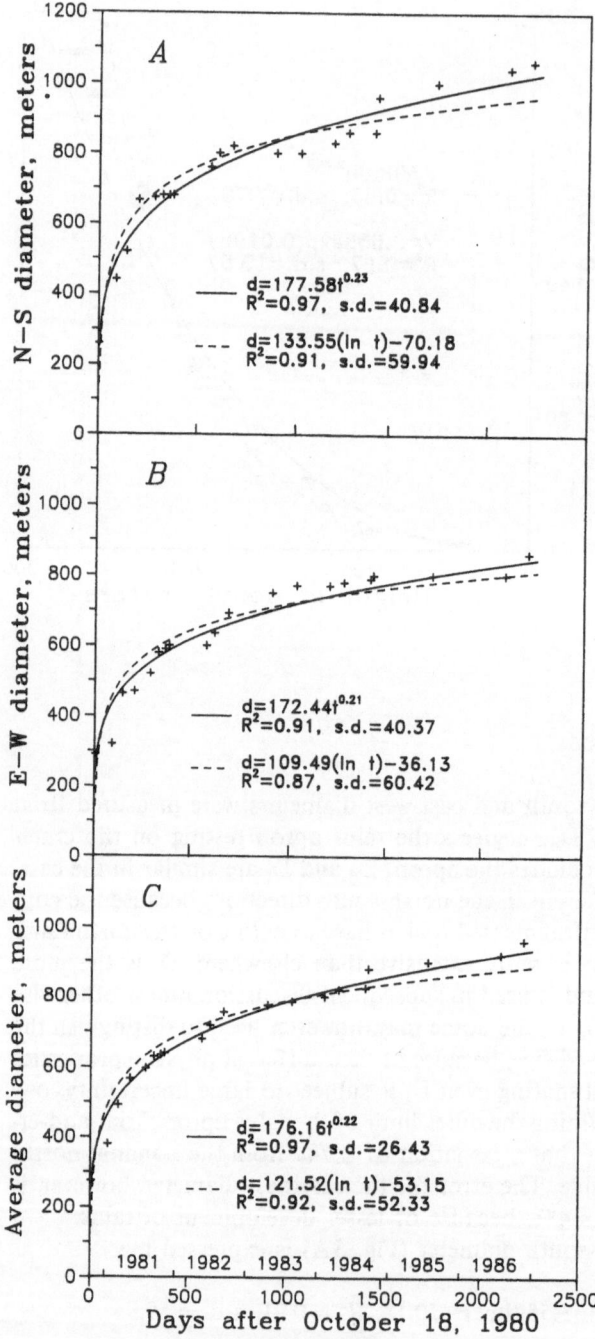

Fig. 5A–C. Variation of diameter (D_t in text) with time for Mount St. Helens dome. Smooth curves are best fits defined by indicated equations. Estimated error for diameter is ±7%. *A* North-south diameter; *B* east-west diameter; *C* average diameter. *d*, diameter in meters

Fig. 6. Variation of dome volume with average diameter of Mount St. Helens dome. *d*, diameter in meters. Standard error *bars* are shown for volumes. Estimated error for diameter is ±7%

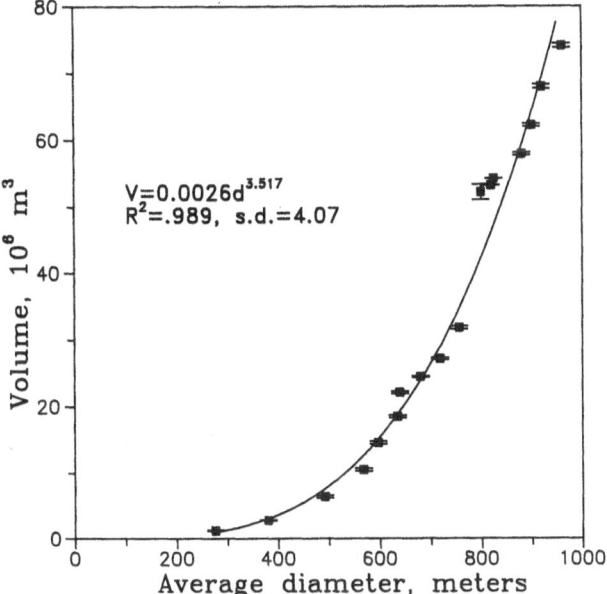

$V = 0.0026 d^{3.517}$
$R^2 = .989$, s.d. = 4.07

Logarithmic fit, $d = 109.49 (\ln t) - 36.13$, $R^2 = 0.87$, s.d. = 60.42

Power fit, $d = 172.44 t^{0.21}$, $R^2 = 0.91$, s.d. = 40.37

The best-fit power curves for both diameters closely resemble one another, and their values for R^2 and standard deviation are better than those for the logarithmic functions. Both diameters lengthened according to nearly the same function despite the pronounced northward slope beneath the northern one-third of the dome.

An average diameter based on the north-south and east-west values was calculated for each interval to facilitate additional comparisons. The variation of the average diameter with time (Fig. 5C) is expressed by:

Logarithmic fit, $d = 121.52 (\ln t) - 53.15$, $R^2 = 0.92$, s.d. = 52.33

Power fit, $d = 176.16 t^{0.22}$, $R^2 = 0.97$, s.d. = 26.43

Again the power fit is superior, as judged from both the statistics and visual comparison of graphs in Fig. 5C.

The average diameter varies smoothly with volume of the dome (Fig. 6) according to the roughly cubic equation $V = 0.003 d^{3.494}$, $R^2 = 0.991$, s.d. = 3.32.

6 Average Shape

To test whether the average shape of the dome changed with time, we set up the general equation, $V = Chd^2$, where C is a shape factor that should remain constant if the shape does not change. The calculated volume and observed values for height and average diameter after each period of dome growth were used to calculate values of C.

Except for the first extrusion in October 1980 (C = 0.4069), the values of C are grouped into two sets, one for shapes before February 1983 and one thereafter. The first set has a mean value of C of 0.2583 (s. d. = 0.0282), and the second set, 0.3341 (s. d. = 0.0196). The increase in C values coincides with the start of the year of continuous growth, which was primarily endogenous. Apparently the endogenous growth, which probably accounted for some 75% of the growth of the dome during the year-long activity, changed the shape of the dome. Thereafter, the dome never resumed its prior shape, although C values for the two 1986 episodes (0.3140 and 0.3009) suggest that it is tending in that direction. The increase in C reflects relative increases in V or decreases in h and d; d is the most likely parameter to have changed, for it influences C most. The average diameters just before and at the end of the year of continuous activity are indeed less than would have been predicted by the general relation of volume to diameter (Fig. 6, the two data points at about 800 m diameter and $52 \times 10^6 \, m^3$).

Another way to evaluate whether the average shape changed with time is to examine the h/d ratio; a constant ratio is a necessary but insufficient condition for constant shape. From December 1980 to the end of 1986, the h/d ratios ranged from 0.227 to 0.292, with a mean of 0.266, median of 0.270, and s. d. of 0.018. The variation about the mean, 6.8%, is small, so that to a first approximation the h/d ratio is consistent with constant shape since June 1981. However, the h/d ratios after the December 1980 and February and April 1981 dome-building episodes are all somewhat lower, 0.227 – 0.245, than those after later episodes (0.248 – 0.292). If these differences are significant, the dome was slightly flatter during its early stages than during later stages. This analysis found no change that correlates with the increase in C values described in the preceding paragraph.

The h/d ratio after the October 1980 extrusion is only 0.142, much lower than that after later extrusions. Such a low ratio indicates a relatively flat shape that may be consistent qualitatively with a thin crust mantling the hot interior of the young, monogenetic dome. A similar explanation could account for the marginally lower h/d ratios after the next three episodes, during which the ratio increased as the skin became thicker, talus cones started to mantle and buttress the flanks of the dome, and the internal structure of the composite dome became more complex. By June 1981 a sufficiently thick skin and talus apron had developed to allow the dome to steepen and reach its final relatively constant h/d ratio. The possible importance of the skin, with or without the buttressing effect of talus, in governing the shape of the dome is also suggested by Iverson (this Vol.) and Denlinger (this Vol.).

7 Discussion

7.1 Volume Predictability

The history of the dome can be divided into three periods of different rates of growth. Within each period, the volume of lava produced during a single growth episode is directly proportional to the time since the previous episode. This pattern is one of the most significant regularities of the Mount St. Helens dome and means that the volume produced during a growth episode can be predicted knowing the length of time since the previous episode. One interpretation of volume predictability is that the volume of erupted lava depends on the degree to which the reservoir and conduit system were pressurized in excess of lithostatic load. If overpressurization occurs at a constant rate, then the volume of material erupted might vary directly with the duration of pressurization. Overpressurization can occur 1) by reduction in confining pressure, as in areas of extensional tectonism, 2) by increase in magmatic pressure, as caused by vesiculation or influx of new magma into the conduit system, or 3) by a combination of the two. The frequency of growth episodes at Mount St. Helens suggests that stress buildup is rapid and hence is more likely related to processes such as vesiculation within the magma body itself than to tectonism.

In two recent papers, Kuntz et al. (1986) argued that basaltic volcanism along the Great Rift, Idaho, was volume predictable and controlled by buildup in magmatic pressure, whereas Bacon (1982) found that bimodal volcanism at Coso, California, was time-predictable and controlled by the rate of regional tectonic extension. Both of these studies dealt with the development of a volcanic field during thousands of years or more, in contrast to the growth of a single dome during a few years, as described here. Probably it is meaningless to compare closely the results of the two types of study, because of the greatly different time scales. In general, however, rates of single relatively short-lived eruptions might more likely be controlled by processes within the parent magma body than by tectonism, whereas the long-term development of an entire field might or might not be controlled by tectonism, depending on regional stress rates and style.

7.2 Changes in Growth Rate Related to Other Events

The growth rate of the Mount St. Helens dome was itself complex and decreased abruptly from one period of constant growth to the next. The rate dropped by about 30% after 1981 (between periods A and B) and by about 50% between February and March 1984 (between periods B and C). These times correlate with other significant events in the history of the dome.

The significant decline from A to B may also correspond with the start of a weakly defined trend of chemical variation, from increasingly more mafic to increasingly more silicic (Fig. 1). Such a correspondence might imply that slightly more silicic magma became available for eruption but at a reduced rate. Another possibility is that fractionation began to produce increasingly more silicic magma at a relatively slow rate. Clearly the chemical trend is weak, however, and its presence or absence does not bear on the definition of the two periods.

The first episode of dome growth in period B began on March 19, 1982, following the longest period of inactivity to that time. This episode was preceded by small, relatively deep (up to 12 km) earthquakes that were deeper than those associated with any of the other dome-building episodes before or since (Weaver et al. 1983). The episode began with a significant magmatic explosion that produced about $0.8 \times 10^6 \, m^3$ of pumice (Swanson et al. 1987; Waitt et al. 1983; Waitt and MacLeod 1987). This event might record the influx of fresh, gas-rich magma into high-level storage beneath the volcano (Weaver et al. 1983).

The change from period B to C coincides with the end of the year of continuous activity between February 1983 and February 1984 and as such reflects a notable change in behavior of the dome. About 75 days of inactivity followed before renewed growth began on March 29, 1984. All subsequent eruptive episodes had a component of endogenous growth greater than those of earlier episodes, except for the year of continuous growth. In addition, most lobes formed during period C had only a thin (March, June, and September 1984) or nearly absent (May 1985, May and October 1986) scoriaceous carapace, a pattern suggesting that the volatile content of the parent magma had slowly decreased with time. SO_2 discharge, as measured by a correlation spectrometer (COSPEC), declined with time during growth of the dome (Casadevall et al. 1983; unpub. data, Cascades Volcano Observatory); this pattern also suggests that the parent magma body was degassing. In addition, rising magma may have degassed more thoroughly during period C than previously. This possibility is suggested by the particularly intense, long-lasting shallow seismicity that preceded most dome-growth events of period C, as if the magma were having a "difficult" time rising the final several hundred meters to the surface and hence had a greater opportunity than did previous batches of magma to degas during prolonged residence at shallow depth. The formation of scoria on the dome is a complex process, however, and its absence may not necessarily reflect simple volatile loss (Anderson and Fink 1986). Despite the uncertainty as to process, the lack of significant scoria clearly distinguishes period C from earlier periods.

The recurrence interval between onsets of dome-growth episodes also changed its pattern each time that the volumetric supply rate declined. The best-fit curve relating episode number to time follows a power law, $t = 4.97 \, E^{2.22}$ ($R^2 = 0.91$, s.d. $= 164.2$), where E is episode number (Fig. 7). Clearly the recurrence interval shows a general increase with time, but the standard deviation of this fit is large. Moreover, scrutiny of Fig. 7 shows major departures from the best-fit curve after episodes 8 and 12. We therefore divided the data into three sets for further analysis.

The intervals between each of the 7 episodes before the end of 1981 (period A) were rather similar (mean = 63 days, median = 64, s.d. = 14.03) and conse-

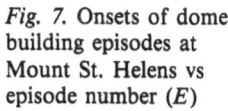

Fig. 7. Onsets of dome-building episodes at Mount St. Helens vs episode number (*E*)

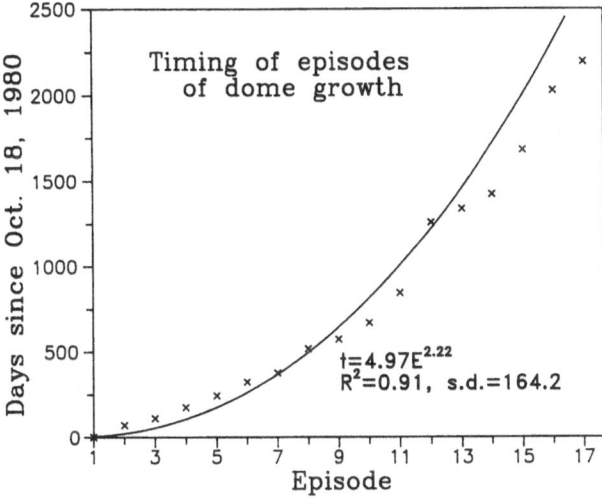

Fig. 8. Onsets of dome-building episodes at Mount St. Helens vs episode number (*E*), divided into three sets of time intervals corresponding to periods *A, B,* and *C* of Fig. 2. *D,* days since October 18, 1980; *E,* episode number

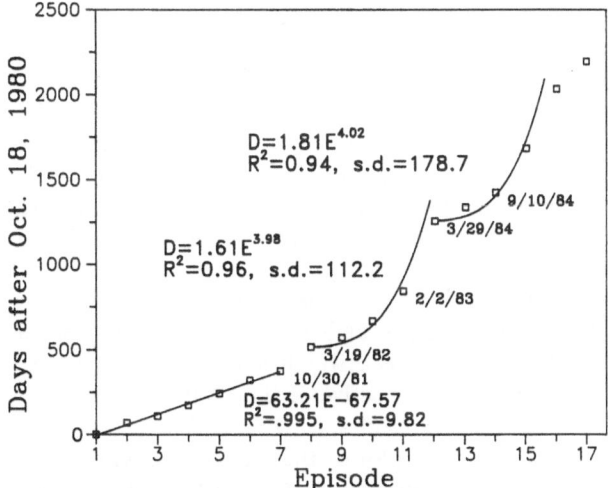

quently can be fit by a straight line in Fig. 8. A period of 140 days separated episodes 7 and 8 (end of period *A* and start of period *B*). Thereafter the recurrence intervals comprise two sets separated by the year of continuous growth in 1983–84 and corresponding to the growth periods *B* and *C*; each set is fit by similar exponential functions (Fig. 8), with D equal to about $1.7E^4$, (R^2 of about 0.95 and s.d. of about 145), where D is number of days between episodes. The last two episodes of 1986 fall slightly off this trend (Fig. 8), just as they depart somewhat from the volumetric growth curve for period *C* (Fig. 2A).

Thus several significant events in the history of the dome correlate in time. The first decline in growth rate corresponds with the possible start of in-

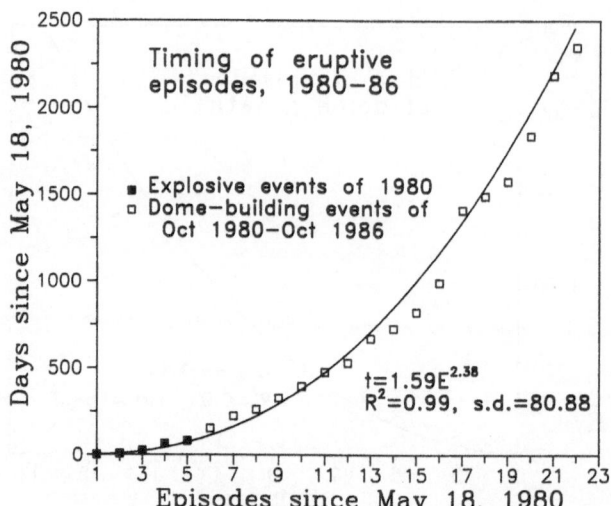

Fig. 9. Recurrence intervals of onsets of all eruptive episodes at Mount St. Helens from May 18, 1980 to October 1986. *E*, episode number

creasing content of SiO_2 and the onset of an unusual period of deep earthquakes and magmatic explosivity after the longest period of inactivity to that time. The second decline in growth rate corresponds with the end of a year of continuous growth, increasingly significant endogenous growth during short-lived eruptive episodes, lack of development of a significant scoriaceous carapace, and more sluggish rise of magma (possibly but not certainly related to declining volatile content). Both declines in growth rate correlated with changes in recurrence interval. We believe it unlikely that these correlations are fortuitous yet see no obvious cause for them. One can imagine several scenarios involving magma recharge to the reservoir system, exponentially-driven fractionation and-or degassing processes, and other possibilities, but testing such ideas is difficult and beyond the scope of this paper.

A broader perspective of the 1980−86 eruption results if the timing of all of the eruptive episodes, starting with that of May 18, 1980, is evaluated as a single data set. A plot of these data (Fig. 9) shows quite regular variation according to the power law, $t = 1.59E^{2.38}$ ($R^2 = 0.99$, s. d. = 80.88). The good statistical fit and relatively small standard deviation suggest that the recurrence interval was *systematically* increasing from the start of the eruption on May 18, 1980, to the end of 1986. In this view, the two excursions from the overall pattern at the end of periods *B* and *C* are second-order effects superposed on the generally declining frequency of activity. A process that accounts for this general trend as well as the excursions should be a focus of future petrologically oriented study.

7.3 Height-Diameter Ratio and Hazards

The *h/d* ratio for the Mount St. Helens dome falls within Moriya's (1978) field for 110 young lava domes in Japan (Fig. 10). Moriya's data suggest that a limit

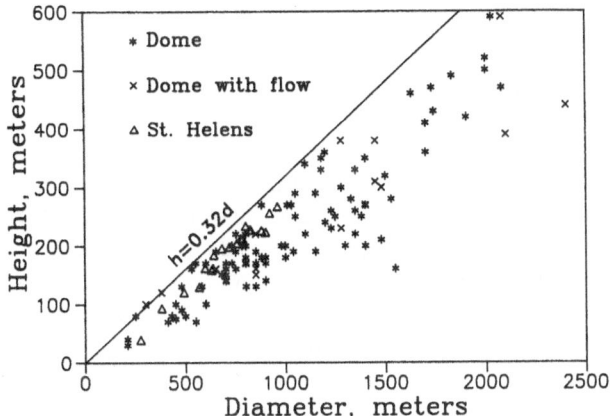

Fig. 10. Relation of heights to diameters for 110 domes in Japan (data from Moriya 1978, Table 1 and Fig. 4) and for 20 stages in growth of Mount St. Helens dome. Seven cryptodomes and 25 flat-topped domes omitted from Moriya's data all have *h/d* ratios less than 0.32. Slope of line defining maximum *h/d* ratio is equivalent to angle of repose of fragmental debris

of about $h = 0.32 d$ exists for domes, and the Mount St. Helens data are consistent with this limit (Fig. 10). Another way to express the limit is $h = 0.32 (2r)$, or as the tangent function, $h/r = 0.64$, where r is the radius of the dome. This ratio is equivalent to a slope of about 33°, near the angle of repose of loose rubble. This relation suggests the unlikely possibility that domes are simply piles of rubble that cannot maintain slopes steeper than the angle of repose. Field evidence, however, shows that domes are *not* piles of rubble; they are capable of maintaining steep and locally overhanging flanks, and old dissected domes have massive, unbroken cores.

Young domes, on the other hand, are typically surrounded by talus aprons. Map measurements of the diameters of such domes must necessarily include at least part of these aprons. Apparently the inclusion of talus in the diameter measurements accounts for the *h/d* limit suggested by Moriya's data. Thus this ratio has little significance rheologically, other than indicating that the angle of repose of talus aprons probably controls the *h/d* ratio.

The *h/d* limit is useful, however, in evaluating how high an unconfined dome can grow and hence its potential for sudden collapse into a rock avalanche or block-and-ash flow. An active dome whose height is far less than $0.32 d$ has potential to grow upward without widening, thereby becoming steeper and increasing its chance of collapse. On the other hand, an *h/d* ratio consistently much lower than 0.32 through several periods of growth suggests that either magma in the conduit has a relatively low head or the strength of the dome is too weak to support a greater height. An active dome whose *h/d* ratio is greater than 0.32 should be considered unstable (or at best metastable) and subject to hazardous collapse at any time.

7.4 Causes of Relatively Constant Shape

The relatively constant shape of the dome since June 1981 suggests some control beyond that supplied simply by the stacking of overlapping lobes. Iverson (this volume) modeled the general shape of the dome as controlled chiefly by its cool skin, a complex unit probably of variable thickness from place to place in accord with the complicated manner in which the dome grew. Another important factor is the presence of a core of relatively low shear strength, which developed and may have enlarged with time as magma entered but did not leave the dome. Still another potentially important element is the pile of flanking talus, which exerts a resistance (buttress) to lateral spreading of the dome if the pile does not oversteepen and fail gravitationally. A major challenge is to understand the mechanical properties of the skin, core, and talus, the response of these elements to gravitational stresses (as reflected in the measured rates and magnitudes of dome deformation), and the interactions and feedbacks among the elements. The models devised by Huppert et al. (1982) for the simple, monogenetic Soufriere (St. Vincent) dome, Murase et al. (1985) for an early stage of the Mount St. Helens dome, and Blake (this volume) for domes in general cannot account for many of the observations of the composite, long-active Mount St. Helens dome. Rheologic studies require more knowledge of the internal nature of the dome before a comprehensive unifying model can be devised. To this end, current work on the dome uses geophysical means to attempt to estimate its internal magnetic and hence thermal structure – vital information for evaluating the viscosity, yield strength, and other physical properties of the core and skin.

The complexity of the dome may seem discouraging to those wishing to understand its rheologic behavior, but the observations of regularities in growth are encouraging to us, because they suggest that certain principles and processes recur and largely control growth patterns. The papers by Iverson (this volume) and Denlinger (this volume) are the first to use some of these observations in a mechanistic study of the dome. This paper will have served its purpose if it entices further such work, despite the complexities alluded to in this paper and spelled out by Swanson et al. (1987).

Acknowledgments. Many staff members of the Cascades Volcano Observatory helped in obtaining data germane to this chapter; in particular, we thank Dan Dzurisin, Dick Iverson, and Roger Denlinger for discussion regarding the shape problem and many other matters. Kathy Cashman calculated the average chemical composition from data in her files. Kazuaki Nakamura provided a copy of Ichio Moriya's interesting paper that, through one of life's bizarre circumstances, Swanson was discussing with Ichio just days before Kazu's untimely death. Jon Fink accompanied us on a "rheologist's climb" of the dome and arranged for the volume calculations to be made. Curtis Manley spent scores of hours digitizing the maps for the volume calculations. Mike Malin graciously donated much time to those calculations, using a state-of-the-art computer routine he wrote. Herbert Huppert spent an instructive evening with us in Reno during a Geological Society of America meeting and somehow focused on the dome rather than other attractions. The chapter benefited from informal reviews by Dan Dzurisin, Rick Hoblitt, and Richard Waitt and formal reviews by Roger Denlinger and Wendell Duffield.

References

Anderson SW, Fink J (1986) Rate-dependent formation of textures in the Mount St. Helens lava dome (abs). Eos 67:1250

Bacon CR (1982) Time-predictable bimodal volcanism in the Coso Range, California. Geology 10:65–69

Carey S, Sigurdsson H (1985) The May 18, 1980 eruption of Mount St. Helens 2. Modeling of dynamics of the plinian phase. J Geophys Res 90:2948–2958

Casadevall T, Rose W, Gerlach T, Greenland LP, Ewert J, Wunderman R, Symonds R (1983) Gas emissions and the eruptions of Mount St. Helens through 1982. Science 221:1383–1385

Cashman KV, Taggart JE (1983) Petrologic monitoring of 1981 and 1982 eruptive products from Mount St. Helens. Science 221:1385–1387

Chadwick WW Jr, Swanson DA, Iwatsubo EY, Heliker CC, Leighley TA (1983) Deformation monitoring at Mount St. Helens in 1981 and 1982. Science 221:1378–1380

Chadwick WW Jr, Archuleta RJ, Swanson DA (1988) The mechanics of ground deformation precursory to dome-building extrusions at Mount St. Helens: 1981–1982. J Geophys Res 93:4351–4366

Dzurisin D, Westphal JA, Johnson DJ (1983) Eruption prediction aided by electronic tiltmeter data at Mount St. Helens. Science 221:1381–1383

Fremont M, Malone SD (1987) High precision relative locations of earthquakes at Mount St. Helens, Washington. J Geophys Res 92:10,223–10,236

Heliker CC (1984) Inclusions in the 1980–1983 dacite of Mount St. Helens, Washington. MS thesis, Western Washington Univ., Bellingham, 185 p

Heliker CC (1985) Inclusions in the 1980–1983 dacite of Mount St. Helens, Washington. Geol Soc Am Abstr with Programs 17:361

Huppert HE, Shepherd JB, Sigurdsson H, Sparks RSJ (1982) On lava dome growth, with application to the 1979 lava extrusion of the Soufriere of St. Vincent. J Volcanol Geotherm Res 14:199–222

Kuntz MA, Champion DE, Spiker EC, Lefebvre RH (1986) Contrasting magma types and steady-state, volume-predictable, basaltic volcanism along the Great Rift, Idaho. Geol Soc Am Bull 97:579–594

Mellors RA, Waitt RB, Swanson DA (1988) Generation of pyroclastic flows and surges by hot-rock avalanches from the dome of Mount St. Helens volcano, USA. Bull Volcanol 50:14–25

Moriya I (1978) Morphology of lava domes. Bull Dept Geography Kanazawa Univ Japan 14:55–69 (in Japanese)

Murase T, McBirney AR, Melson WG (1985) Viscosity of the dome of Mount St. Helens. J Volcanol Geotherm Res 24:193–204

Scandone R, Malone SD (1985) Magma supply, magma discharge and readjustment of the feeding system of Mount St. Helens during 1980. J Volcanol Geotherm Res 23:239–262

Swanson DA (1985) Graben formation, thrust faulting, and growth of the Mount St. Helens dacite dome. EOS 66:852

Swanson DA, Casadevall TJ, Dzurisin D, Malone SD, Newhall CG, Weaver CS (1983) Predicting eruptions at Mount St. Helens, June 1980 through December 1982. Science 221:1369–1376

Swanson DA, Dzurisin D, Holcomb RT, Iwatsubo EY, Chadwick WW Jr, Casadevall TJ, Ewert JW, Heliker CC (1987) Growth of the lava dome at Mount St. Helens, Washington (USA), 1981–1983. In: Fink J (ed) The emplacement of silicic domes and lava flows. Geol Soc Am Spec Pap 212:1–16

Waitt RB, Pierson TC, MacLeod NS, Janda RJ, Voight B, Holcomb RT (1983) Eruption-triggered avalanche, flood, and lahar at Mount St. Helens – effects of winter snowpack. Science 221:1394–1397

Waitt RB, MacLeod NS (1987) Minor explosive eruptions at Mount St. Helens dramatically interacting with winter snowpack in March–April 1982. In: Schuster JE (ed) Selected papers on the geology of Washington. Wash Div Geol Earth Resources Bull 77:355–379

Weaver CS, Zollweg JE, Malone SD (1983) Deep earthquakes beneath Mount St. Helens: evidence for magmatic gas transport? Science 221:1391–1394

The Development and Distribution of Surface Textures at the Mount St. Helens Dome

S. W. ANDERSON and J. H. FINK

Abstract

Recently acquired detailed topographic data covering the 6-year emplacement of the Mount St. Helens dacite lava dome, and hydrogen isotopic analyses of lava samples, have revealed relationships among lava surface texture, underlying slope, volatile content, and repose period. Two principal types of surface texture, smooth and scoriaceous, have formed on the Mount St. Helens dome during two different types of extrusive episodes. Type I extrusions begin with the emplacement of a small amount of smooth lava followed by a larger amount of scoriaceous lava, resulting in a predominantly scoriaceous lobe. Type II extrusions are dominated by large, smooth fractures, called crease structures. Type II lobes can be further subdivided into II-A types, in which the initially smooth crease structure becomes largely scoriaceous away from the vent during emplacement, and II-B types, whose surfaces remain entirely smooth.

The underlying slope and water content of the extruding lava appear to play the dominant roles in determining extrusion type and associated textural pattern. Crease structures, which form on shallow slopes, induce rapid cooling of the crack tip region, resulting in the formation of smooth lava. Volatile contents of around 0.3 to 0.4 wt% allow vesiculation of lava at the surface or slightly beneath the initially smooth crease structure walls, forming surface scoria. This vesiculation causes volatile contents to drop to around 0.1 wt%. Type I lobes are predominantly scoriaceous, possibly because large crease structures do not form on slopes of greater than 20°. Type II lobes form on flatter areas near the top of the dome or on the crater floor. Type II-A lobes occur when higher water contents cause the lava to slowly expand beneath and break apart the overlying smooth surface, resulting in a distal increase in scoriaceous lava. Type II-B lobes have lower water contents and fail to exhibit appreciable scoria at the surface.

The observed long-term increase in the percentage of smooth lava on the dome surface may be related to more thorough degassing of magma during ascent and emplacement, rather than to drying out of the parent magma body.

Introductory note. Silicic lava commonly exhibits a variety of vesicular (e.g. Fink and Manley 1987), petrologic (e.g. Sampson 1987), and devitrification textures (e.g. Lofgren 1971). The term "lava texture" is often ambiguous due to the many textural types in each category and should only be used if defined in a specific manner. Because this chapter addresses the formation of different lava surface morphologies resulting from variations in vesicularity, the terms "textures", "lava textures", and "surface textures" will only refer to differences in the macroscopic appearance of lava caused by variations in the shape and distribution of bubbles.

1 Introduction

Studies of surface textures on silicic lava flows and domes have previously focused on prehistoric flows (Loney 1968; Fink 1979, 1980a, b, 1983; Fink and Manley 1987; Manley 1986; Eichelberger and Westrich 1984; Eichelberger et al. 1986). Since the flows were not observed during extrusion, attempts to explain relationships between surface morphology and conditions of emplacement have been based primarily on structural and chemical studies. A major problem encountered in these studies has been the lack of pre- and syn-emplacement data needed to test models of texture formation. In particular, the lack of data on extrusion rates, preeruption topography, and repose periods have hindered the application of models for degassing and textural development derived from such studies to currently active volcanoes.

In October 1980, a small dacite lava dome extruded onto the crater floor at Mount St. Helens. Since then, nearly 20 separate lava flows, or lobes, have piled upon this initial extrusion to form a composite lava dome which (in 1989) is approximately 250 m high and 1000 m wide at the base (Swanson and Holcomb, this Vol.). The growth of this lava dome has been more thoroughly documented than any other in history (see Moore et al. 1981; Chadwick et al. 1983; Swanson et al. 1987; Swanson and Holcomb, this Vol.). Detailed topographic maps of the dome compiled after almost every extrusion provide a basis for accurately calculating eruptive volumes and pre-eruption slopes. The availability of these data, combined with the extensive observational record of the growing dome and the uniform petrology of the dacitic lava (Cashman and Taggart 1983), makes the dome at Mount St. Helens an ideal location to study the formation of surface textures and their relationship to emplacement conditions.

In this chapter, we present detailed observations of texture development during the emplacement of many lobes, and simple geologic maps showing the final distribution of textures. Then we combine this information with isotopic and topographic data in order to define the factors which control the development and distribution of lava textures at the Mount St. Helens dome. Finally we discuss some of the ways that the surface appearance of silicic extrusions may be used to interpret volcanic hazards.

2 Growth of the Mount St. Helens Composite Dome

Following the climactic eruption of Mount St. Helens on May 18, 1980, dacitic lava domes were emplaced onto the crater floor during the waning stages of two smaller explosive eruptions in June and August of 1980. These domes, which were subsequently destroyed by ensuing eruptions, were eventually replaced by the extrusion which formed in October 1980.

Since October 1980, the lava dome has experienced three major periods of growth. From October 1980 through the end of 1982, the dome grew through an episodic series of extrusions, erupted primarily from vents near the central top of the dome (Swanson and Holcomb, this Vol.). Individual extrusions produced short, thick lava flows and usually lasted a few days, with periods of repose lasting several weeks or months. A fairly complete record of the growth of the Mount St. Helens dome from 1980 through 1983 can be found in Moore et al. (1981) and Swanson et al. (1987).

Beginning abruptly in February 1983, and continuing for a 12-month period, the typical, episodic growth pattern changed to slow, continual intrusion (endogenous growth) and extrusion (exogenous growth). Many features, such as spines, mounds, and a "piggy-back" lava lobe (an actively extruding lobe situated on top of another active lobe), characterized the surface of the 1983 lobe. In March 1984, episodic growth resumed and continued through October 1986. To date (January 1989) no extrusions have occurred since October 1986. Because of the unusual development of lava morphologies during the continuous intrusive and extrusive period of dome growth (see Swanson et al. 1987), this chapter will focus primarily on the more typical, episodic growth of the Mount St. Helens dome from 1980–1982 and 1984–1986.

The lava which comprises the Mount St. Helens dome is a porphyritic dacite with a silica content of approximately 62 to 63%. The lava contains phenocrysts of plagioclase, orthopyroxene, hornblende, magnetite-ilmenite, and clinopyroxene (Cashman and Taggart 1983). The chemical composition of the dacite has not changed appreciably since October 1980 (Swanson and

Table 1. Eruption variables calculated for lobes of the Mount St. Helens Dome

Extrusion date	Lobe volume $(10^6 \, m^3)$	Exogenous/ endogenous ratio	Underlying slope (degrees)	Extrusion rate $(10^6 \, m^3 \, d^{-1})$	Repose period (days)	Extrusion type
Oct 1980	1.2	∞	5.0	1.2	–	II
Dec 1980	1.92	39.80	18.8	0.27	71	II
Feb 1981	2.19	4.31	34.0	1.10	32	I
Apr 1981	1.87	3.66	28.0	0.93	61	I
Jun 1981	3.49	9.35	25.1	3.49	66	I
Sep 1981	1.78	1.88	33.9	0.36	78	I
Oct 1981	1.64	1.60	35.9	0.55	50	I
Mar 1982	–	–	36.0	–	137	I
May 1982	–	–	24.4	–	56	I
Aug 1982	2.51	2.22	31.3	0.50	89	I
Mar 1984	0.62	1.26	3.5	0.12	47	II
Jun 1984	0.94	11.80	23.7	0.06	72	I
Sep 1984	1.53	0.71	5.7	0.76	69	II
May 1985	0.83	0.24	19.0	0.28	258	II
May 1986	–	–	10.0	–	354	II
Oct 1986	–	–	–	–	160	II

Volumes for individual lobes, underlying slopes, and exogenous/endogenous ratios were measured using digital topographic data sets, except Oct 1980 volume which was taken from Swanson and Holcomb (this volume). Digital data for individual lobes erupted in March 1982, May 1982, May 1986, and October 1986 were not available

Holcomb, this Vol.). The crystallinity of the lava (approximately 50%) has also remained essentially constant due to crystal growth rates of less than 10^{-11} cm s^{-1}, which would only result in approximately 0.02 m of growth during the 6-year history of the dome (Cashman and Marsh 1986).

Topographic data sets for almost every extrusion and observations made by members of the U. S. Geological Survey (USGS) have been used to calculate various eruptive parameters, such as overall extrusion rate, underlying slope, the ratio of extrusive to intrusive growth, and repose periods between eruptions (Table 1). The topographic data sets were constructed with a 2-m contour interval by the National Mapping Division of the USGS and then digitized and manipulated at Arizona State University's Image Processing Facility in order to calculate eruptive volumes and underlying slope for each extrusion. This information was then combined with observational data regarding the timing of eruptive events to compute extrusion rates and repose periods. The exact methods used to calculate these parameters can be found in Anderson (1988).

3 Lava Textures and Crease Structures at Mount St. Helens

3.1 Texture Types and Proportions of Textures on Lobe Surfaces

Two different lava textures have been observed on the surface of the Mount St. Helens dome: smooth and scoriaceous (Fig. 1). Scoriaceous lava occurs as a 1- to 2-m thick carapace over a denser, flow banded interior (Cashman and Taggart 1983). The scoriaceous carapace is very irregular when viewed in profile owing to the presence of many small mounds and spires of vesicular material. The vesicles in the scoriaceous lava are highly irregular in shape and are commonly elongate perpendicular to the flow surface (Fig. 2). The shapes of the vesicles are largely dependent upon the proximity of phenocrysts, with very thin films of glass forming boundaries between the vesicles and crystals. The vesicles can occupy over 50% of the volume of a sample. The second surface texture is a smooth, relatively nonvesicular lava (Fig. 1). When viewed in profile, the areas of smooth lava lack the spires and mounds found with the scoriaceous lava. Smooth lava usually contains less than 15% vesicles by volume which are irregular but not elongate in shape (Fig. 3).

A prominent structural feature found on many of the active Mount St. Helens lobes is a fracture with outwardly convex walls, called a crease structure (Anderson and Fink 1987, 1988). These features, which have not received much attention in the past, are always associated with smooth lava and appear to play a major role in determining both the sequence and patterns of textures.

Geologic maps showing the final distribution of textures on each lobe (Fig. 4, from Holcomb and Colony 1987) have been digitized in order to calculate the proportions of surface textures. As Table 2 indicates, the relative proportions of the two textures vary considerably from lobe to lobe. Geologic maps are not yet available for the June 1980, August 1980, May 1986, and October

Fig. 1. June 1981 lobe showing the two different lava textures. In the center of the lobe is a prominent structural feature called a crease structure, which consists of smooth, relatively nonvesicular lava. The rest of the flow is mantled by scoriaceous lava which forms a 1–2 m thick carapace over a more dense, flow-banded interior. (Photo courtesy of the USGS)

1986 lobes. Therefore, proportions of textures on these extrusions have been estimated from photographs or field inspection.

3.2 Temporal Sequences and Final Patterns of Textures

Because of the ongoing monitoring of Mount St. Helens by the USGS, several observations of texture development on growing lobes have been made. According to Cashman and Taggart (1983), lava less than 24-h-old generally displays a smooth surface texture, becoming more fragmented and scoriaceous as it moves downslope. They suggested that this fragmentation was probably due to an increase in extrusion rate and downslope velocity. Swanson et al. (1987) also noticed changes from smooth to scoriaceous lava during individual extrusions. Although they were not able to delineate the process responsible for this change, they suggested that degassing may have played an important role. Finally, Swanson and Holcomb (this Vol.) noted that younger lobes showed greater proportions of smooth-textured lava, and they speculated that this change might reflect progressive degassing of the magma source.

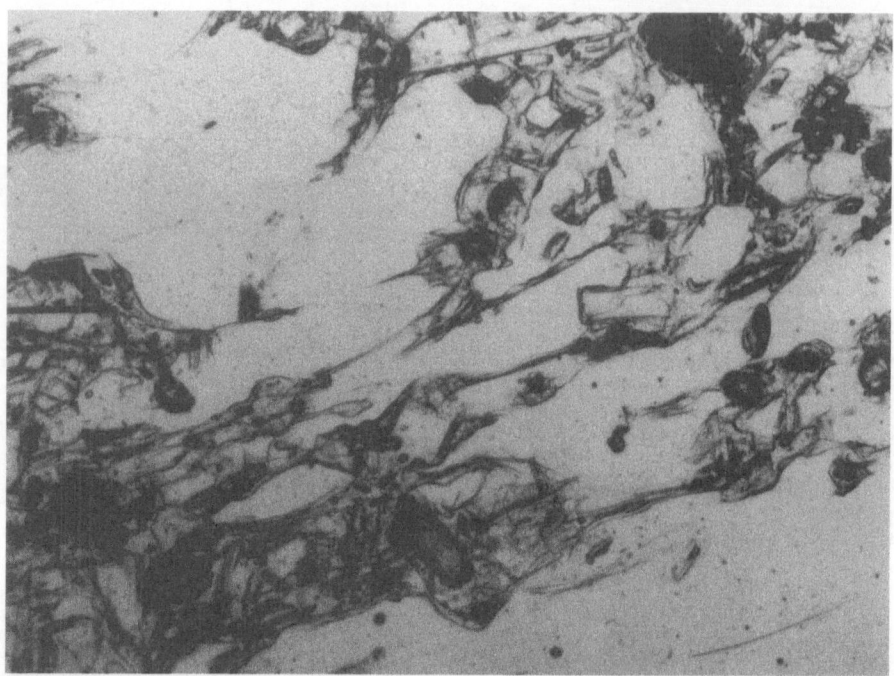

Fig. 2. Photomicrograph of scoriaceous dacite. Note the filaments of glass and large volume of void space. Magnification is 25×

In the following section, observations of texture development on the most thoroughly documented extrusions will be summarized. Although Moore et al. (1981) and Swanson et al. (1987) include some observations of texture development, their descriptions are generally insufficient to base a thorough discussion of texture formation. For this reason, we used slide collections provided by USGS personnel and U.S. Forest Service aerial photographs as a basis for defining the distribution of textures during different stages of the various extrusions that form the dome. The two main textures are quite distinctive, even in air photos taken at approximately a 1 : 15 000 scale.

October 1980. Extrusion of the October 1980 lobe began approximately 50 minutes after the conclusion of a series of explosions during October 16 – 18 (Moore et al. 1981). When first viewed, the dome, which was 5 m high and 25 m wide, was located on the flat crater floor and had a relatively smooth surface texture. Inspection of the crater on October 19 revealed a much larger dome, 185 m wide and 50 m high (Moore et al. 1981). The dome was still mostly smooth and was dominated by a large, lobe-bisecting crease structure (Fig. 5). A small amount of scoria can be seen faintly in the central valley of the crease structure (Fig. 5). Extrusion of the lobe probably ceased sometime during October 19, but slow spreading and sagging under its own weight con-

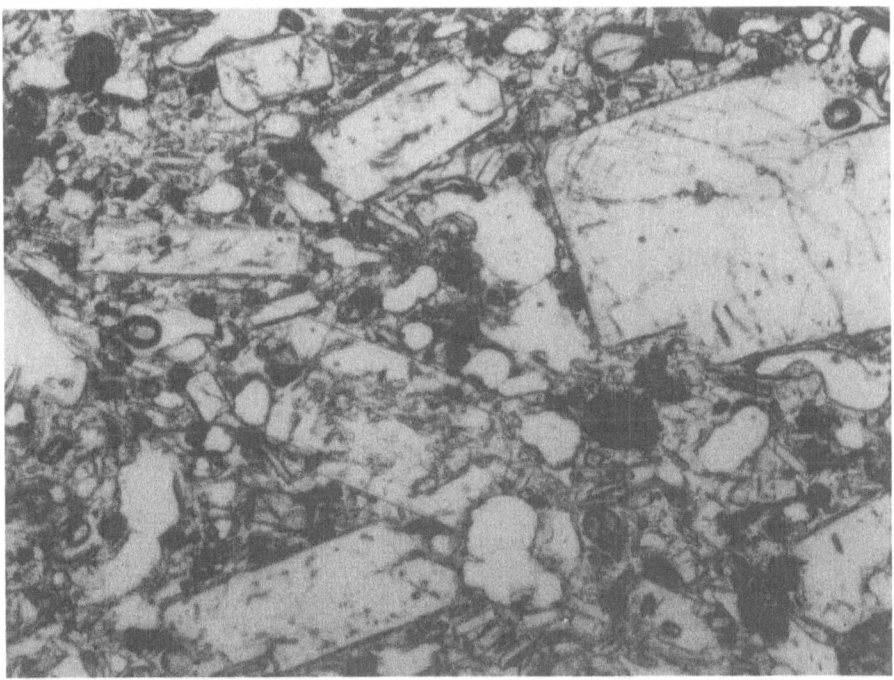

Fig. 3. Photomicrograph of smooth dacite. Magnification is 25×

Table 2. Surface areas of texture types

Extrusion date	Scoria (m²)	Smooth (m²)	Scoria (%)	Extrusion type
Oct 1980	33463	96580	26	II
Dec 1980	92512	32515	74	II
Feb 1981	84032	11487	88	I
Apr 1981	102441	18942	84	I
Jun 1981	117982	19094	86	I
Sep 1981	96776	3962	96	I
Oct 1981	122984	14442	89	I
Mar 1982	53385	4963	91	I
May 1982	66791	35504	65	I
Aug 1982	118927	34337	78	I
1983	–	–	–	Continuous
Mar 1984	21080	29371	42	II
Jun 1984	63546	1007	98	I
Sep 1984	25528	64173	28	II
May 1985	–	–	0*	II
May 1986	–	–	10*	II
Oct 1986	–	–	0*	II

Percentage of dome area covered by scoriaceous and smooth lava, calculated from digital versions of geologic maps compiled by Holcomb and Colony (1987), along with lobe type. *indicates values estimated in the field.

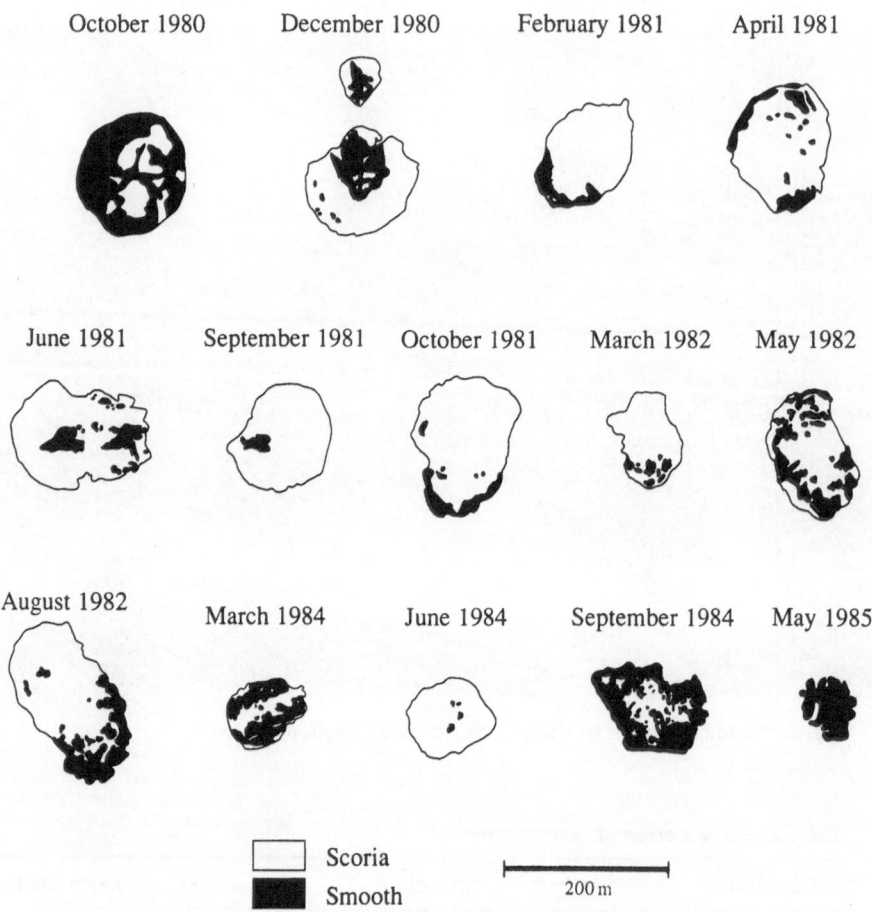

Fig. 4. Sketch maps of textural distributions on the various Mount St. Helens lobes. Sketch maps for May and October 1986 are not available but lobes are almost entirely smooth

tinued for the next few days (Moore et al. 1981). The completely emplaced lobe was predominantly smooth, with a few small, localized areas of scoria near the center of the lobe (Fig. 6). The crease structure observed during emplacement was not present on the fully emplaced lobe.

December 1980. The beginning of extrusion of the two December 1980 lobes was not witnessed, but is believed to have occurred on December 27 (Moore et al. 1981). When observed from an aircraft at about 2.45 P. M. on December 28, the SE lobe displayed a predominantly scoriaceous carapace, although the central region was dominated by a smooth crease structure which became more scoriaceous outward toward the margins of the flow. During the final stages of extrusion on January 2, 1981, the crestal region of the SE lobe consisted of several slab-like, smooth areas surrounded by highly scoriaceous lava

Fig. 5. Growing October 1980 lobe on the 19th. Note the large crease structure that bisects the lobe. Lobe diameter is approximately 185 meters. (Photo courtesy of the USGS)

(Fig. 7). This represented the final textural pattern of the fully emplaced December lobe.

Observations of the smaller, NW lobe are incomplete. When first viewed on December 28, the NW lobe was entirely scoriaceous. The next available photograph was taken after extrusion ceased and shows a somewhat linear outcrop of smooth lava extending along a deep crack from the vent to the toe of the mostly scoriaceous lobe.

September 1981. Extrusion of the September 1981 lobe was perhaps the best-documented photographically from start to finish. The first moments of extrusion were captured on film the evening of September 6 from a fixed-wing aircraft. These photos show the presence of a smooth, relatively small (approximately 30 m in diameter) crease structure. On the morning of September 7, the dome was predominantly scoriaceous, had grown considerably, and lacked any crease structure. One small outcrop of smooth lava at the toe of the flow (Fig. 8) may have represented the remains of the crease structure seen during the evening of September 6.

The lobe continued to descend the steep flank of the dome through September 10. It remained mostly scoriaceous throughout extrusion, with the only smooth area located at the toe. On September 11 a small crease structure was

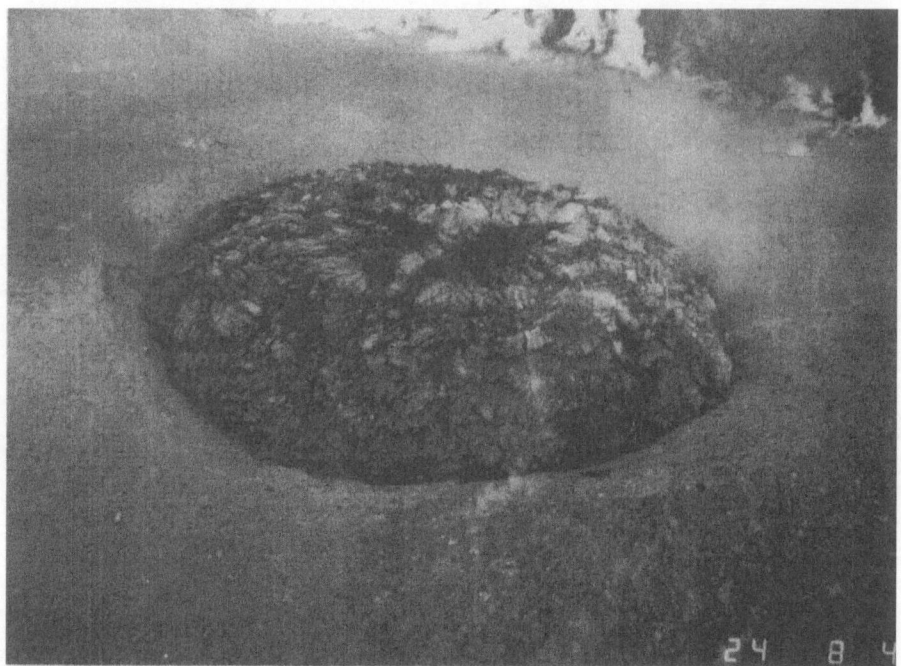

Fig. 6. Completely-emplaced October 1980 lobe. The surface is entirely smooth except for a few localized areas of scoria. Lobe diameter is approximately 225 meters. (Photo courtesy of the USGS)

seen over the vent area (Fig. 9), which was not present in photographs taken between September 7–10. When extrusion ceased on or about September 11 (Swanson et al. 1987), the surface of the lobe was predominantly scoriaceous except for two small areas of smooth lava: one near the vent and one at the toe.

October 1981. Extrusion of the October 1981 lobe probably began on October 30, but was obscured by poor weather (Swanson et al. 1987). When first viewed on the morning of October 31, the new lobe had a scoriaceous central area and smooth, blocky margins, suggesting the earliest material was probably smooth lava that was pushed to the flow margins as new, scoriaceous material was extruded. Emplacement of scoriaceous material continued through the end of the extrusion on November 4. The fully emplaced lobe was situated on a relatively steep slope and was predominantly scoriaceous, with smooth areas only at the flow margins and near the vent area.

May 1982. Extrusion of the May 1982 lobe began on May 15 with emplacement of smooth, blocky lava on the steep margin of the dome. On May 16, a small outcrop of scoriaceous lava was present near the vent, and smooth, blocky lava was concentrated near the flow front and flow margins. The sur-

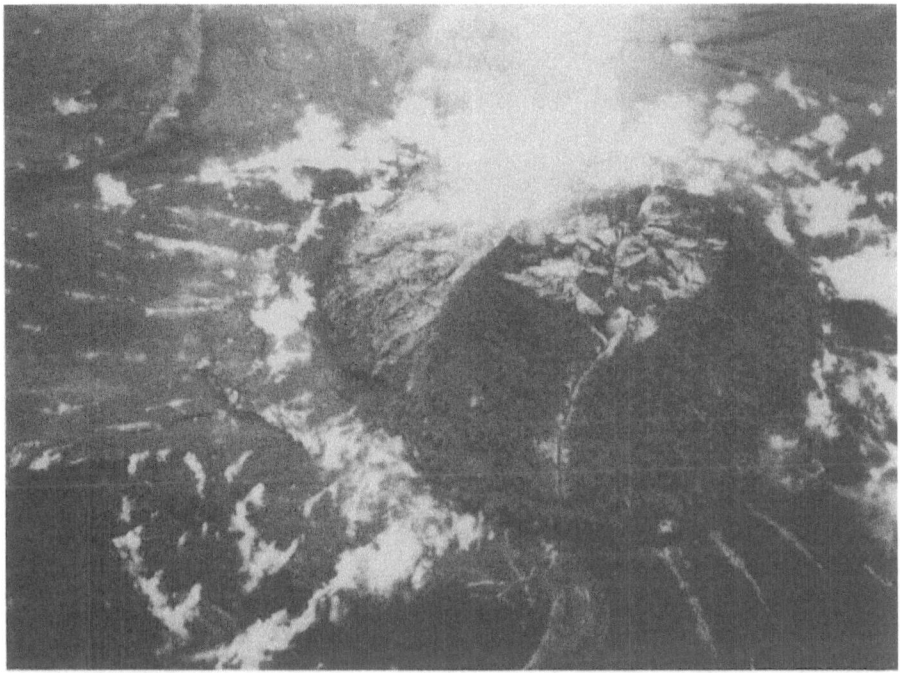

Fig. 7. December 1980 lobe showing a smooth crestal region surrounded by scoriaceous outer margins. (Photo courtesy of the USGS)

face of the lobe did not change appreciably between May 16 and 19, except that the scoriaceous area near the vent had greatly enlarged. Extrusion probably ceased on or about May 20. The fully emplaced lobe had a smooth, blocky area comprising approximately 35% of the surface near the flow front and along margins close to the vent. Scoria made up the remainder of the surface.

September 1984. The texture of the September 1984 lobe was obscured by steam and poor lighting from the 10th to the 11th. The first clear view, on the 12th, revealed a smooth surface texture and an apical crease structure. The lobe was situated on the flat, crestal portion of the dome and had grown to nearly its maximum extent at this point. The surface of the fully emplaced lobe was mostly smooth, although the central portion consisted of a wide, linear strip of scoriaceous lava that apparently formed during the final stages of extrusion.

October 1986. A time-lapse Super-8 movie camera installed on the dome prior to extrusion of the October 1986 lobe enabled observers to document a profile view of a large crease structure which was present throughout the course of the extrusion. The movie clearly reveals that the crease structure walls were smooth during all stages of development. Examination of air photos taken later also showed that the October 1986 lobe surface remained entirely smooth through-

Fig. 8. Extrusion of lava with a scoriaceous carapace on September 7, 1981. Note smooth toe from early-phase crease structure. (Photo courtesy of the USGS)

out emplacement on the flat crest of the dome, although the large surfaces of the crease structure broke up into several smooth slabs at some point during or after extrusion.

3.3 Classification of Extrusion Types

Many extrusions show striking similarities both in their sequence of texture formation and in their final distribution of textures on the fully emplaced lobe. Thus, we present a classification of lobes based on both their sequence and final pattern of textures. A scheme of this type allows the classification of all lobes, including those which were not observed during emplacement. This classification provides a basis for comparing various eruptive parameters with the different kinds of extrusions. For those lobes not observed during extrusion, characterization is based upon the final distribution of surface textures.

Type I Extrusions. Initial extrusion of a small amount of smooth lava (generally less than 15% of the total surface area), followed by a larger amount of scoriaceous lava. In some cases, small (less than 30 m wide), smooth crease structures are observed in the vent area at both the outset and end of extrusion.

Fig. 9. Small crease structure over the vent of the predominantly scoriaceous September 1981 lobe. (Photo taken on September 11 by the USGS)

The fully emplaced lobe is almost completely covered by scoriaceous lava and may have a small, smooth crease structure over the vent. Type I extrusions form on slopes of greater than 20° (Table 1).

Type II Extrusions. (Endmember A) Distinguished by a large, smooth crease structure which bisects the entire lobe (often greater than 150 m in diameter). The crease structure walls remain smooth throughout their lateral extent, resulting in an entirely smooth lobe. The central valley of the crease structure is commonly evident on the fully emplaced lobe.

(Endmember B) Characterized by a scoriaceous surface resulting from the breakup of a large, smooth-surfaced crease structure as the walls migrate outward from the central valley. The fully emplaced lobe is entirely scoriaceous.
The February 1981, April 1981, June 1981, September 1981, October 1981, May 1982, August 1982, and June 1984 lobes form on steep slopes, are predominantly scoriaceous, and are classified as Type I lobes. The June 1980, August 1980, May 1985, and October 1986 lobes are completely smooth and are examples of the II-A endmember. Although the October 1980, February 1984, March 1984, September 1984, and May 1986 lobes display both smooth and scoriaceous surface areas, they are primarily smooth and thus are also classified as Type II-A. The mostly scoriaceous surface of the SE December

1980 lobe typifies the class B endmember. Table 2 lists the various lobes according to extrusion type and shows the percentages of smooth and scoriaceous lave exposed on their surfaces.

4 Water Contents of the Mount St. Helens Dome Lavas

4.1 Previous Studies

Several recent studies of the distribution of volatiles within silicic lava flows have indicated that volatile content is a key factor in determining both the mode of eruption (explosive or effusive) and the formation of glassy and vesicular textures. Friedman and Smith (1958) demonstrated that fresh volcanic glass can retain some magmatic water. More recently, Eichelberger and Westrich (1984) suggested that the water contents and isotopic signatures of relatively nonvesicular volcanic glass reflect comparable properties in the preeruption magma. Taylor et al. (1983) measured water contents, $\delta^{18}O$ and δD values of obsidian samples from several different eruptive sequences in the western United States. They found that $\delta^{18}O$ values varied only from +7.0 to +8.0 per mil, which they interpreted to be primary magmatic values. In contrast, large variations in the water content and hydrogen isotope composition observed within individual eruptive sequences were related to the mode of eruption. Glassy tephra samples had higher water contents and were richer in deuterium than obsidian from flows. They suggested that water contents were lower in flow obsidian samples because this material underwent greater degassing during ascent. Such degassing causes large variations in deuterium and small variations in $\delta^{18}O$ values because the fractionation effect between magma and vapor for oxygen isotopes is much smaller than for hydrogen isotopes.

Little previous work has been published on the water contents of products from the 1980–1986 eruptions of Mount St. Helens. Although Rutherford et al. (1985) estimate that magma erupted on May 18, 1980 initially contained approximately 4.6 weight per cent water, no data have been published on the water contents of the effusive Mount St. Helens dome products.

4.2 Procedure

Water content and isotope analyses were performed in the Stable Isotope Laboratory at Arizona State University. A hydrogen extraction line was employed to derive δD and water content values. Crushed whole-rock samples were heated to 1000°C for 3 h after initial drying overnight. Hydrogen was produced by reaction of expelled water with uranium. A manometer was employed to measure the H_2O content. Stepwise heating of selected samples showed that most of the water was given off between 400° and 600°C. There is a 2

Table 3. Water content and isotope data

Extrusion date	Sample No.[a]	Texture	Location[b]	Water content (%)	δD (‰)	Extrusion type
May 1986	5HCS786	Smooth	F	0.111	−43	II
May 1986	6HCS786	Smooth	M	0.187	−71	II
May 1986	9HCS786	Smooth	M	0.271	−67	II
May 1986	14VC786	Smooth	M	0.337	−40	II
May 1986	11HCS786	Scoria	V	0.149	−46	II
May 1985	12HCS85	Smooth	F	0.210	−56	II
May 1985	6HCS85	Smooth	M	0.200	−59	II
May 1985	2HCS85	Smooth	V	0.230	−71	II
Jun 1981	SH91	Scoria	F	0.126	−71	I
Jun 1981	SH108	Smooth	V	0.194	−95	I
Feb 1981	SH82	Smooth	F	0.081	−82	I
Feb 1981	SH69	Scoria	M	0.105	−71	I
Feb 1981	SH73	Smooth	V	0.212	−90	I
Dec 1980	SH61	Scoria	F	0.127	−73	II
Dec 1980	SH66	Smooth	V	0.165	−87	II
Oct 1980	SH53	Smooth	F	0.109	−69	II
Oct 1980	SH33	Scoria	V	0.161	−60	II

[a] Samples collected by USGS personnel from earlier lobes are designated by "SH" prefixes. For a given lobe, samples are listed in order of decreasing distance from the vent. Deuterium values for meteoric water from Mount St. Helens area have been reported at around −90 per mil (Barnes 1984).
[b] F denotes samples from the flow front; V denotes samples from the vent area; M denotes samples between the vent and the flow front

per mil error for the δD analyses and a 0.002% error for the water contents (S. Roberts, personal communication, 1988). The analyses were not corrected for crystal content or chemistry because the crystallinity and petrology were essentially constant for all samples.

Samples from the surfaces of the May 1986 and May 1985 lobes were collected along lines extending from the vent to the flow front (Table 3). Although more than 200 samples of lobes from 1980–1984 were collected by members of the USGS, only a few proved useful for this study because the exact locations of most samples were not well-documented.

4.3 Results

Water content and isotope values of Mount St. Helens dome samples are shown in Table 3. The δD values range between −40 and −95 per mil. Measured water contents vary between 0.081 and 0.337 wt%. None of the samples are believed to have been contaminated by meteoric water because (1) most samples were collected during or shortly after extrusion, (2) the hydrogen

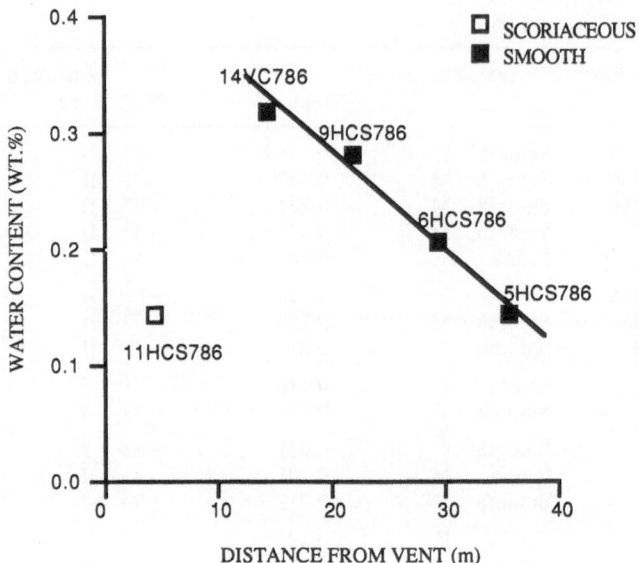

Fig. 10. Water contents of May 1986 lava samples plotted against distance from the vent. All samples are of smooth lava except for 11HCS786

isotope values fall within the "normal" range of magmatic waters (Taylor 1974), and (3) the scoriaceous samples, which should be most susceptible to hydration, are generally the most enriched in deuterium (see Table 3). If hydration occurred after extrusion, the δD values of the scoriaceous material would tend towards the δD of local meteoric water. Meteoric waters that emerge from springs in volcanic rocks in the Mount St. Helens area have δD values ranging from -82.3 to -111.4 per mil, and a sample of water from a pond located in the summit crater of Mount St. Helens immediately prior to the May 18 eruption yielded a value of -89.8 per mil (Barnes 1984).

One of the most striking trends in this data set is that most extrusions show an increase in water content during extrusion. Analyses of five samples from the May 1986 lobe, which were collected systematically from the flow front to the vent area, show this most dramatically. The lava is all smooth except for a scoriaceous zone near the vent (11HCS786). Water contents of the four smooth samples show a clear linear increase from the flow front toward the vent (Fig. 10). Since the lava at the flow front was the first erupted, the trend in Fig. 10 indicates a progressive increase in water content of magma as it emerged during the course of extrusion. It is unlikely that this linear increase in water content is due to degassing after the lava reached the surface because (1) there is no increased vesicularity in distal samples, (2) sampled lava quenches rapidly at surface conditions, and (3) degassing at the surface would require the unlikely condition of vent lava quenching more rapidly than the earlier erupted flow front lava.

5 Discussion

5.1 Interpretation of the Water Content Data

The observed distribution of textures and water contents appears to indicate that volatile content increased during extrusion of some individual lobes (e.g., May 1986). Although some lobes show little or no vent-ward increase in water contents, analyses of pairs of samples from the vents and flow fronts of the February 1981 and May 1985 lobes also indicate an increase in water during extrusion. No examples of water content decreases during extrusion have been found for any of the Mount St. Helens lobes we have examined to date. Patterns of textures similar to that of May 1986 on several other extrusions suggest progressive water content increases may be common at Mount St. Helens. In particular, the October 1980, May 1982, and September 1984 lobes have smooth margins surrounding a strip of scoriaceous material in the vent region. It seems likely that these lobes have patterns of water content similar to those of the May 1986 lobe: low near the margins and gradually increasing until the material vesiculates near the center.

Such an increase could reflect degassing of the upper part of the magma column prior to eruption (Chadwick et al. 1988), with volatiles escaping through the fractured host rock around the conduit or within the dome. Smooth lava represents magma that degasses thoroughly enroute to the surface and is unable to vesiculate at surface pressures before quenching occurs. Flow front smooth lava is more thoroughly degassed than near-vent smooth lava because it represents the top of a rising magma body which had ample opportunity to degas through fractures in the conduit and dome as it slowly broke a path to the surface. Near-vent smooth lava is more water-rich because it had less time to degas as it rose more rapidly through the already opened conduit. If the magma retains enough water to vesiculate before quenching occurs, then the scoriaceous carapace will form as the bubbles grow and release gas at the surface.

The low water value of scoriaceous sample 11 HCS 786 illustrates the change that occurs when lava vesiculates at the surface. The adjacent sample, 14 VC 786, was taken at the contact between the smooth and the scoriaceous lava, approximately 13 meters from the vent, whereas 11 HCS 786 is from scoriaceous lava located 3 meters from the vent. By projecting the trend defined by the four distal samples, we infer that this near-vent lava was erupted with a water content of between 0.4 and 0.5 wt%. The resulting vesiculation released approximately 0.3 wt%. The scoriaceous lavas, in some cases, have higher water content values than the smooth samples from the flow front. We suggest that the flow front smooth lava is able to degas more thoroughly than the scoriaceous lava because it rises more slowly and has more time to lose volatiles during ascent, whereas the scoriaceous material may not be able to thoroughly equilibrate at surface pressures before quenching occurs.

5.2 The Role of Crease Structures in Texture Formation

We have compared extrusion type with eruptive parameters, such as lobe volume, extrusion rate, ratio of exogenous to endogenous growth, pre-eruptive slope, and repose period, and found that only slope shows a clear relationship (Table 1). Type I extrusions tend to form on the steep portions of the dome where slopes are greater than about 20°, whereas Type II extrusions form on the flatter, crestal portions of the dome where the slope is less than 20°. A similar relationship is seen if one compares the proportion of a lobe's surface that is covered by scoria (Table 2) with slope. Lobe volume, extrusion rate, ratio of exogenous to endogenous growth, and repose period appear to have little or no influence on extrusion type.

We suggest that slope influences texture development at Mount St. Helens by controlling the development of crease structures. Anderson and Fink (1987, 1988) proposed qualitative and quantitative models of crease structure development based on observations of active flows at Mount St. Helens and field measurements of flow surfaces at the Little Glass Mountain, Crater Glass, and Medicine Lake Dacite flows of the Medicine Lake Highland volcano in northern California. At Mount St. Helens, crease structures form throughout the extrusion of lobes situated on shallow slopes (Type II extrusions), but only at the very beginning or end of extrusion of lobes situated on steep slopes (Type I extrusions). These temporal observations suggest that crease structures form when lava is forced to spread laterally, either as the flow advances over a flat area or as downslope movement ceases near the close of extrusion.

Two lines of evidence suggest crease structures form by a cyclic fracture mechanism. First, crease structures on the various flows at Medicine Lake Highland are commonly covered by subhorizontal "steps" or striations with spacings ranging from a few millimeters to several tens of centimeters, similar to features found on the faces of basalt columns. Ryan and Sammis (1981) and DeGraff and Aydin (1987) showed that each striation on a basalt column forms during a single fracture event. Ryan and Sammis (1981) suggested that striae widths depend on the temperature gradient near the crack tip; high gradients result in rapid cooling and accumulation of elastic strain energy, allowing the fracture to propagate by numerous small crack advances, whereas smaller gradients result in fewer, larger cracks.

The second argument is based on evaluation of simple conductive cooling and fracture models for Mount St. Helens lobes. Here we have assumed that fracture requires lava to cool below the glass transition temperature, T_g (Ryan and Sammis 1981), and that cooling does not involve crystallization (Turcotte and Schubert 1982). The time necessary for a lobe to cool below T_g to a depth of 15 meters and fracture, calculated using values of 725 °C for T_g (Ryan and Sammis 1981), 980 °C for the emplacement temperature (Friedman et al. 1981), and 0.005 $cm^2 s^{-1}$ for the thermal diffusivity (Friedman et al. 1981), is over 5 years. However, a 15-m-wide crease structure formed on the surface of the September 1981 lobe in less than 24 h. This discrepancy can be resolved if we assume that the fracture propagated incrementally, with each crack advance exposing the tip to the atmosphere (Anderson and Fink 1987, 1988). Such interrupted growth will result in much more rapid cooling of the flow surface. Exposure of a crack tip is evident in several photographs taken during formation of the December 1980 crease structure, where incandescent lava can be seen along the central axis.

5.3 Factors Controlling Texture Development

Since the main difference between the two texture types at Mount St. Helens lies in their near-surface vesicularity, the factors which control bubble growth should also determine the textural makeup of a given lobe. Laboratory and theoretical studies (e.g., Murase and McBirney 1973) show that bubble growth in silicic magmas is promoted by relatively high water contents and high temperatures. Water has two effects: it increases the volatile pressure and lowers the lava viscosity. High temperatures also serve to lower the viscosity, and they increase the diffusivity of volatiles in the magma. Thus, conditions which cause higher volatile contents or temperatures should favor the production of scoriaceous lava.

The distribution of volatiles in an extruding lobe may reflect both the preeruption arrangement and processes that occur as the lava is being emplaced. Stratification of volatiles within a source magma body should be indicated by the textures produced when that magma erupts, unless effusion rates are high enough to allow selective withdrawal of deeper material (e.g., Blake 1981). Volatiles may also become concentrated during the flow of lava on the surface. Fink and Manley (1987) proposed that volatiles migrate upward within active glassy rhyolite flows through microcracks that form in response to shear stresses. Such a process could also occur in the Mount St. Helens dacite, especially where the lava was moving rapidly. Thus the association of scoriaceous Type I lobes with steep slopes could indicate that volatiles were being redistributed within the active lava.

Bubble growth in lavas is inhibited by cooling. Crease structures, which develop when lava spreads laterally on shallow slopes, induce rapid cooling of the crack tip area and prevent the formation of scoria. For those lobes in which volatile content is relatively high, either due to initial concentration or redistribution during flow, bubbles may grow and expand beneath the smooth crease structure walls, eventually breaking them apart and exposing a near-surface scoriaceous zone. This process may lead to the formation of scoria-rich Type II B lobes on shallow slopes in which initially smooth crease structures become scoriaceous during the course of emplacement. Type II A lobes, whose surfaces remain smooth, apparently form from relatively dry magma.

Volatile content data lend support to this model. A distal scoria sample (SH 61) from the Type II-B December 1980 lobe has a water content of 0.127%. If we assume that near-surface vesiculation of this lava caused it to lose an amount of water (~0.30%) comparable to that assumed for the near-vent May 1986 scoria sample (11 HCS 786), then its original water value (~0.43%)

would have been unusually high for distal smooth lava samples, which commonly have values around 0.10–0.20% in Type II-A lobes (e.g., May 1985). Thus its high water content may have caused this lobe to experience late stage vesiculation which broke apart the surface of its crease structure.

The observed distribution of textures and water contents appears to indicate that volatile content increasing during extrusion of some individual lobes (e.g. May (1986). On other lobes, the textural pattern seems to reverse at the end of extrusion, with smooth lava occurring near the vent as well as at the flow front. This late-stage smooth lava may result from the presence of a crease structure. Alternatively, these cases may show evidence for volatile reduction in the last-erupted magma. Thus the observed textural patterns may be indicating that the parent magma body exhibits varying degrees of volatile stratification over time.

Other workers (e.g., Cashman and Taggart 1983; Swanson and Holcomb, this Vol.) have suggested that the progressive decrease in scoria production observed throughout the history of the dome (Table 2) indicates that the parent magma body is drying out with time. Our work, however, suggests the decrease in scoria with time may be due to greater amounts of degassing during ascent and emplacement, which could be related to a higher degree of fracturing and deformation around the conduit. We also suggest that the volatile content of the parent magma body cannot be directly inferred from lava texture distribution, In fact, the highest water contents we measured were in the predominantly smooth May 1985 and May 1986 lobes.

The apparent increase in water content during extrusion of several lobes has important implications for volcanic hazards. It is widely accepted that hazards from explosions are relatively high prior to an extrusive episode, and are significantly reduced once lava appears at the surface. Our work, however, indicates that the potential for an explosion may actually increase as extrusion proceeds. We suggest that late stage scoria development may be an indicator of increased explosive hazards due to higher water contents.

Acknowledgments. We thank Don Swanson, Dan Dzurisin, and Bobbie Myers of the USGS for their cooperation and logistical support, as well as Robin Holcomb and Wayne Colony for graciously providing early drafts of their geologic maps. Water content and isotope analyses were cheerfully performed by Sarah Kealy in ASU's Stable Isotope Laboratory. Special thanks to Mike Malin for lending his computer software and expertise. The manuscript benefited greatly from reviews by Kathy Cashman, Hank Westrich, and Dan Dzurisin. This reasearch was supported by National Science Foundation Grant EAR-8618365 and Arizona State University Grant CGIA-494653.

References

Anderson SW (1988) The development and distribution of lava textures at the Mount St. Helens dome (MS Thesis). Arizona State University, Tempe, 187 p

Anderson SW, Fink JH (1988) Crease structures as lava emplacement rate indicators (abs). EOS 69:1486

Anderson SW, Fink JH (1987) Modeling crease structures on silicic lava flows (abs). EOS 68:1545

Barnes I (1984) Volatiles of Mount St. Helens and their origins. J Volcanol Geotherm Res 22:133–146

Blake S (1989) Viscoplastic models of lava domes. IAVCEI Proc Volcanol 2:88–126

Cashman KV, Marsh BD (1986) Use of crystal size distributions to place constraints on magmatic crystallization models – Mount St. Helens as an example (abs). International Volcanological Congress, Auckland, 232

Cashman KV, Taggart JE (1983) Petrologic monitoring of the 1981 and 1982 eruptive products from Mount St. Helens. Science 221:1385–1387

Chadwick WW, Swanson DA, Iwatsubo EY, Heliker CC, Leighley TA (1983) Deformation monitoring at Mount St. Helens in 1981 and 1982. Science 221:1378–1380

Chadwick WW Jr, Archuleta RJ, Swanson DA (1988) The mechanics of ground deformation precursory to dome-building extrusions at Mount St. Helens 1981–1982. J Geophys Res 93:4351–4366

DeGraff JM, Aydin A (1987) Surface features of columnar joints and their significance to mechanics and direction of joint growth. Geol Soc America Bull 99:605–617

Eichelberger JC, Westrich HR (1984) Degassing of magma in an obsidian flow and inferred degassing at depth. In: Proceedings of Workshop XIX: Active tectonic and magmatic processes beneath Long Valley Caldera, CA, Volume 1: U.S. Geological Survey Open File Report 84-939:147–150

Eichelberger JC, Carrigan HR, Westrich HR, Price RH (1986) Non-explosive silicic volcanism. Nature 323:598–602

Fink JH (1979) Surface structures on obsidian flows (PhD dissertation). Stanford University, 164 p

Fink JH (1980a) Gravity instability in the Holocene Big and Little Glass Mountain rhyolitic obsidian flows, northern California. Tectonophysics 66:147–166

Fink JH (1980b) Surface folding and viscosity of rhyolite flows. Geology 8:250–254

Fink JH (1983) Structure and emplacement of a rhyolitic obsidian flow: Little Glass Mountain, Medicine Lake Highland, northern California. Geol Soc America Bull 94:362–380

Friedman I, Smith RL (1958) The deuterium content of water in some volcanic glasses. Geochem Cosmochem Acta 15:218–228

Friedman JD, Olhoeft GR, Johnson GR, Frank D (1981) Heat content and thermal energy of the June dacite dome in relation to total energy yield, May–October 1980. In: Lipman PW, Mullineaux DR (eds) The 1980 eruptions of Mount St. Helens, Washington, US Geological Survey Professional Paper 1250:557–568

Holcomb RT, Colony WE (1987) Large-scale maps of a growing lava dome, Mount St. Helens, Washington (abs), IUGG Assembly XIX abstracts, 2:417

Lofgren GE (1971) Spherulitic textures in glassy and crystalline rocks. J Geophys Res 76:5635–5648

Loney RA (1968) Flow structures and composition of the Southern Coulee, Mono Craters, California – A pumiceous rhyolite flow. Geol Soc America Memoir 116:415–440

Manley CR (1986) Textural studies of young rhyolite flows (MS Thesis). Arizona State University, Tempe, 121 p

Manley CR, Fink JH (1987) Internal textures of rhyolite flows as revealed by research drilling. Geology 15:549–552

Moore JG, Lipman PW, Swanson DA, Alpha TR (1981) Growth of lava domes in the crater, June 1980–January 1981. In: Lipman PW, Mullineaux DR (eds) The 1980 eruptions of Mount St. Helens, Washington: US Geological Survey Professional Paper 1250: 541–548

Murase T, McBirney AR (1973) Properties of some common igneous rocks and their melts at high temperatures. Geol Soc Amer Bull 84:3563–3592

Murase T, McBirney AR, Melson WG (1985) Viscosity of the dome of Mount St. Helens. J Volcanol Geotherm Res 24:193–204

Rutherford MJ, Sigurdsson H, Carey S, Davis A (1985) The May 18, 1980 eruption of Mount St. Helens, 1. Melt composition and experimental phase equilibria. J Geophys Res 90:2929–2947

Ryan MP, Sammis CG (1981) The glass transition in basalt. J Geophys Res 86:9519–9535

Sampson DE (1987) Textural heterogeneities and vent area structures in the 600 year old lavas of the Inyo volcanic chain, eastern California. In: Fink JH (ed) The emplacement of silicic domes and lava flows. Geol Soc America Special Paper 212:89–102

Swanson DA, Holcomb RT (1989) Regularities in growth of the Mount St. Helens dacite dome, 1980–1985 (this volume)

Swanson DA, Dzurisin D, Holcomb RT, Iwatsubo EY, Chadwick WW Jr, Casadevall TJ, Ewert JW, Heliker CC (1987) Growth of the lava dome at Mount St. Helens, Washington. In: Fink JH (ed) The emplacement of silicic domes and lava flows. Geol Soc America Special Paper 212:1–16

Taylor BE, Eichelberger JC, Westrich HR (1983) Hydrogen isotopic evidence of rhyolitic degassing during shallow intrusion and eruption. Nature 306:541–545

Taylor HP (1974) The application of oxygen and hydrogen isotope studies to problems of hydrothermal alteration and ore deposition. Economic Geology 69:843–883

Turcotte DL, Schubert G (1982) Geodynamics. John Wiley and Sons, New York, 450 p

Lava Domes Modeled as Brittle Shells that Enclose Pressurized Magma, with Application to Mount St. Helens

R. M. IVERSON

Abstract

Lava domes can be modeled mathematically as brittle shells that enclose pressurized magma. This chapter describes a static, brittle-shell model that is conceptually distinct from previous models of lava domes. The governing equations of the brittle-shell model embody several simplifying assumptions, none of which restricts the rheology of lava-dome constituents. The single morphologic assumption is that lava domes are axially symmetric. The most important mechanical assumption is that, in domes which are growing slowly and endogenously, stresses are in quasi-static equilibrium. This equilibrium may be disrupted by extrusions or explosions, which reflect transient adjustments that lead to a new equilibrium of the dome. The mechanical parameters included in the model are the thickness and tensile strength of the dome's outer shell and the unit weight and excess pressure of the enclosed magma and gas. These four parameters combine to form a single dimensionless number, D, which completely governs the equilibrium dome shape. The value of D is about one for the Mount St. Helens dome. Morphologic measurements on the Mount St. Helens dome show that its growth has been nearly self-similar since May 1981, and field observations show that failure of the dome's outer shell, accompanied by extrusions, has been an important growth process (Swanson and Holcomb, this Vol.). Theoretical predictions based on the brittle-shell model are consistent with both of these phenomena.

1 Introduction – The Conceptual Model

A lava dome is a steep-sided, rounded extrusion that forms a dome-shaped or bulbous mass of congealed lava above and around a volcanic vent (Bates and Jackson 1980). Quantifying and testing hypotheses about the physical processes that control the shape and growth of lava domes can improve understanding of eruptive mechanisms and hazards at volcanoes where domes exist.

This chapter describes formulation and testing of quantitative predictions based on the hypothesis that lava domes are essentially two-component systems in static or quasi-static mechanical equilibrium. One component of the system is an internal body of ductile magma, and the other component is an external shell or carapace of solid, brittle rock (Fig. 1). When forces in the two components of the system are balanced in static equilibrium, the system resides in a state of minimum free energy. Nonequilibrium states of the system, which occur during eruptive lava-dome growth, are transient phenomena that restore the system to a new state of equilibrium. Lava-dome growth consequently is regarded as a succession of static or quasi-static equilibrium states, which are punctuated by intermittent eruptions.

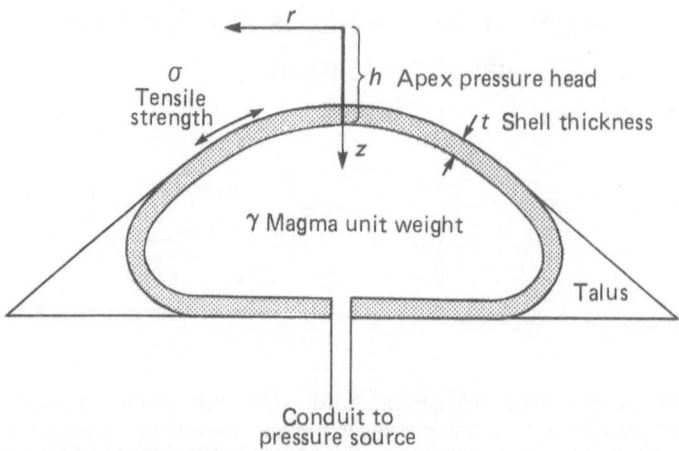

Fig. 1. Conceptual model of a lava dome viewed in cross-section. The parameters that play a role in the brittle-shell dome model are identified

Another component that might be considered part of the lava-dome system is a skirt or apron of talus around the dome (Fig. 1). Here the talus is assumed to be mechanically decoupled from the dome system. Thus, the talus can contribute to the external morphology of the dome, but it cannot contribute strength that affects the equilibrium of the dome. Although this assumption is not completely realistic, its use nevertheless appears warranted, owing both to its role in simplifying predictions and to the observation that talus surfaces generally slope at the angle of repose. Thus, any outward force that steepens the talus slopes will cause them to fail, and they will provide relatively little support for the growing dome carapace.

It also seems reasonable to assume that the dome's solid carapace and internal magma each are isothermal and homogeneous, and that the dome is axially symmetric. These assumptions are reasonable not because they are likely to apply exactly to lava domes, but because they lead to simple, unambiguous predictions that can be readily compared and contrasted to the behavior of lava domes.

Under the assumptions described above, the problem of predicting equilibrium dome configurations requires that only four physical parameters be considered. These are the tensile strength of the dome's solid carapace, σ; the thickness of the carapace, t; the unit weight of the fluid magma, γ; and the pressure head, $h(=\text{pressure}/\gamma)$ of the magma at the apex of the dome (Fig. 1). The analysis that follows this discussion shows that these four parameters combine to form a single dimensionless number, D, that governs equilibrium dome configurations.

None of the parameters that combine to form D embodies assumptions about rheology. The ductile magma within the dome can be linearly or nonlinearly viscous, and the solid rock carapace can be elastic or plastic, as long as the stress state departs insignificantly from equilibrium. Assumptions about rheology are unnecessary because the problem of finding equilibrium dome configurations is statically determinate.

Equilibrium is regarded as a critical limiting state, and its role might be clarified by a simple thought experiment. At equilibrium the dome carapace is uniformly stressed; its strength just suffices to contain the magma pressure exerted from within. In other words, if σ, t, and γ are, for the moment, assumed to be constants, then an increase in h will be accommodated either by a change in equilibrium dome shape or by a non-equilibrium response such as tensile failure of the dome carapace, which would be accompanied by an extrusion or explosion. An extrusion or explosion changes the magma pressure and/or the properties of the dome carapace, resulting in a new state of equilibrium. Alternatively, quasi-equilibrium dome growth can occur continuously and endogenously if, for example, the thickness and/or strength of the solid carapace increases with time. The *rate* at which endogenous dome growth or eruptions occur depends on the dome-rock rheology and the subterranean magma pressure and supply rate, which are analyzed in an accompanying chapter (Denlinger 1987, and this Vol.).

Before considering the brittle-shell analysis in detail, it is worthwhile to note that the rheology of lava-dome constituents probably varies more or less continuously with distance from the center of the dome. Near the center, the magma may exist at temperatures high enough to ensure fully fluid behavior; complete stress relaxation probably occurs in tens of seconds (cf. Goetze 1971), and the magma probably flows in response to any deviatoric stress. Toward the dome exterior, the dome rock becomes progressively cooler and more solid. At the exterior surface it has considerable rigidity, and it can support some deviatoric stress almost indefinitely. The brittle-shell model idealizes this continuously varying system by treating it as a two-layer system. This idealization, although imperfect, can provide insights to dome behavior that are not provided by models that idealize lava domes as homogeneous, single-phase fluids or solids (e.g., Huppert et al. 1982).

2 Analysis

The analysis described here follows and builds upon the analysis of drop-shaped fluid storage tanks described by Flügge (1967, p. 1–45), and the reader is referred to Flügge's exhaustive treatment for further details. Drop-shaped storage tanks provide a close mechanical analog to the brittle-shell concept of lava domes, because in each case the tensile strength of an exterior carapace or shell constrains the shape of the pressurized fluid mass within.

Referring to the curvilinear, orthogonal coordinate system imposed on the curved shell segment pictured in Fig. 2, consider the system of stresses and stress resultants (which are forces per unit length of the surface upon which they act) in the shell. Normal- and shear-stress resultants that act in the plane of the shell are designated by N, and stress resultants that act transverse to the

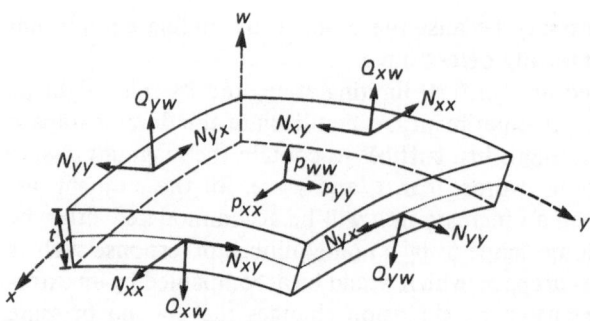

Fig. 2. A segment of
the curved shell of a
lava dome, showing the
coordinate system and
stress components

plane of the shell are designated by Q. The first subscript of each resultant in
Fig. 2 designates the coordinate direction of the normal to the plane upon
which it acts, and the second subscript designates its direction of action. In addition to the stress resultants N and Q, there is a stress, p, that acts through
the centroid of the shell. This stress is caused by internal magma pressure and
by the weight of the rock in the shell.

As described in detail by Flügge (1967), a remarkable simplification of the
stress system shown in Fig. 2 is possible if some tenable assumptions are made.
Assuming that no moment flexes the shell out of its equilibrium curvature, the
stress resultants N_{xx} and N_{yy} must vary slightly with w just so as to preserve
equilibrium. Similarly, assuming that no moment twists the shell, N_{yx} and N_{xy}
must also vary with w just so as to preserve equilibrium. These assumptions
appear reasonable, given that internal magma pressure is the only "external"
force available to flex or twist the shell. At equilibrium, internal magma
pressure varies linearly with depth and imposes no loads that distort the shell.
The assumption of insignificant bending and twisting moments therefore appears good, and it is reasonable to neglect such moments.

A further simplification of the stress system shown in Fig. 2 arises from
consideration of the moment equilibrium of surfaces in the x-y plane. First,
there is equilibrium of such surfaces with respect to torsion about the w-axis
only if $N_{xy} = N_{yx}$. Second, equilibrium of such surfaces with respect to torsion about the y-axis exists only if $Q_{xw} = 0$, and equilibrium with respect to
torsion about the x-axis exists only if $Q_{yw} = 0$. This means that only the
resultants designated by N and the stresses designated by p affect the balance
of forces in the shell, and the analysis reduces to that of "membrane" forces
(Flügge 1967).

In an analysis of membrane forces, the shell is replaced, in effect, by an in-
finitesimally thick membrane. Despite the fact that the shell has a finite
thickness, t (Figs. 1 and 2), it can be modeled as a membrane (Fig. 3) with little
loss in accuracy.

For convenience, now consider angular coordinates of the membrane,
designated ϕ in the meridional direction and θ in the parallel direction (Fig.
3). The radial coordinate of the membrane, r, varies as some unknown function
of ϕ and θ, and finding this function is the essence of the problem.

Fig. 3. The membrane
model of the dome
shell, showing the coor-
dinate system and stress
resultants. r_1 is the
radius of curvature in
the meridional direc-
tion; it is perpendicular
to the shell surface but
does not necessarily
terminate on the axis
of symmetry. r_2 is the
radius of curvature in
the parallel direction; it
is perpendicular to the
shell surface and ter-
minates on the axis of
symmetry

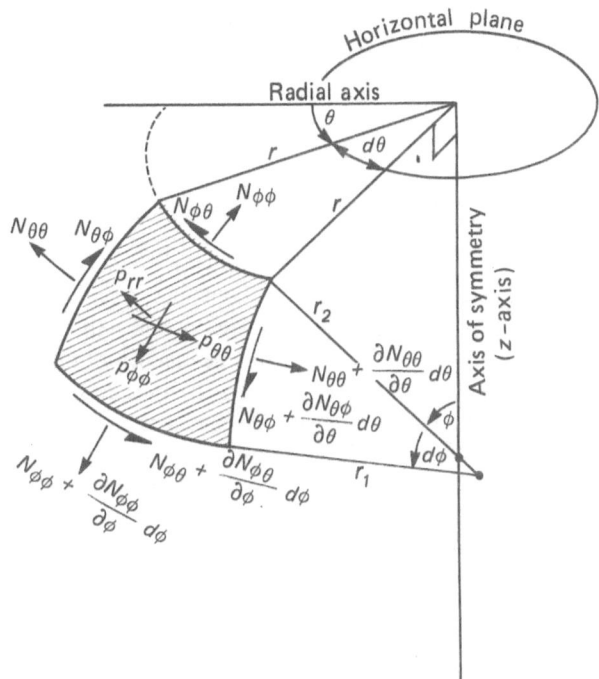

Balancing the force components in the membrane leads to force-
equilibrium equations for the ϕ, θ, and r directions. For the meridional (ϕ)
direction, equilibrium is expressed by

$$\frac{\partial}{\partial \phi}(rN_{\phi\phi}) + r_1 \frac{\partial N_{\theta\phi}}{\partial \theta} - r_1 N_{\theta\theta}\cos\phi + p_{\phi\phi}rr_1 = 0 ; \qquad (1\,\text{a})$$

and in the parallel (θ) direction, by

$$\frac{\partial}{\partial \phi}(rN_{\phi\theta}) + r_1 \frac{\partial N_{\theta\theta}}{\partial \theta} + r_1 N_{\phi\theta}\cos\phi + p_{\theta\theta}rr_1 = 0 ; \qquad (1\,\text{b})$$

and in the radial (r) direction, by

$$\frac{N_{\phi\phi}}{r_1} + \frac{N_{\theta\theta}}{r_2} = p_{rr} . \qquad (1\,\text{c})$$

For the somewhat lengthy derivation of Eqs. (1), the reader is referred to
Flügge (1967). Definitions of the variables in these equations are illustrated in
Fig. 3.

Because the dome is axially symmetric, all derivatives with respect to the parallel coordinate, θ, must vanish. Equations (1a) and (1b) thus reduce to

$$\frac{d}{d\phi}(rN_{\phi\phi}) - r_1 N_{\theta\theta} \cos \phi + p_{\phi\phi} r r_1 = 0 \; ; \tag{2a}$$

$$\frac{d}{d\phi}(rN_{\phi\theta}) + r_1 N_{\phi\theta} \cos \phi + p_{\theta\theta} r r_1 = 0 \; , \tag{2b}$$

where Eq. (2b) contains only the shear-stress resultant ($N_{\phi\theta}$) and is independent of Eqs. (2a) and (1c), which specify the normal-stress resultants ($N_{\phi\phi}$ and $N_{\theta\theta}$). Thus, Eq. (2b) is not needed to obtain the $r(\phi)$ or $\phi(r)$ solution.

The external load imposed on the membrane by the magma pressure acts normal to the membrane surface, because the liquid magma supports no shear stress. Moreover, the membrane approximation assumes that the body force caused by the weight of the rigid shell is negligible compared to the external (magma) load and its attendant reaction stresses. The stress p consequently has no component in the ϕ- or θ-directions, and is given simply by

$$p = p_{rr} = \gamma z \; , \tag{3}$$

where γ is the magma's unit weight, and z is a vertical coordinate with its origin ($z = 0$) at a height h above the apex of the dome. The coordinate z is reckoned positive downward (Fig. 1).

For an equilibrium shell of constant strength, σ, and thickness, t, the stress resultants $N_{\phi\phi}$ and $N_{\theta\theta}$ must by definition each equal σt. Using this fact and Eq. (3), Eq. (1c) reduces to

$$\sigma t (1/r_1 + 1/r_2) = \gamma z \; ; \tag{4}$$

and Eq. (2a) reduces to a simple geometric relation:

$$\sigma t (dr/d\phi - r_1 \cos \phi) = 0 \; . \tag{5}$$

Equations (4) and (5) contain only the variables ϕ, r, and z. However, they also contain the geometric parameters r_1 and r_2, which need to be eliminated before solutions can be obtained.

The parameter r_2 is eliminated by expressing it in terms of the geometry of Fig. 3:

$$r_2 = r/\sin \phi \; , \tag{6}$$

and r_1 is eliminated by expressing it in terms of Eq. (5):

$$\frac{1}{r_1} = (\cos \phi) \frac{d\phi}{dr} = \frac{d}{dr}(\sin \phi) \; . \tag{7}$$

Substituting Eqs. (6) and (7) into Eq. (4) yields a differential equation that governs the shape of a meridian on the dome surface:

$$\frac{d(\sin\phi)}{dr}+\frac{\sin\phi}{r}=\frac{\gamma z}{\sigma t} . \tag{8}$$

This equation, however, contains both ϕ and z as undetermined functions of r. Thus, another differential equation, which is deduced from the geometry of Fig. 3, provides closure to Eq. (8) by relating ϕ, z, and r along a meridian:

$$\tan\phi = dz/dr . \tag{9}$$

Simultaneous solution of Eqs. (8) and (9) therefore yields ϕ and z as functions of r along a meridian; this relation completely determines the shape of the dome. The nonlinear system of Eqs. (8) and (9) is not tractable analytically, but is readily solved by elementary numerical methods.

To maximize the applicability of numerical solutions, it is useful to normalize or "scale" Eqs. (8) and (9) so that all their variables are dimensionless. To normalize the equations three dimensionless variables are introduced:

$$\varrho = r/\sqrt{\frac{\sigma t}{\gamma}} ; \quad \xi = z/\sqrt{\frac{\sigma t}{\gamma}} ; \quad \eta = \sin\phi . \tag{10a,b,c}$$

Substituting Eqs. (10) into Eq. (8) then yields

$$\frac{d\eta}{d\varrho} = \xi - \frac{\eta}{\varrho} ; \tag{11}$$

and substituting Eqs. (10) into Eq. (9) yields

$$\frac{d\xi}{d\varrho} = \frac{\eta}{\sqrt{1-\eta^2}} , \tag{12}$$

so that there is a nonlinearly coupled pair of first-order, ordinary differential equations (11 and 12) to solve for the dependent variables η and ξ.

Note that the parameter $\sqrt{\sigma t/\gamma}$ appears prominently as the natural length scale in Eqs. (10), (11), and (12). Scaled against this reference length, points on the dome surface have the coordinates (ϱ,ξ), where ϱ is the normalized radial distance from the dome axis and ξ is the normalized vertical distance below an origin that lies a distance h above the dome apex (Fig. 1). The distance h also can be scaled relative to the reference length $\sqrt{\sigma t/\gamma}$, yielding the dimensionless number

$$D = \sqrt{\frac{\sigma t}{\gamma}}/h . \tag{13}$$

This dimensionless number incorporates all the physical parameters that affect the shape of the dome, and it therefore governs the mathematical solution completely. That is, the solutions of Eqs. (11) and (12) comprise an orderly family, the members of which are distinguished only by their value of D.

The boundary conditions used to solve Eqs. (11) and (12) are found at the dome apex, where $r = 0$ and, in terms of the normalized variables, $\varrho = 0$. The boundary conditions are

$$\eta(\varrho = 0) = 0 \; ; \quad \xi(\varrho = 0) = h / \sqrt{\frac{\sigma t}{\gamma}} = 1/D \; . \tag{14a, b}$$

The first of these conditions is derived from the requirement that the dome-surface slope must be zero at the dome apex, and the second is derived from the requirement that, by definition, $z = h$ at the apex (Fig. 1).

Solving the system (11, 12, 14a, b) numerically requires relatively simple but rather unique tactics. The uniqueness arises out of difficulty in beginning the computation stably and out of a necessity to switch computational algorithms depending on whether $\sin \phi$ or $\cos \phi$ is in the neighborhood of zero. For details of these tactics, the reader is referred to the discussion by Flügge (1967). It is easy to implement Flügge's suggestions computationally and to solve the system of equations by using a numerical integration procedure. A user-friendly BASIC program called "Halfdome" (Appendix 1) solves the equations using a Runge-Kutta algorithm (Kreyszig 1979, p. 797) and plots the cross-sectional shape of half of an axially symmetric lava dome.

3 Solutions

Figure 4 shows a series of solutions generated by using "Halfdome". The solutions illustrate how the equilibrium shape and relative size of the dome change as a function of D, and additional solutions show the same trend over a much wider range of D. The talus aprons shown on the solutions in Fig. 4 adjoin the dome surface arbitrarily at the point where the slope angle is 55°, but the talus aprons themselves slope at an arbitrary angle of 35°. These angles are adjustable parameters that do not affect the solution of the equilibrium equations in "Halfdome" (Appendix 1).

4 Discussion

The model results described above constitute an hypothesis aimed at explaining the morphology and style of growth of lava domes. Here this hypothesis is compared with other quantitative hypotheses for lava-dome growth and with data collected at Mount St. Helens.

Fig. 4. A family of lava-dome cross-sectional profiles computed using "Halfdome". Profiles are shown for different values of *D*. The *shaded* part of each profile represents the dome shell and interior, and the *unshaded* part represents the bordering talus apron. The talus apron adjoins the shell at an arbitrary angle of 55° but slopes at an angle of 35°

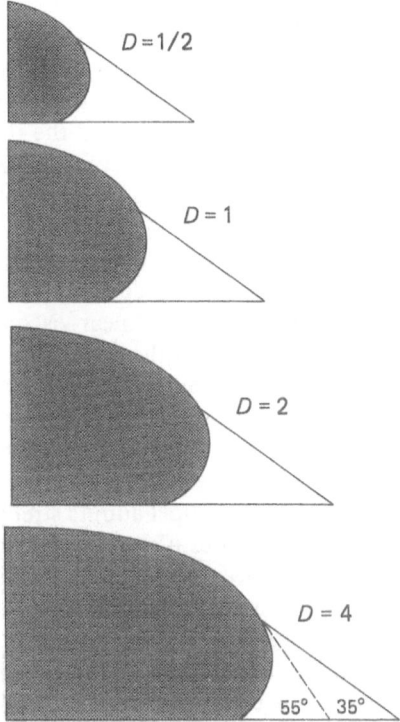

4.1 Comparison with Other Hypotheses

Two quantitative hypotheses for the mechanical controls of lava-dome growth and morphology have been advanced previously. The first of these, by Huppert et al. (1982), postulates that lava domes behave like a pile of linearly viscous fluid. Fluid is continuously injected through the base of the pile to maintain the rotund shape of the lava dome. Thus, the lava dome cannot achieve a state of static equilibrium. It must constantly expand volumetrically; otherwise it sags into a flatter and flatter viscous pile. Measurable sagging of this type occurs between eruptions on the Mount St. Helens dome (D. A. Swanson, U. S. Geological Survey, personal communication), but the dome does not continue to sag indefinitely.

Another important feature of the Huppert et al. (1982) linear-viscous model is that fluid pressures within the dome are hydrostatic; there can be no buildup of excess pressure that could potentially cause an explosion, and there is no means of generating localized extrusions.

The key parameter in the linear-viscous model is the fluid or lava viscosity, which determines the shape of the lava dome and its rate of spreading. In order to achieve reasonable fits to the observed lava-dome shape at the Soufriere of St. Vincent volcano, Huppert et al. (1982) needed to use a lava viscosity of

2×10^{12} poise, which is several orders of magnitude larger than typical measured lava viscosities. Huppert et al. attributed their discrepant theoretical viscosity to the presence of a high-viscosity "skin" they observed on the outer surface of the dome but could not simulate with their uniform-viscosity model. The skin constrained the flow of the magma within the dome to such an extent that it effectively dominated the dome behavior.

A second hypothesis for the controls of lava-dome growth and morphology has been advocated by Blake (1987, and this Vol.), who models a dome as a pile of Bingham (viscoplastic) material. This material has a finite shear strength as well as viscosity, and Blake's model consequently appears to be more realistic than the linear-viscous model. However, it uses two key concepts that are identical to those of the linear-viscous model: (1) that the dome is a pile of homogeneous material, with properties in the outer shell that are identical to those in the dome interior; and (2) that fluid pressure in the dome is hydrostatic. The Bingham model is appealing in its ability to represent static, equilibrium dome states (this is rendered possible by the yield strength in the model), but the model admits strength only in shear, and not in tension. Blake obtained estimates of lava-dome yield strengths by using an equilibrium relationship between dome height, radius, density, and yield strength that he derived from small-scale model experiments using kaolin slurries. The relationship shows that the dome height is always proportional to the square root of its radius if the dome is composed of Bingham material. Such domes must therefore become relatively flatter as they enlarge. Moreover, a plot of dome height as a function of the square root of its radius must be a straight line that intersects the origin. The Bingham hypothesis consequently can be tested with height-radius data obtained during growth of a dome (cf. Fig. 9).

The major conceptual contrast between the model proposed here and the viscous and viscoplastic models is that the brittle-shell model treats the outer, cooled carapace of a lava dome as a material that is distinct from the ductile magma within the dome. Because the dome carapace is assumed to be under uniform tension, a magma (or gas) pressure in excess of hydrostatic pressure exists within the dome. This excess pressure balances the tensile stress in the carapace. Higher excess pressures require greater curvature, strength, or thickness of the carapace in order to balance the stresses. Consequently, if all other factors are constant, domes that are more nearly spherical can contain greater excess pressures and thus would be more prone to explosive failure than would relatively flatter domes. Quiescent failures (extrusions) would be more likely in flatter domes that have less surface curvature. Thus, the brittle-shell model is fundamentally different from previous models: it supposes a distinct mechanical behavior of the carapace, and it provides an explicit mechanism (tensile failure of the carapace) for eruptive dome growth, including both explosions and extrusions.

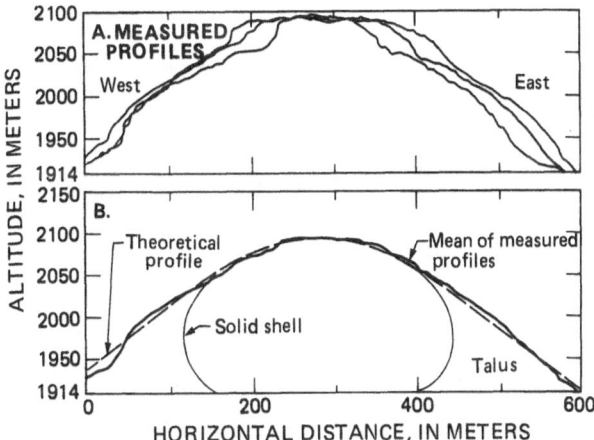

Fig. 5 A, B. Comparison of computed and measured topographic profiles of the Mount St. Helens lava dome. *A* Three profiles measured from a photogrammetric topographic map made from air photos taken on November 4, 1981. The three profiles are all roughly east-west, having azimuths of 73°, 91°, and 116° from due north, and each passes through the apex of the dome. The location of the upper margin of the talus apron is very difficult or impossible to distinguish in the field and is not useful as a modeling constraint. *B* The mean of the three measured profiles compared with a profile computed using "Halfdome". A *D* value of 1 and talus slopes of 36° were used in the computation

4.2 Comparison with Morphologic Data from Mount St. Helens

One means of testing model predictions with field data is to compare calculated dome profiles with those measured at Mount St. Helens (e.g., Swanson et al. 1987; Swanson and Holcomb 1987, and this Vol.). Such comparisons show excellent correspondence between predicted and measured cross-sectional profiles, particularly if the value of the single model parameter, *D*, is selected to maximize the goodness of fit (e.g., Fig. 5). However, it is important that the value of *D* is not selected through an exercise in unrestrained curve fitting. Table 1 shows that plausible values of *D* for the Mount St. Helens dome are of the order of one. The use of a *D* value of one to fit the curve of Fig. 5 is therefore physically reasonable.

As shown in Fig. 5, the correspondence between predicted and measured dome profiles is improved if several of the irregular, measured profiles are averaged to obtain a mean profile. For Mount St. Helens these mean profiles are best obtained by averaging several east-west profiles, because north-south profiles are asymmetrical owing to a pronounced south to north slope of the crater floor.

The close agreement between predicted and measured topographic profiles is perhaps surprising, in light of the complexity of dome growth as documented by Swanson et al. (1987). This documentation shows that the dome has grown partly by endogenous growth and partly by localized extrusions. Local extrusions impose local loads on the dome surface, which might be expected

Table 1. Plausible values of *D* and its constituent parameters for the Mount St. Helens dome

σ (Pa)[a]	t (m)[b]	γ (N/m³)[c]	h (m)	D
1×10^7	10	26000	62	1
1×10^7	20	26000	44	2
1×10^7	30	26000	27	4
1×10^6	10	26000	20	1
1×10^6	20	26000	14	2
1×10^6	30	26000	11	3
1×10^6	10	26000	39	0.5
1×10^6	20	26000	55	0.5
1×10^6	30	26000	68	0.5

[a] Estimates of the tensile strength of dome-carapace rock are based on typical crustal rock tensile strengths tabulated by Jaeger and Cook (1979, p. 190 – 191). Such tensile strengths vary by about one order of magnitude.
[b] Estimates of the dome carapace thickness are based on preliminary analyses of magnetic data for the Mount St. Helens dome (personal communication, D. Dzurisin and R. P. Denlinger, U.S. Geological Survey).
[c] The estimate of dome-rock unit weight is based on hundreds of measurements of Mount St. Helens dacite erupted in 1980 (personal communication, R. P. Hoblitt, U.S. Geological Survey).

to cause asymmetrical dome growth. However, the rock added to the dome surface during extrusions probably has about the same density as the magma inside the dome, and if the dome shell acts as a membrane, it should comply to the new load over a period of time so as to preserve the equilibrium dome configuration. Furthermore, the thickened dome shell that results from an extrusion will strengthen that part of the dome so that subsequent extrusions will tend to occur elsewhere. (This type of behavior has been documented at Mount St. Helens.) Thus, in an average sense, the equilibrium shape of the dome can be maintained even under the influence of local extrusions.

4.3 Comparison with Dome-Growth Data from Mount St. Helens

The brittle-shell model does not include explicit time dependence, and consequently it cannot predict rates of dome growth without independent knowledge of the rate of change of the parameters that compose *D*. Similar independent knowledge (of the rate of magma entry, for example) is necessary for *any* model to predict rates of dome growth. However, even without this independent knowledge, the brittle-shell model *can* predict how the dome shape and volume change during growth. That is, it can predict scenarios for how the ratios of dome height to volume, diameter to volume, and height to diameter change as the dome grows. These predictions can be compared with detailed data collected over a 6-year period at Mount St. Helens.

A complication in making theoretical predictions of height-volume and similar relations is that it is not known a priori which physical parameters change and which, if any, stay nearly the same during dome growth. That is, among the parameters that compose D (i.e., σ, t, γ, and h), one or more must change for dome growth to occur. For fixed values of these parameters, the dome can exist in only *one* equilibrium size and shape.

The simplest scenario for dome growth is that it occurs through compensating increases in σ, t, γ, and h, which cause D to remain roughly constant. Dome growth then occurs as self-similar expansion, and the dome shape remains roughly the same. During this self-similar growth, t and h are perhaps the parameters most likely to change, whereas σ and γ are more likely to remain almost constant.

Swanson and Holcomb (1987, and this Vol.) report that growth of the dacite dome at Mount St. Helens was approximately self-similar after May 1981. They measured dome heights and diameters on a series of detailed topographic maps and calculated volumes from digitized versions of the maps. The base of the dome was fixed at 1914 m elevation for all measurements and calculations. Swanson and Holcomb's best-fit, nonlinear regression equation for the height-volume data is

$$V = 657\,700\, e^{0.019H} \quad (r^2 = 0.97) \; , \tag{15}$$

where V is the volume of the dome measured in cubic meters, and H is the height of the dome measured in meters. Their best-fit, nonlinear regression for the diameter-volume data is

$$V = 0.003\, d^{3.494} \quad (r^2 = 0.99) \; , \tag{16}$$

where d is the diameter of the dome base measured in meters. Plots of these equations, along with the field data and theoretical predictions, are shown in Figs. 6 and 7.

Swanson and Holcomb (this Vol.) describe the errors implicit in the data plotted in Figs. 6 and 7. The errors in computed dome volumes arise principally from imperfect digital representation of the irregular dome surface, and may be as large as 10%. These errors must be borne in mind when deciding how many refinements of theory are justified in pursuing a good fit to the data.

The theoretical predictions of dome height-volume and diameter-volume relations shown in Figs. 6 and 7 are based on several scenarios for self-similar growth. Different scenarios employ different values of D and/or different assumptions about the extent of the talus apron that mantles the lower parts of the dome. For each scenario, however, the basic strategies used to calculate the theoretical dome volume are the same.

Two methods exist for calculating the theoretical volume enclosed by the brittle dome carapace. The first method relies on "Halfdome" and uses numerical integration to compute the volume above any selected horizontal plane that transects the dome (Appendix 1). The second method is useful only

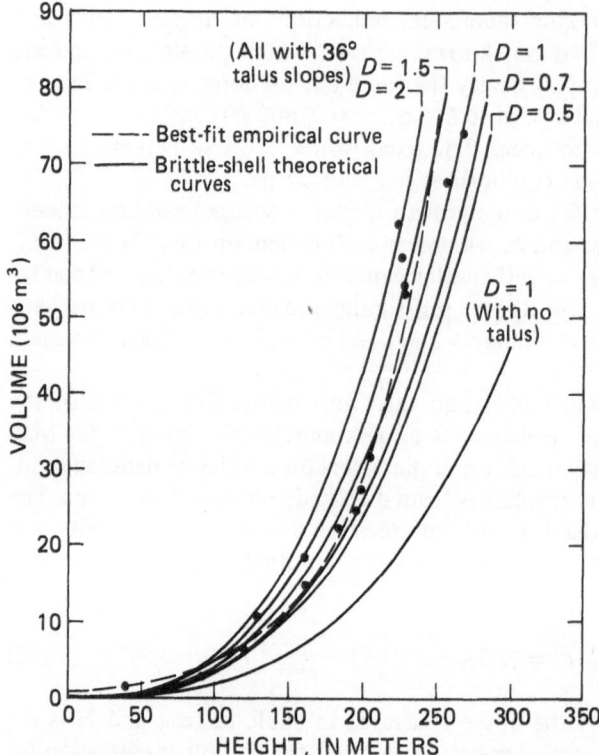

Fig. 6. Lava dome volume as a function of height. Data points are from measurements of the Mount St. Helens dome by Swanson and Holcomb (this Vol.), and the *dashed line* is the empirical, exponential curve that best fits the data. The *solid lines* represent theoretical results for five self-similar dome growth scenarios. Swanson and Holcomb (this Vol.) report that dome growth was approximately self-similar after May 1981, when the dome was about 130 m high. The theoretical curves were obtained from numerical simulations using "Halfdome"

for determining the *total* volume enclosed by the carapace. This method exploits the balance of forces that act on the base of the carapace, which has a radius R_2 (Fig. 8). The magma pressure on the base is equal to $(h+H)\gamma$, where h is the pressure head at the apex of the dome and H is the height of the apex above the base (Fig. 8). Multiplying this magma pressure by the area of the base, πR_2^2, gives an expression for the total vertical force on the base. An additional expression for the total vertical force is given by the product of the dome volume, V, and the unit weight of the dome, γ. Equating these two expressions for the vertical force and dividing each by γ gives an equation by which to find the dome volume enclosed by the brittle carapace:

$$V = \pi R_2^2 (H+h) \ . \tag{17}$$

This equation, which fortuitously has the same, simple form as the equation for a right, circular cylinder, provides a means for checking the accuracy of the numerical calculation of the dome's volume. However, if a talus apron is to be included in the calculation of the total dome volume, Eq. (17) serves no useful purpose.

To include bordering talus in the volume calculation, the volume of the part of the dome above the talus apron and having thickness H_1 (Fig. 8) is ob-

Fig. 7. Lava dome volume as a function of the base diameter. Data points are from measurements of the Mount St. Helens dome by Swanson and Holcomb (this Vol.), and the *dashed line* is the empirical, power-law curve that best fits the data. The *solid lines* represent theoretical results for five self-similar dome growth scenarios. (Two scenarios produce nearly identical curves.) The theoretical curves were obtained from numerical simulations using "Halfdome"

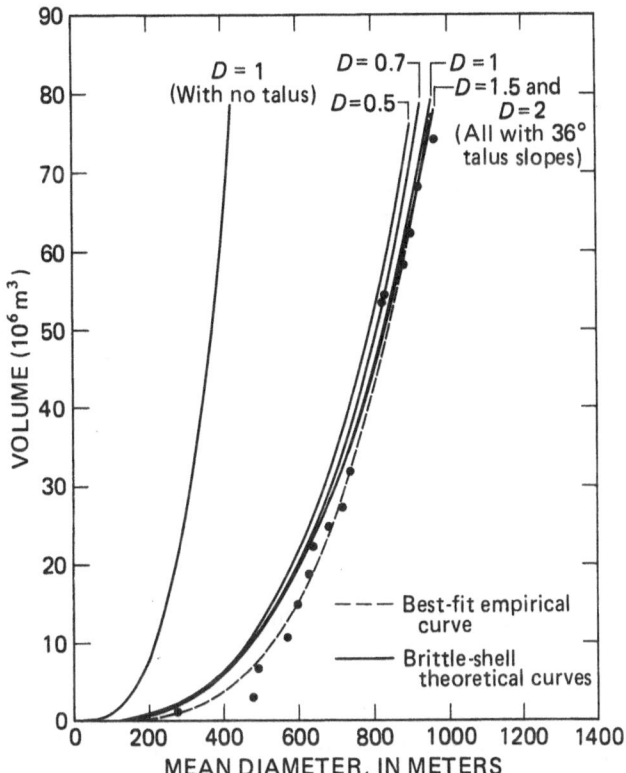

Fig. 8. Definition sketch of a cross-section of half a lava dome, showing the geometric parameters used in computing theoretical dome volumes

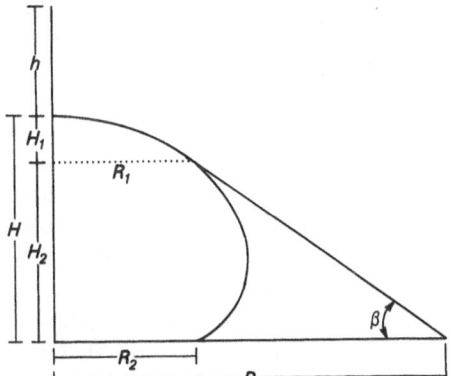

tained from numerical integration using "Halfdome." Designate this volume V_1. Below H_1 the volume of the dome is approximated as that of a right conic frustrum. The sloping surface of the conic frustrum is the surface of the talus apron, and the volume of the frustrum is given by (Tuma 1979, p. 27)

$$V_2 = \frac{\pi H_2}{3} (R_1{}^2 + RR_1 + R^2) \ ,$$ (18)

where V_2 is the volume of the part of the dome below the upper margin of the talus, H_2 is the height of the talus apron above the dome base (i.e., the height of the frustrum), R_1 is the radius of the dome at height H_2, and R is the radius of the dome base (Fig. 8). If the surface of the talus apron slopes at a uniform angle β, then H_2, R, and R_1 are related by the simple geometric equation

$$H_2 = (R - R_1) \tan \beta \ .$$ (19)

The total volume of the dome plus talus apron is found by adding Eq. (18) to V_1, making use of Eq. (19) if desired:

$$V = V_1 + V_2 = V_1 + \frac{\pi (R - R_1) \tan \beta}{3} (R_1{}^2 + RR_1 + R^2) \ .$$ (20)

Inserting appropriate values for the geometric parameters in Eq. (20) allows theoretical predictions of dome height-volume and diameter-volume relations to be compared with field data. Each point on the theoretical curves of Figs. 6 and 7 was obtained by extracting the necessary values from numerical simulation results generated using "Halfdome". For the curves that include talus aprons in the computed dome volumes, the angle of the talus slopes was fixed at 36°. On the basis of several measurements made on each of seven topographic maps of the Mount St. Helens dome (R. T. Holcomb and D. A. Swanson, U. S. Geological Survey, unpublished data), 36° ± 5° is a reasonable estimate of the mean talus slope angle.

Figures 6 and 7 show that the best match of the theoretical curves to the field data is generated if the D value is about one and the 36° talus apron is included in the dome volume. This scenario does not provide the best match to both the volume-height and volume-diameter data, but it provides the best comproprise in simultanously fitting both types of data. The deviation of the best-fit theoretical curves from most of the field data is negligible if the possibility of 10% error in the data is taken into account.

For dome diameters less than about 750 m, however, the deviation of the theoretical curves from the diameter-volume data is systematic and significant (Fig. 7). In such instances the theory overestimates the volume of the dome. This systematic deviation reflects at least two phenomena. First, during early stages of growth of the Mount St. Helens dome, the dome was somewhat more asymmetrical than during later stages (Swanson and Holcomb, this Vol.). The theory uses the mean dome diameter as the basis for volume calculations, and consequently it systematically overestimates the volume when the dome is significantly asymmetrical. Second, the dome did not grow in a self-similar fashion until its mean diameter reached nearly 600 m (Swanson and Holcomb,

Fig. 9. Comparison of
Bingham models and
brittle-shell models for
predicting dome height
as a function of the
square root of radius.
Data points are the
same as those plotted
in Figs. 6 and 7

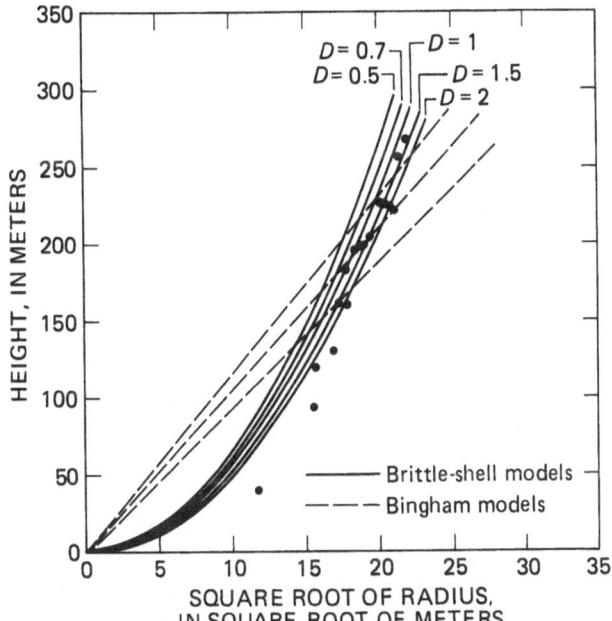

this Vol.). Thus, the self-similar growth curves cannot be expected to predict the data for smaller domes with great accuracy.

Another means of assessing self-similar growth predictions entails plotting dome height as a function of the square root of its radius. This permits a straightforward comparison of the Bingham and brittle-shell hypotheses. Predictions based on the brittle-shell theory are plotted on Fig. 9, along with the pertinent data. The Bingham-model prediction described by Blake (1987, and this Vol.) plots as a straight line that passes through the origin of Fig. 9. The slope of the line is determined by the yield strength and density of the dome's lava. Neither the Bingham nor the brittle-shell theory fits all the data, but the inaccuracy of the brittle-shell theory is negligible except for cases in which the dome was less than 150 m in height. For these cases growth of the dome was not self-similar.

Some alternative hypotheses for lava-dome growth make use of the brittle-shell model but not of the self-similar growth concept. Instead, such hypotheses suppose that D changes during dome growth. A scenario of this type that seems quite plausible is one in which the thickness of the dome carapace, t, increases during growth, while the other parameters that contribute to D remain more or less constant. In this case the basic dome-shape equation (17) can be scaled against the value of h, which is assumed constant, and Eq. (17) can be reduced to a dimensionless form:

$$\frac{V}{h^3} = \pi \frac{[R_2(D)]^2}{h^2} [H(D)/h + 1] \ . \tag{21}$$

Here, R_2 and H are both functions of D that need to be obtained from numerical solutions using "Halfdome". The disadvantage of this type of hypothesis is that it involves more than one free parameter. That is, it allows the dome shape to change as the dome enlarges, with no independent constraints imposed by physical data. Thus, in the absence of additional information, matching the theory to the data becomes a relatively unrestrained exercise in curve-fitting. It therefore yields a rather small increment of physical insight, and the topic is not pursued here.

5 Conclusions and Unresolved Questions

This chapter describes a brittle-shell model for lava domes that is conceptually simple, quantitatively explicit, and computationally economical. The model results constitute a testable hypothesis for the mechanical controls of lava-dome morphology and growth.

The brittle-shell model differs significantly from previous mechanical models of lava domes. The primary difference is the presumption that the solid rock composing the outer carapace of the dome has mechanical properties that differ markedly from those of the molten rock within the dome. The important properties of the outer carapace are its tensile strength and thickness, whereas the important properties of the molten interior are its pressurization and weight. A single dimensionless number, D, represents the combined effect of these four properties, and its value determines the shape of the dome. Talus slopes that mantle the lower parts of the dome are assumed to contribute only cosmetically, and not mechanically, to its shape.

An important feature of the brittle-shell model is that it provides conceptually for a means of eruptive lava-dome growth. The dome can grow through tensile failure of its outer shell. Alternatively, endogenous dome growth can occur if changes in the dome-shell thickness or strength keep pace with pressurization in the feeding conduit, which causes intrusion of new magma into the dome interior.

The brittle-shell model also adequately matches morphologic data obtained for the growing Mount St. Helens dome. Plots of dome volume as a function of its diameter and height and of height as a function of radius reflect the shape of the dome and how it might change during growth. The match between theoretical predictions and the data is best if a scenario of self-similar-dome growth is assumed with D always equal to about one. A D value of the order of one is commensurable with estimates of the strength, shell thickness, unit weight, and pressurization of the Mount St. Helens dome. Moreover, a D value of one produces a good match of theoretical and measured topographic profiles for the Mount St. Helens dome.

A more rigorous test of the brittle-shell hypothesis would require independent measurement of each of the parameters that contribute to D. The most poorly constrained parameters appear to be the thickness of the solid dome carapace and the pressure of the fluid phase within the dome. Future research that included drilling into the dome's interior therefore could provide valuable data to test the applicability of the brittle-shell hypothesis.

Other future research could be directed toward making the brittle-shell model more realistic by allowing for tensile stress and strength heterogeneities in the mathematical formulation. Moreover, research on the rate-dependent processes by which the dome shell may fracture and grow expands the ramifications of the model and addresses the processes that act during disequilibrium states. Such research is described by Denlinger (this Vol.).

Acknowledgments. I thank the many CVO staff members, and particularly Don Swanson, who collected the extraordinary data set on the Mount St. Helens dome. Chuyler Freeman assisted with construction of topographic profiles, and Rick LaHusen provided advice on making "Halfdome" user-friendly. Don Swanson, Roger Denlinger, and David Pollard provided thoughtful critiques of the manuscript.

Appendix 1: Computer Program Halfdome

The computer program "Halfdome" takes about a minute to compute and plot a solution of the system (11, 12, 14a, b) if an IBM XT[1]-type personal computer with CGA graphics is used and the program is accessed through the hard disk. An option when using "Halfdome" is to include a bordering talus apron adjacent to the side of the dome. The presence of the talus apron does not affect the procedure for solving the equations; the talus only contributes cosmetically to the shape of the dome.

A file output by "Halfdome", which has a default name of "Dome.dat", lists the cumulative volume computed as the numerical integration proceeds from the apex to the base of the dome carapace. The partial dome volume lying above a horizontal plane that transects the carapace at an arbitrary elevation may therefore be read directly from "Dome.dat". The final entry in "Dome.dat" is the total volume enclosed by the dome carapace.

```
10  REM************************************************************************
20  REM                        HALFDOME.Bas
30  REM                     (R.M. Iverson 7/6/87)
40  REM************************************************************************
50  REM This program computes the profile of half of a lava dome by numerical
60  REM solution of two simultaneous, nonlinear, ordinary differential
65  REM equations that have their basis in the following physical assumptions:
70  REM  1. At any moment in time, the dome is in static mechanical equilibrium.
80  REM  2. The dome consists of a pressurized magma body enclosed by a carapace
90  REM     of solid rock. The carapace is assumed to be of uniform thickness
100 REM     and tensile strength. The carapce, in turn, is surrounded along its
110 REM     lower margins by an apron of talus that is assumed to be in plastic
120 REM     equilibrium.  Thus the talus contributes no strength to help support
130 REM     the dome.
140 REM  3. The physical parameters that determine the shape of the dome are
150 REM     the magma pressure (expressed as head) at the apex of the dome,
```

---

[1] The use of brand and trade names in this chapter is for identification purposes only and does not constitute endorsement by the U.S. Geological Survey.

```
160 REM     the magma density (expressed as unit weight), which is assumed to be
170 REM     constant, the tensile strength of the solid rock carapace,
175 REM     and the thickness of the carapace.  These parameters are combined
180 REM     into one dimensionless parameter, D, that characterizes the forces
190 REM     that determine the shape of the dome.  D is the only free parameter
200 REM     that affects the numerical solution of the governing equations.
210 REM*****************************************************************************
215 REM
220 REM To run the program, enter the BASIC interpreter program, type load
230 REM "Halfdome", and then type run.  The program will first prompt for the
240 REM name of a file into which numeric output data are placed, and then will
245 REM prompt for the desired value of D, for the desired angle of the
250 REM bordering talus slope, and for a scale factor that determines the size
260 REM of the graphic output on the monitor screen.
270 REM
280 REM*****************************************************************************
290 REM
300 REM DEFINITION OF VARIABLES:
310 REM D - A dimensionless number that incorporates the effects of magma pres-
320 REM     sure and density, and of dome-carapace thickness and tensile strength.
325 REM Rho - A dimensionless horizontal coordinate of a point on the dome
330 REM     surface.  The origin for Rho is at the axis of the dome.
340 REM Xi - A dimensionless vertical coordinate of a point on the dome surface.
345 REM     The origin for Xi is at a distance h above the dome apex, where
350 REM     h is the pressure head at the apex.
360 REM Eta - A number equal to the sine of the dome-surface slope.
370 REM Mu - A number equal to the cosine of the dome-surface slope.
380 REM COUNT - A counter to keep track of which of three segments of the
390 REM     dome profile is being calculated.
400 REM N and NN - Iterative indices for loop procedures.
410 REM STP - Size of the space step used in the numerical solution
420 REM SCALE - A number which gives the linear scale of the solution plotted
430 REM     on the graphics monitor.  The number is relative only and has no
440 REM     particular physical meaning or importance.
450 REM RHOPLOT - Value of rho scaled appropriately for graphic plotting.
460 REM XIPLOT - Value of XI scaled appropriately for graphic plotting.
470 REM OLDRHO - Value of Rho saved from previous numerical step
480 REM OLDXI - Value of XI saved from previous numerical step
490 REM ORHOPLO - Value of OLDRHO scaled for graphics use
500 REM OXIPLO - Value of OLDXI scaled for graphics use
510 REM TALUSLOPE - User-specified slope angle (in degrees) of the talus apron
520 REM     surrounding the base of the dome
530 REM RHOTALUS - Value of RHO where the talus apron adjoins the dome
540 REM XITALUS - Value of XI where the talus apron adjoins the dome
550 REM RHOTALUSPLO - Value of RHOTALUS scaled for graphics use
560 REM XITALUSPLO - Value of XITALUS scaled for graphics use
570 REM TALUSADJOIN - Slope angle (in degrees) at the spot on the dome where
580 REM     the adjoining talus lope contacts the dome.  If no talus slope is
590 REM     specified, then this parameter defines the slope of the dome at the
600 REM     point where the dome contacts the horizontal dome base.
610 REM TALUSAPRON - If this parameter = 0, then no talus slope is specified
620 REM     to apron the dome.  If this parameter = 1, then a talus apron exists
630 REM TESTANGLE - This angle is used to test the calculated dome slope
640 REM     against the angle specified for TALUSADJOIN
650 REM VOLSUM - Cumulative sum of the dome volume above current Xi
660 REM*****************************************************************************
670 REM
680 REM****** BEGIN EXECUTABLE PART OF PROGRAM**********************************
685 REM
690 REM*****************************************************************************
700 REM PROMPT USER FOR NAME OF NUMERIC OUTPUT FILE
710     CLS:PRINT:PRINT:PRINT"DOME PROFILE PROGRAM"
720     PRINT:INPUT"Enter filename for numeric output (Default = Dome.dat)";NF$
730     IF NF$="" THEN NF$="DOME.DAT"
740     OPEN NF$ FOR OUTPUT AS 1
750 REM SPECIFY N AND C AS INTEGER-VARIABLE MARKERS
760     DEFINT N,C
770 REM PROMPT FOR THE VALUE OF THE PARAMETER D (=a/h)
780     PRINT:PRINT "The D parameter should be0.5 < D <20"
790     INPUT "Enter the value of the parameter D";D
800 REM PROMPT FOR USER SELECTION (YES OR NO) OF TALUS APRON AROUND DOME
810     PRINT " "
820     PRINT "Is a talus apron around the dome desired? Enter 0 (zero) if
830     INPUT "no apron is desired and 1 if it is desired"; TALUSAPRON
```

```
840 REM PROMPT FOR USER SELECTION OF TALUSADJOIN ANGLE
850     PRINT:PRINT "If no talus apron was specified, now enter as TALUSADJOIN"
860     PRINT "the angle of the dome surface where it contacts the dome base."
870     PRINT "Otherwise, TALUSADJOIN specifies the slope angle (in degrees)"
880     PRINT "of the point on the dome where talus adjoins the dome surface."
890     INPUT "Enter TALUSADJOIN value in degrees (between 15&90)";TALUSADJOIN
900 REM PROMPT USER FOR THE ANGLE OF THE TALUS SLOPE, IF SLOPE IS SPECIFIED
910     PRINT:PRINT "The TALUSLOPE parameter specifies the angle of the talus"
920     PRINT "apron adjoining the dome, ranging from 15 to 90 degrees"
930     INPUT "Enter TALUSLOPE value in degrees (Default is 35)";TALUSLOPE
940 REM SET DEFAULT VALUES OF PARAMETERS
950     IF TALUSLOPE=0 THEN TALUSLOPE=35
960     IF TALUSADJOIN=0, THEN TALUSADJOIN=35
970     TALUSLOPE=TALUSLOPE*.0174533
980     TALUSADJOIN=TALUSADJOIN*.0174533
990     IF D=0 THEN D=4
1000 REM SPECIFY THE SIZE OF THE SPACE STEP FOR THE NUMERICAL SOLUTION
1010    STP=.01
1020 REM PROMPT FOR THE "SCALE" FACTOR TO DETERMINE THE SIZE OF THE PLOT
1030    PRINT:PRINT "The scale of the plotted ouput is set by specifying SCALE"
1040    PRINT "The default value of SCALE is 50"
1050    INPUT "Enter the value of SCALE"; SCALE
1060    IF SCALE=0, THEN SCALE=50
1070 REM FIX VALUES OF Rho, Eta, Xi, AND Volsum AT THE APEX OF THE DOME
1080    RHO=0
1090    ETA=0
1100    VOLSUM=0
1110    XI=1/D
1120 REM PRINT HEADINGS AND THE DATA VALUES FOR THE DOME APEX TO OUTPUT FILE
1130    PRINT#1, "Rho          Xi            Eta          Volume Sum"
1140    PRINT #1, RHO,XI,ETA,VOLSUM
1150 REM CLEAR THE MONITOR AND PREPARE FOR GRAPHICS OUTPUT
1160    CLS
1170    SCREEN 2
1180    KEY OFF
1190 REM DRAW A VERTICAL LINE OF LENGTH h/a (=1/D) ATOP THE DOME APEX
1200    RHOPLOT=0
1210    XIPLOT=XI*SCALE+20
1220    LINE(0,20)-(RHOPLOT,XIPLOT)
1230 REM RETAIN THE CURRENT RHO VALUE FOR FUTURE USE
1240    OLDRHO=RHO
1250 REM****************************************************************
1260 REM
1270 REM*********** BEGIN COMPUTATION OF DOME PROFILE *****************
1280 REM (THIS ROUTINE IS USED TWICE TO COMPUTE THE FLATTER PARTS OF THE
1290 REM DOME PROFILE.  Rho IS THE INDEPENDENT VARIABLE IN THIS CALCULATION)
1300 REM****************************************************************
1310 REM
1320    COUNT=0
1330 REM ANALYTICALLY CALCULATE THE VALUE OF Eta ONE STEP FROM THE ORIGIN
1340    RHO=RHO+STP
1350    ETA=.5 * (1/D) * RHO
1360 REM BEGIN LOOP TO CALCULATE THE REST OF THE DOME PROFILE, STEPPING
1370 REM OUTWARD AND DOWNWARD FROM THE APEX OF THE DOME
1380    FOR N = 1 TO 500
1390    IF COUNT=1,THEN IF N=1, THEN GOTO 1530
1400    OLDXI=XI
1410    IF COUNT=0, THEN XI=XI+STP*(ETA/SQR(1-ETA*ETA))
1420    IF COUNT=1, THEN XI=XI-STP*(ETA/SQR(1-ETA*ETA))
1430    VOLSUM=VOLSUM + (XI-OLDXI)*3.14159*((RHO+OLDRHO)/2)^2
1440 REM PRINT CURRENT VALUES OF RHO, XI, ETA, AND VOLSUM TO OUTPUT FILE
1450    PRINT #1, RHO,XI,ETA,VOLSUM
1460 REM PLOT A LINE CONNECTING TWO ADJACENT POINTS ON THE DOME PROFILE
1470    ORHOPLO=2.02*OLDRHO*SCALE
1480    OXIPLO=OLDXI*SCALE+20
1490    RHOPLOT=2.02*RHO*SCALE
1500    XIPLOT=XI*SCALE+20
1510    LINE(ORHOPLO,OXIPLO)-(RHOPLOT,XIPLOT)
1520 REM USE RUNGE-KUTTA ALGORITHM TO CALCULATE Eta FOR THE NEXT STEP
1530    AN=STP*(XI-ETA/RHO)
1540    BN=STP*(XI-(ETA+AN/2)/(RHO+STP/2))
1550    CN=STP*(XI-(ETA+BN/2)/(RHO+STP/2))
1560    DN=STP*(XI-(ETA+CN)/(RHO+STP))
1570    ETA=ETA+(1/6)*(AN+2*(BN+CN)+DN)
```

```
1580 REM STORE AND INCREMENT THE VALUE OF Rho
1590     OLDRHO=RHO
1600     RHO=RHO+STP
1610 REM TEST SLOPE ANGLE FOR MATCH WITH TALUSLOPE ANGLE
1620     TESTANGLE = ETA - SIN(TALUSADJOIN)
1630     IF COUNT=0, THEN IF ETA>.25, THEN IF ABS(TESTANGLE)<=.01,THEN GOTO 1650
1640     GOTO 1700
1650     RHOTALUS=RHO
1660     XITALUS=XI
1670 REM IF THERE IS NO TALUS APRON AND THE BASE SLOPE IS REACHED, GOTO END
1680     IF TALUSAPRON=0, THEN GOTO 2370
1690 REM TEST FOR 50-DEGREE SLOPE TO ENACT ALGORITHM SWITCH
1700     IF ETA>=.776, THEN GOTO 1830
1710     IF COUNT=1, THEN IF ETA<.01, THEN COUNT=2
1720     IF COUNT=2, THEN GOTO 2370
1730     NEXT N
1740     IF COUNT=2 GOTO 1080
1750     IF N=500, THEN STOP
1760 REM*******************************************************************
1770 REM
1780 REM BEGIN CALCULATION OF THE STEEP PART OF THE DOME PROFILE,
1790 REM SWITCHING TO AN ALGORITHM USING Xi AS THE INDEPENDENT VARIABLE
1800 REM AND INTRODUCING Mu AS A DEPENDENT VARIABLE
1810 REM
1820 REM*******************************************************************
1830     MU=SQR(1-ETA*ETA)
1840     OLDRHO=RHO
1850     OLDXI=XI
1860 REM BEGIN LOOP TO CALCULATE THE STEEP PART OF THE DOME PROFILE
1870     FOR NN = 1 TO 500
1880 REM USE RUNGE-KUTTA ALGORITHM TO CALCULATE Mu FOR THE NEXT STEP
1890     AN=STP*((SQR(1-MU*MU)/RHO)-XI)
1900     BN=STP*((SQR(1-(MU+.5*AN)*(MU+.5*AN))/RHO)-(XI+.5*STP))
1910     CN=STP*((SQR(1-(MU+.5*BN)*(MU+.5*BN))/RHO)-(XI+.5*STP))
1920     DN=STP*((SQR(1-(MU+.5*CN)*(MU+.5*CN))/RHO)-(XI+STP))
1930     MU=MU+(1/6)*(AN+2*(BN+CN)+DN)
1940 REM CALCULATE Rho FOR THIS STEP
1950     RHO=RHO+STP*(MU/SQR(1-MU*MU))
1960 REM CALCULATE Eta
1970     ETA=SQR(1-MU*MU)
1980 REM*******************************************************************
1990 REM THE NEXT NINE LINES LOCATE THE POSITION OF THE TOP OF THE TALUS SLOPE
1995 REM*******************************************************************
2000 REM TEST FOR SLOPE ANGLE MATCH WITH TALUSADJOIN ANGLE
2010     IF RHO<=OLDRHO, THEN GOTO 2100
2020     TESTANGLE=ETA-SIN(TALUSADJOIN)
2030     IF ABS(TESTANGLE)<=.01, THEN GOTO 2050
2040     GOTO 2100
2050     RHOTALUS=RHO
2060     XITALUS=XI
2070     IF TALUSAPRON=0, THEN GOTO 2370
2080 REM*******************************************************************
2090 REM INCREMENT THE VALUE OF Xi
2100     XI = XI + STP
2110     VOLSUM = VOLSUM + (XI-OLDXI)*3.14159*((RHO+OLDRHO)/2)^2
2120 REM PRINT NEW VALUES OF Rho, Xi, Eta, AND Volsum
2130     PRINT #1, RHO,XI,ETA,VOLSUM
2140 REM PLOT Rho AND Xi
2150     ORHOPLO=2.02*RHO*SCALE
2160     OXIPLO=OLDXI*SCALE+20
2170     RHOPLOT=2.02*RHO*SCALE
2180     XIPLOT=XI*SCALE+20
2190     LINE(ORHOPLO,OXIPLO)-(RHOPLOT,XIPLOT)
2200 REM STORE THE CURRENT VALUES OF Rho AND Xi FOR LATER USE
2210     OLDRHO=RHO
2220     OLDXI=XI
2230 REM TEST FOR 140-DEGREE SLOPE TO ENACT ALGORITHM SWITCH
2240     IF NN=500, THEN STOP
2250     IF MU<=-.766, THEN GOTO 2280
2260     NEXT NN
2270 REM EXIT THIS ROUTINE AND RESUME CALCULATION WITH Rho INDEP. VAR.
2280     STP=-STP
2290     COUNT=1
2300     GOTO 1380
```

```
2310 REM***********************************************************************
2320 REM
2330 REM************** PERFORM CALCULATIONS TO FINISH THE PLOT **************
2340 REM              BY ADDING THE TALUS SLOPE AND DOME BASE
2350 REM***********************************************************************
2360 REM CALCULATE LOCATIONS OF TOP AND BASE OF THE MARGINAL TALUS SLOPE
2370    RHOTALUSPLO=RHOTALUS*2.02*SCALE
2380    XITALUSPLO=XITALUS*SCALE+20
2390    XPLOT = (XIPLOT-XITALUSPLO)/(TAN(TALUSLOPE)/2.02)+RHOTALUSPLO
2400 REM DRAW LINES TO DEMARCATE THE MARGINAL TALUS SLOPE
2410    LINE(RHOTALUSPLO,XITALUSPLO)-(XPLOT,XIPLOT)
2420    LINE(XPLOT,XIPLOT)-(0,XIPLOT)
```

References

Bates RL, Jackson JA (eds) (1980) Glossary of geology, Falls Church, Va. American Geological Institute

Blake S (1987) Modeling lava domes as Bingham fluids (abs.). International Union of Geodesy and Geophysics, XIX General Assembly, Abstracts vol 2

Denlinger RP (1987) An analysis of dome-building eruptions as creep-rupture phenomena (abs.). International Union of Geodesy and Geophysics, XIX General Assembly, Abstracts vol 2:413

Flügge W (1967) Stresses in shells. Springer, Berlin Heidelberg New York

Goetze C (1971) High temperature rheology of Westerly granite. J Geophys Res 76: 1223–1230

Huppert HE, Shepherd JB, Sigurdsson H, Sparks RSJ (1982) On lava dome growth, with application to the 1979 lava extrusion of the Soufriere of St. Vincent. J Volcanol Geotherm Res 14:199–222

Jaeger JC, Cook NWG (1979) Fundamentals of rock mechanics, 3rd edn. Chapman and Hall, London

Kreyszig E (1979) Advanced engineering mathematics, 4th edn. Wiley, New York

Swanson DA, Holcomb RT (1987) General consistencies in the growth of the Mount St. Helens lava dome, 1980–86 (abs.). International Union of Geodesy and Geophysics, XIX General Assembly, Abstracts vol 2:412

Swanson DA, Dzurisin D, Holcomb RT, Iwatsubo EY, Chadwick WW, Casadevall TJ, Ewert JW, Heliker CC (1987) Growth of the lava dome at Mount St. Helens, Washington, (USA) In: Fink JH (ed) GSA Spec Pap 212:1–16

Tuma JJ (1979) Engineering mathematics handbook, 2nd edn. McGraw Hill, New York

A Model for Dome Eruptions at Mount St. Helens, Washington Based on Subcritical Crack Growth

R. P. DENLINGER

Abstract

A model that idealizes the Mt. St. Helens lava dome as a brittle shell enclosing a pressurized, ductile interior (Iverson, this Vol.) is utilized here to interpret accelerating deformation during episodic dome growth. This approach assumes that the brittle exterior of the dome represents the most significant obstacle to lava extrusion, and that lava extrusion begins when the shell ruptures. With this model, the accelerating growth of the dome prior to each extrusion in 1981 and 1982 may be interpreted to be due either to a nonlinear increase in pressure within the dome with time or to slow crack growth that progressively weakens the domes' brittle shell. Comparison with the deformation data for the May 14, 1982, extrusion suggests that slow crack growth within the dome is the most likely factor contributing to the deformation. If this conclusion is supported by future studies, then the increase in the likelihood of an eruption with increasing time between eruptions may be calculated based upon the rate of pressurization of the dome and the crack growth characteristics of its brittle exterior.

1 Introduction

Interpretation of volcano deformation involves testing of alternative hypotheses. The growth of the lava dome at Mount St. Helens results from the addition of material (dacite) during discrete eruptive episodes, as shown in Fig. 1 of Swanson and Holcomb (this Vol.). In this chapter I develop the hypothesis that deformation of the dome during eruptions is due to the progressive growth of cracks that weaken the domes' rigid outer carapace. The theory is then compared to the changes in dome shape during swelling of the dome during a single eruption in 1982.

Accelerating deformation at Mount St. Helens (MSH) has been a premonitory feature of all dome building eruptions since 1980, and is similar to reported premonitory eruptive behavior at other volcanoes (Tokarev 1983; Kirsanov 1982). In 1981 it became apparent that the onset of extrusion of lava onto the dome could be predicted (Swanson et al. 1983) using a premonitory pattern of accelerating deformation of the dome and crater floor (Fig. 1; Chadwick et al. 1983, 1988), accelerating tilt of the crater floor (Fig. 2; Dzurisin et al. 1983), and exponentially increasing seismic energy release (Malone et al. 1983).

Extrusion of lava was usually heralded by an abrupt decrease of crater floor tilt, suggesting a drop in conduit pressure (Dzurisin et al. 1983), and by

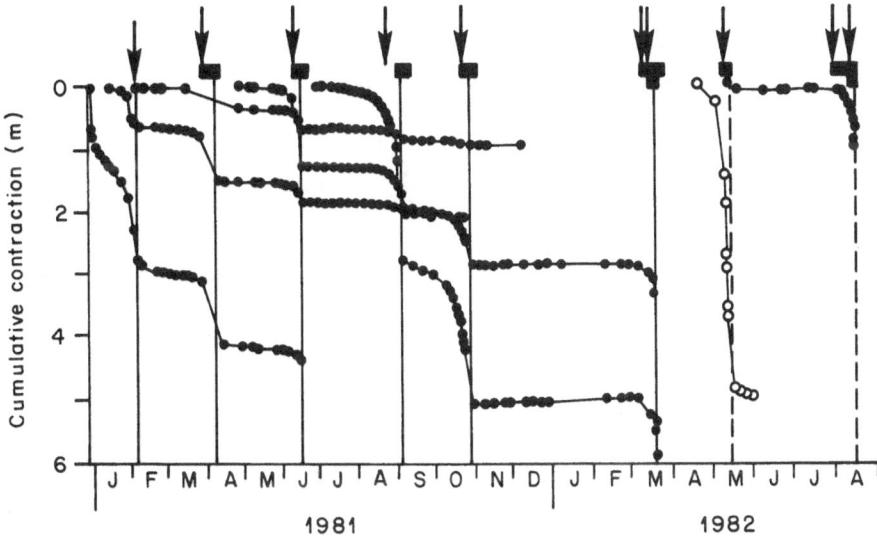

Fig. 1. Crater floor thrust fault displacement (*closed circles*) and dome flank displacement (*open circles*) for dome eruptions in 1981 and 1982. Note the predictable, repetitive pattern that allowed members of the U.S. Geological Survey to predict each eruption days to weeks in advance (Swanson et al. 1983). The *arrows* note when a prediction was made, the *rectangular bar* is the time interval in which the eruption is expected to occur, and the *vertical line* indicates when the eruption occurred. Similar patterns have been observed preceding eruptions of Bezymianny (Tokarev 1983), Mt. Augustine (J. Powers, personal communication), and in the failure of brittle materials (Varnes 1983)

a rapid drop in subsurface seismicity (Malone et al. 1983). The most rapid volumetric growth was exogenous in 1981 and 1982 (Swanson and Holcomb, this Vol.), and culminated a precursory period of accelerating endogenous growth. The pattern of accelerating endogenous growth prior to extrusion for the 1981–1982 eruptions is consistent with descriptions of the delayed failure of brittle material by tertiary creep (Varnes 1983), also known as creep-rupture.

Hypotheses for the tertiary creep behavior observed during eruptive episodes at Mount St. Helens are not limited to brittle failure of the dome (Denlinger 1987). Thermo-mechanical coupling within the conduit magma (i.e. shear heating) could produce accelerating flow of magma into the dome, as could expulsion of a stiff, degassed magma from the conduit system at the onset of an eruptive episode (Chadwick et al. 1988). Development of shear zones in the magma (Cashman and Taggart 1983) and faulting or deformation of the crater floor (Chadwick et al. 1988) also are likely possibilities. Some of these mechanisms are quite complex, involving a detailed knowledge of the rheology of the magma in the conduit and dome system, and more than one mechanism may be active during an eruptive episode. The change in dome behavior between 1980 and 1988 is consistent with the idea that different processes affected eruptive behavior as the dome evolved with time.

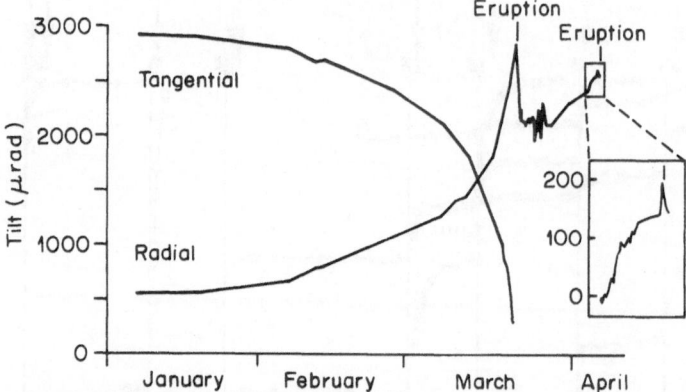

Fig. 2. Accelerating radial tilt of the crater floor in a direction away from the dome prior to the eruption of March 19, 1982 (Dzurisin et al. 1983). The abrupt decrease in radially outward tilt signaled the onset of rapid exogenous growth. This pattern was characteristic of eruptions in 1981 and 1982, and is suggestive of an abrupt pressure drop in the conduit due to rupture of the dome

In order for the brittle dome carapace to control dome growth, it must present the most significant resistance to volumetric growth of the dome. Here the concept that the dome can be idealized as a brittle shell enclosing a pressurized fluid interior (Iverson, this Vol.) is combined with the theory of slow crack growth to construct a model for dome deformation and rupture.

2 Conceptual Model

In the dome model of Iverson (this Vol.) there is a static balance between the pressurized fluid interior and the rigid exterior of the dome. This balance determinees the change, if any, in the static equilibrium shape of the dome as its volume increases. His critical shape parameter for static loading is

$$D = \frac{\sqrt{\bar{\sigma}\tau/\gamma}}{h} \tag{1}$$

where $\bar{\sigma}$ is the tensional strength of the shell, τ is its thickness, γ is the unit weight of the magma, and h is the amount of overpressure, expressed in head in excess of lithostatic pressure.

The rock strength is expected to rapidly decrease with increasing temperature as is observed in lab studies (Kirby 1983; Goetze 1971), where total relaxation of shear stresses may occur in minutes in rock at temperatures exceeding 700 °C. The existence of a cool outer shell enclosing a much hotter interior is supported by modeling studies of magnetic profiles over the dome from 1984 to 1988 (Dzurisin et al. in press). These magnetic studies show that the dome possesses an outer carapace 10 to 30 m thick below 550 °C. A model of a rigid carapace enclosing a pressurized fluid interior is the simplest means of reconciling these data.

The balance between shell strength and internal pressure is maintained indefinitely under conditions of static equilibrium. Delayed failure of the dome can be viewed in this model as either due to a decrease in tensional strength under a constant pressure $\gamma(H+h)$, where H is dome height, or an increase in pressure $\gamma(H+h+\Delta h)$ beyond the strength limit of the shell. Here Δh is the perturbation or increment in the equilibrium magma head.

The process of delayed brittle failure of the dome shell is seen here as one in which growth of cracks at subcritical velocities (velocities less than those required to cause failure) leads eventually to growth of a tensional fissure through the shell. Subcritical crack growth, or slow crack growth, is considered to be an important factor in brittle failure of dome material (dacite), as it is in the delayed failure of ceramics (Wiederhorn et al. 1976) and rocks (Costin 1987; Atkinson 1982; P. L. Swanson 1984) where slow crack growth is affected by environmental factors such as water vapor or corrosive gases.

Here, slow crack growth is assumed to be the mechanism by which there is gradual yielding of the brittle dome exterior, as cracks grow, branch out, and sometimes link to form longer cracks. Under a constant load, crack growth will reduce the tensional strength of the dome with time. On the other hand, crack growth rates increase exponentially with increases in an applied load. Both scenarios predict acceleration of deformation (extension of the shell) with time. The effects of a constant load as well as an increasing load are considered in the analysis below.

3 Model Description

Delayed failure of the brittle exterior of the dome by slow crack growth is estimated here using the pressurized shell model of Iverson (this Vol.). The pressure within the dome increases during the time between eruptions, culminating in rupture of the dome and the next extrusion. Slow crack growth within the brittle shell is assumed to occur by stress corrosion (Atkinson 1982). In the following analysis, an empirical relationship between crack velocity and stress is combined with a modified statistical theory for rock strength to estimate the time-dependent strength of the dome shell. The analysis of time dependent strength closely parallels the work of Wiederhorn and Bolz (1970) and Costin (1987).

The stresses within the brittle shell of the pressurized dome model of Iverson (this Vol.) depend upon the fluid pressure contained by the shell. Within the plane of an axisymmetric membrane, the shear stress is independent of the two tensional stresses $N_{\theta\theta}/r$ and $N_{\phi\phi}/r$ within the plane of the shell. Instantaneous failure of the shell due to these tensional stresses is prevented by the tensional strength of the shell, defined here as the magnitude of uniaxial tensional stress that ruptures the material. This strength, a factor in the equilibrium shape of the dome through the parameter D (Eq. 1), is decreased by the growth of cracks within the dome, resulting in a change in the shape of the dome with time.

Two significant features of slow crack growth in rocks (and stress corrosion theory) affect the growth of cracks in this model. One involves the presence of crack growth limits relative to the stress intensity factor. The stress intensity factor, K_I, is the elastic stress concentration factor for an infinitely sharp crack (Sneddon and Lowengrub 1969). For tensile loading (mode I):

$$K_I = Y \sigma_a \sqrt{c} \tag{2}$$

where σ_a is the regional tensile stress, c is crack length, and Y contains geometric factors. There is a lower limit on K_I below which cracks will not grow (not observed in rocks), and an upper limit, K_{Ic}, at which a crack will propagate unstably so that material failure occurs. Slow crack growth occurs between these two limits, usually at crack velocities less than 10^{-1} m/s (Costin 1987).

The other significant feature of stress corrosion is a temperature-dependent relationship between the stress intensity, K_I, and the crack velocity, v. Atkinson (1982) and Swanson (1984) have shown that between the crack growth limits, the relationship between crack velocity and stress intensity for rocks may be fit by the empirical equation:

$$v/v_0 = A e^{-E/RT} K_I^n \tag{3}$$

where v_0 is initial crack velocity, E is an empirically-determined energy barrier for tensile crack growth, T is temperature (°K), R is the gas constant, and n is a factor determined by the slope of a log v versus log K_I curve. Values for n in rocks range from 20 to 180 (Atkinson 1982; P. L. Swanson 1984). Note that since K_I is a function of stress, the rate of crack growth either increases with tensile stress or with temperature.

Following Wiederhorn and Bolz (1970), the empirical relationship between crack growth and stress concentration may be used to determine the time at which unstable crack propagation begins, leading to material failure. As cracks lengthen (or as the load increases), the stress intensity factor increases to its threshold value for unstable crack propagation, K_{Ic}, at which time the material ruptures. (Note that the stress intensity factor will depend upon the loading conditions.)

The following analysis assumes that a constant tensional load causes tensile failure. Writing the crack velocity as a change in crack length with time

$$v = \frac{dc}{dt} \tag{4}$$

and combining this with the change in stress intensity with crack length obtained by differentiating Eq. (2) for a constant stress

$$dc = \frac{2 K_I dK_I}{Y^2 \sigma_a^2} \tag{5}$$

the time to failure may be obtained by integration to yield

$$t_f = \frac{2}{A} \int_{K_{Ii}}^{K_{Ic}} \frac{K_I \, dK_I}{v Y^2 \sigma_a^2}$$

where K_{Ii} = initial stress intensity factor. Using the relationship between crack velocity and stress intensity factor in Eq. (3), then the time to failure may be written (Costin 1987)

$$t_f = \frac{2 e^{E/RT}}{A Y^2 \sigma_a^2 v_0} \frac{K_{Ic}^{2-n} - K_{Ii}^{2-n}}{(2-n)} . \tag{7a}$$

For the largest crack in the material (the weakest link), the ratio of stress intensity at failure, K_{Ic}, to the initial stress intensity, K_{Ii}, can be shown to be

$$K_{Ic}/K_{Ii} = \sigma_c/\sigma_i \tag{7b}$$

where σ_c = stress required to cause immediate failure and σ_i = the initial applied stress = σ_a (Costin 1987). Costin also shows theoretically that the time to failure is proportional to the applied stress, and that Eq. (7a) provides a very good fit to the limited data on static tensile failure of rock once the relationship between K_I and crack velocity is known (i.e. once a factor of n in Eq. (3) is found empirically).

However, the stress applied to the dome shell is not constant, since a constant stress is inconsistent with the history of dome growth. The volume-predictability of the dome growth at Mount St. Helens (Swanson and Holcomb, this Vol.) requires some rate of volumetric production from the conduit. Since the volumetric production is effectively zero between eruptive episodes, pressure must increase within the conduit with time. The simplest explanation for a constant time-averaged rate of supply involves a constant magma compressibility combined with a constant rate of pressure increase with time. This condition will be assumed here in estimating crack growth within the shell. The rate of loading of the shell varies linearly with the rate of pressurization of the dome in the dome shell model (Iverson, this Vol.), and this depends upon the rate of pressurization of the conduit with time.

For dynamic loading of the shell at a constant stress rate,

$$\frac{dc}{dt} = \left(\frac{dc}{d\sigma}\right)_{K_I} \left(\frac{d\sigma}{dt}\right) \tag{8}$$

since the stress intensity factor is a unique function of the crack velocity (Costin 1987). Differentiating the change in stress intensity factor with crack length yields:

$$d\frac{K_I}{dc} = \frac{d\sigma}{dc} Y \sqrt{c} + \frac{\sigma Y}{2\sqrt{c}} . \tag{9}$$

Let $\sigma = \sigma_a = \sigma_0(1+ft)$, then

$$dc = \frac{dK_I}{[f/(1+ft)](K_I/v) + \sigma^2 Y^2/2K_I} \tag{10}$$

and the time to failure at a constant stress rate is

$$t_f = \int_{K_{Ii}}^{K_{Ic}} \frac{K_I^{-1} dK_I}{(f/(1+ft))+(v/2c)} \ . \tag{11}$$

For $\left(\dfrac{v}{2c}\right) \geqslant \left(\dfrac{f}{1+ft}\right)$ the time to failure is well approximated by the expression for constant stress (Eq. 6).

Applying this analysis to Mount St. Helens, the effect of dome pressurization on crack growth will depend upon the crack size distribution and the crack growth behavior under load. For example, if failure is induced by small cracks (small c) that grow very fast (high v), then very small changes in pressure will greatly accelerate the rupture of the dome. On the other hand, if failure occurs by the slow growth (low v) of large cracks (large c), then very large changes in dome pressure are required to cause rupture of the carapace. Thus if the dome behaves as a brittle shell enclosing a pressurized interior, then its dynamic behavior critically depends upon the crack growth characteristics of the dacite comprising the dome shell.

The crack size distribution in rocks determines the stress intensity K_{Ii} in Eq. (6) and hence determines the initial strength of the material when crack growth begins. When this distribution is combined with the $K_I - V$ relationship as is done in Eq. (7a), the time to failure may then be estimated since both a rate and one end point are determined. Statistical theories for crack size distributions in rocks have proven successful for evaluating rapid, brittle failure (Vardar and Finnie 1975), and represent an intuitively satisfying way of estimating the range of flaws present in the rock.

In statistical theories, tensile strengh or failure stress is not unique, but varies statistically with the spatial distribution of flaws that can cause failure. Weibull (1939) considered that volumes of rock are linked serially, such that the probability of breakage was determined by the probability of finding a critical flaw in a series of sample volumes. For a random distribution of flaws in each sample volume, Weibull (1939) showed that the probability of failure $P(\sigma_c)$ at a stress σ_c is

$$P(\sigma_c) = 1 - \exp\{-\int (\sigma_c/\sigma_0)^m \, dV\} \ . \tag{12}$$

Here σ_c is the tensile strength in the absence of slow crack growth, m is constant for a given material and σ_0 is a constant that depends upon loading conditions and sample volume (Jaeger and Cook 1979, p. 196–201). As the applied stress increases to infinity, the proportion of unstable cracks in the total population increases so that P approaches unity.

Using Weibull statistics, the variation of tensional strength with volume, V, for rapid crack growth under the same loading conditions may be expressed as

$$\sigma_c V^{1/m} = \text{constant} \ . \tag{13}$$

Table 1. Ratio of tensional strength of stressed dome shell to its uniaxial tensional strength for two different combinations of the shell stresses, using Weibull statistics. S1 = stress in circumferential direction; S2 = stress in meridional direction

m:	4	6	8
S1/S2 = 1:	0.72	0.78	0.82
S1/S2 = −1:	1.06	1.04	1.03

Vardar and Finnie (1975) found values for m of 10−12 for Brazilian tests (defined in Jaeger and Cook 1979) on granodiorite and limestone, and these values will be assumed here.

For the failure probability defined by Eq. (12), Weibull (1939) showed that the tensional strength of material for rapid fracture is given by

$$\bar{\sigma} = \int_{0}^{\infty} (1 - P) \, d\sigma_c \qquad (14)$$

under the condition that $\lim_{\sigma_c \to \infty} (1 - P)\bar{\sigma} = 0$ and this value is different than the breaking strength of material in uniaxial tension σ_c because of the difference in geometry and stress condition (Jaeger and Cook 1979). The ratio of the actual breaking strength to the uniaxial strength varies as $(\sigma_c / \sigma_0)^m$ and may be calculated once the stress conditions and geometry are known.

This ratio was calculated for the two limiting stress conditions in the dome shell model of Iverson (this Vol.) and these values are listed in Table 1. One condition exists at the apex of the shell, where the meridianal and hoop normal stresses $N_{\phi\phi}/r$ and $N_{\theta\theta}/r$ are tensional, equal, and proportional to the excess internal pressure. I will assume that lower down on the flank of the shell, $N_{\phi\phi}/r$ decreases and becomes compressional whereas the circumferential stress $N_{\theta\theta}/r$ remains tensional.

For this case the breaking strength of the shell will be considered for the condition $N_{\phi\phi}/r = -N_{\theta\theta}/r$. This latter stress condition is a variation of the stress condition obtained by Iverson (this Vol.), and is never achieved with a positive value for h, but rather exists as a theoretical limit. However it illustrates the lower probability of tensional failure using weakest link statistics when one of the principal stresses is compressional.

For these "weakest link" statistics, the tensional strength of material on top of the dome is significantly less than the strength of the same material on the flanks of the dome (see Table 1). Thus, all other factors being equal, the probability of tensional failure is greatest near the apex of the dome and least on the flanks. With the two layer model of Iverson (this Vol.) the apex of the dome is more likely to fail in tension than any other location because of the biaxial tensional stresses acting on it, favorably loading a larger distribution of cracks in the same sample volume.

For slow crack growth at subcritical velocities, the Weibull theory as stated above is inappropriate, but may be modified using stress corrosion theory to provide a good estimate of the relationship between failure stress, applied stress, and time to failure (Costin 1987; Wilkins 1980). A proper estimate requires two sets of laboratory measurements; the first set determines the rapid fracture Weibull statistics, the second set relates low stresses and increased failure times to the Weibull distribution from the first set. This procedure is described in detail by Costin (1987) and Wilkins (1980).

In the first set of experiments, a number of rock samples (enough to form a good statistical sample, usually about 30) are loaded to failure. The failure stresses at zero time are then ranked, and the probability of failure for stresses within the range of stress that caused failure is

$$P(\sigma_c(j)) = \frac{j}{N+1} \tag{15}$$

where j is the rank within the population of N samples. A plot of $\log[-\log[1-j/(N+1)]]$ versus $\log(\sigma_c)$ gives a linear dependence with a slope of m.

In the second set of experiments, a number of samples are loaded to some stress within the range of failure stress determined in the first set. A certain number of samples will fail immediately and the remainder will fail after varying lengths of time. The failure times from this second set of experiments are ranked and paired with the ranked values of stress from the first experiment. If these data are used to define the ratio of applied stress σ_a to zero-time failure stress σ_c for each ranked failure time, then a plot of $\log(\sigma_a/\sigma_c)$ to $\log t_f$ will yield a straight line with a slope n that is the same n in the $K_I - v$ relationship (Eq. 7a), (Costin 1987). This observation justifies the use of a known $K_I - v$ relationship to determine a $t_f - (\sigma_a/\sigma_c)$ relationship and hence predict a distribution of failure times for a given loading condition.

In the example that follows, I assume that the dome model of Iverson (this Vol.) applies to the Mount St. Helens' lava dome. The loading condition is represented by the equilibrium parameter D that balances the resistance of the shell to breaking with the pressure in the magma enclosed by the shell. Changes in D during swelling of the dome are used to examine the balance between magmatic pressurization, changes in shell thickness, and progressive crack growth.

4 Application to a Dome Building Eruption at Mount St. Helens

Detailed deformation data during an eruptive episode and strength and fabric tests on dome dacite are needed to test this dome failure model. Deformation data during dome eruptions are difficult and dangerous to obtain, and sufficiently complete data for the dome are not available until mid-1982. The May

Table 2. Dimensions of the dome prior to the May
14, 1982 eruption

---

h_0 = 166 m
w_0 = 262 m
l_0 = 352 m
r_0 = $(w_0 + l_0)/2$ = 307 m
D_0 = 1.0

14, 1982 eruption was chosen since adequate displacement data are available (Cascades Volcano Observatory, unpublished data) to determine dome shape and the dome was still structurally simple enough to be idealized as a single ellipsoidal shell.

The May 14, 1982 dome-building eruption may be analyzed in terms of the delayed failure statistics presented earlier if the change in the equilibrium shape of the dome with time is due to crack growth within the brittle exterior of the dome. The change in D with time is either due to a change in tensional strength with time, a change in the shell thickness with time, or a change in internal dome pressure with time prior to rupture of the shell. Magma density is assumed to be constant. In the absence of laboratory measurements, the tensional strength of the dome dacite is estimated here using an average crack growth n value of 40 estimated from strength tests on granite and andesite (Atkinson and Meredith 1987).

The initial and final dimensions of the dome for the May 14, 1982 extrusion are given in Table 2 (from Chadwick et al. 1988) and these dimensions are consistent with an initial value of D (Eq. 1) of about 1.0 when the horizontal dimensions are averaged. As shown by Iverson (this Vol.), the dimensions of the dome uniquely determine D. In particular, if r = (w+l)/2 and H = height, then

$$D = B(r/H)^b \quad \text{and} \tag{16a}$$

$$D = D_0[1 + \Delta (r/H)]^b \tag{16b}$$

where w = width, l = length, the subscript 0 denotes initial values and B and b are empirical coefficients. Note that the exact value of D is not critical, only the changes in D (and hence dome equilibrium) are important as determined from deformation.

Changes in D over a 30 to 40 day period of dome expansion are affected by changes in the product of shell thickness and tensional strength, or by changes in overpressure h. The magnetic studies of Dzurisin et al. (in press) have shown that the 550 °C isotherm moves toward the inside of the dome at rates between 2 cm/day (winter) and 2 mm/day (summer) unless the carapace is severely fractured, but may be arrested altogether during active periods of swelling. This study suggests that the mechanically rigid dome shell, defined as rock at temperatures less than 550 °C, was about 10 to 11 m thick in May 1982.

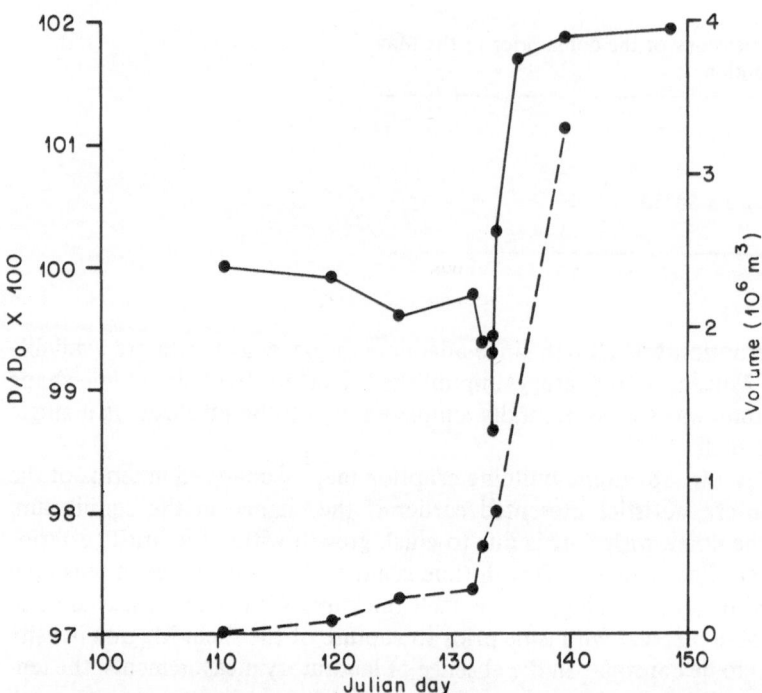

Fig. 3. Change in the equilibrium shape of the dome as expressed by the factor D (*solid line*) relative to dome volume (*dashed line*) during the May 14, 1982 eruptive episode. The factor D expresses a balance between brittle shell strength and the internal pressure contained by the shell (Iverson, this Vol.), and a decrease in D corresponds to an increase in the ratio of dome height to dome radius as the shell weakens or pressure increases. An abrupt decrease in the outward tilt of the crater floor was noted at 0000 hours on May 14th and an extrusion was documented about 4 h later. May 14th = Julian day 134

A 10-m-thick shell that thickens at a rate of 2 cm/day will increase in thickness by 0.6 m in 30 days of swelling unless the dome is heated during endogenous growth. The magnetic studies of Dzurisin et al. (in press) show that the rate of increase in shell thickness may stop 2–4 days prior to extrusion. Their study prompts the examination of two limiting conditions: one in which changes in D occur at a constant shell thickness, and one in which changes in D occur when the shell is assumed to thicken at a rate of 2 cm/day.

Besides shell thickness τ, the conduit overpressure h is also assumed to increase linearly during the eruptive episode. Linear rates of increase for h are determined from the observed decrease in D (Fig. 3) for successively increasing time increments in Table 3a, 3b. In Table 3a a constant shell thickness is assumed, and in Table 3b the dome shell is allowed to thicken linearly with time.

The change in the shape of the dome during the period of swelling prior to the May 14, 1982 extrusion indicates an accelerating decrease in D (Fig. 3) that cannot be explained solely with linear increases in shell thickness or dome overpressure. Here the dome shell model is applicable until deformation rates

Table 3 a. Change in the linear rate of increase in h to fit D, with shell thickness constant. Using $D = \dfrac{\sqrt{\bar{\sigma}\tau/\gamma}}{h}$ and $h = h_0(1 + ft)$:

Date	Time (t)	D/D_0	τ (m)	$f\,(\times 10^{-4})$
April 12	0.0	1.0	10	0
April 29	17.23	0.9991	10	0.5
May 5	23.04	0.9960	10	1.7
May 11	29.25	0.9977	10	0.8
May 12	30.06	0.9937	10	2.1
May 13	30.858	0.9943	10	1.9
May 13	30.861	0.9929	10	2.3
May 13	30.876	0.9866	10	4.4

Table 3 b. Change in the linear rate of increase in h to fit D, with shell thickness given by $\tau = \tau_0(1 + 0.02\ \text{m/day}\cdot\text{time})$, using $D = \dfrac{\sqrt{\bar{\sigma}\tau/\gamma}}{h}$ and $h = h_0(1 + ft)$:

Date	Time (t)	D/D_0	τ (m)	$f\,(\times 10^{-4})$
April 12	0	1.0	10.0	–
April 29	17.23	0.9991	10.34	1.0
May 5	23.04	0.9960	10.46	1.16
May 11	29.25	0.9977	10.59	1.07
May 12	30.06	0.9937	10.60	1.20
May 13	30.858	0.9943	10.62	1.18
May 13	30.861	0.9929	10.62	1.22
May 13	30.876	0.9866	10.62	1.44

are high enough to sustain shear stresses within the ductile magma in the domes' interior (about 10^{-2}/s as shown in Appendix 2).

The overall shape of the dome flattened due to this extrusion (a net increase in D), and the curves in Fig. 3 are drawn through the extrusive episode to provide continuity.

The accelerating decrease in D during the period of dome swelling prior to extrusion could be either due to an increase in internal pressure or due to a decrease in tensional strength because of slow crack growth (Table 3). The decrease in D may be produced solely with a nonlinear rate of increase in dome overpressure h (Table 3 a). Alternatively, the decrease in D can be ascribed to a constant shell growth rate of 2 cm/day as well as nonlinear changes in h (Table 3 b). In both cases the changes in the equilibrium shape of the dome during swelling must be accommodated either by a nonlinear rate of increase in dome overpressure or by a nonlinear decrease in tensional strengh $\bar{\sigma}$.

Within the context of this work three possibilities exist:

1. Changes in the equilibrium shape of the dome during swelling occur as the dome resists increases in conduit pressure. All of this resistance is provided

by a dome carapace or shell that is at temperatures less than 550 °C. Changes in dome shape are affected by slow crack growth within, and rate of thickening of, this shell.

2. Conditions in (1) apply except that the tensile strength of the shell is constant until rupture occurs by rapid brittle fracture (i.e. slow crack growth is absent). In this case the pressure increase within the dome must be nonlinear in time to explain the deformation data.

3. Deformation of the dome is not controlled by the carapace, but is a passive response to the mass flux of magma through the conduit and into the dome.

It will be possible to rigorously distinguish between these possibilities with more complete lab and field data. However, several lines of evidence, considered together, favor the first hypothesis for this eruption. The geophysical evidence referenced in the introduction shows that rupture of the dome exterior coincides both with an abrupt deflation of a bulge in the crater floor surrounding the dome and with the onset of rapid growth. These observations are not consistent with hypothesis (3). Of the two remaining hypotheses, the second hypothesis is not consistent with the nearly linear, volume-predictable nature of dome growth over periods of 1 to 2 years (Swanson and Holcomb, this Vol.) if the magma in the conduit system has a constant volume compressibility. Also, hypothesis (2) is qualitatively eliminated by the ubiquitous slow growth of surface cracks during swelling of the dome, and the often observed growth of large fissures due to swelling in the vicinity of the eventual site of lava extrusion. Continuous magnetic measurements up until the point of extrusion could constrain changes in shell thickness, and values of n, m, and σ_c could be determined with lab tests of dome dacite. With these measurements it would be possible to determine whether the slow growth of cracks in the brittle exterior of the dome controls the rate of swelling prior to extrusion.

An estimate of the relationship between time to rupture and conduit pressurization may be made by developing hypothesis (1). Combining Eqs. (7) and (12), a relationship for the time to failure that incorporates slow crack growth is obtained:

$$t_f = \frac{2}{n-2} \left[\frac{K_{Ic}^2}{Y^2 \sigma_c^2 v_c^2} \right] \left(\frac{\sigma_a}{\sigma_c} \right)^{-n} \left[\ln \left(\frac{1}{1-P} \right) \right]^{(n-2)/m}. \qquad (17)$$

Since n values range between 20 and 150, small changes in the applied stress σ_a result in large changes in t_f.

The exact values of these constants are not available for Mount St. Helens dome dacite, and measurements from other rocks are used here to make an estimate of the relationship between time to failure and applied load. The crack growth in Yugawa andesite, the only volcanic rock of intermediate composition that was available, gave a value for n of 31 (Waza et al. 1980). The measurements of Vardar and Finnie (1975) on granite and limestone were used for the fatal flaw distribution parameter m of 8. The ratio

Table 4. Parameters used for Mount St. Helens' dome

$\bar{\sigma}$	$= 1\,\text{MPa}$
σ_c	$= 1\,\text{MPa}$
τ_0	$= 10\,\text{m}$
h_0	$= 20\,\text{m}$
γ	$= 26\,000\,\text{Nt/m}^3$
D	$= 1$
n	$= 31$ (Yugawa andesite, Waza et al. 1980)
m	$= 8$ (granite, Vardar and Finnie (1975)
$\dfrac{K_{Ic}^2}{Y^2\sigma_c^2v_c^2}$	$= 25\,\text{s}$ (Atkinson and Meredith 1987)
h	$= h_0(1 + 10^{-3}/\text{day}\cdot\text{time in days since April 12 at noon})$

$$\frac{K_{Ic}^2}{Y^2\sigma_c^2v_c^2} \tag{18}$$

was estimated from the data of Atkinson and Meredith (1987) for crack growth in basalts, granite, and tuff. These values are listed in Table 4, and used in Eq. (17) to construct Fig. 4 for constant values of probability P.

The probability of a dome-shell rupture, and hence an extrusion, strongly depends on the rate of dome and conduit pressurization relative to the overpressure at the start of the eruptive episode. For the parameters in Table 4, I estimate that the dome overpressure at the end of the April 1982 eruption was about 0.5 MPa, and the shell tensional strength was about 1 MPa. These values are consistent with the average dome shape in 1982 (Iverson, this Vol.).

Assuming that the dome strength is about equal to the critical stress for rupture, σ_c, the stress history for the May 1982 eruption may be plotted on Fig. 4. A constant stress ratio is shown as a dotted line, and the linear rate of stress increase (determined from the apparent rate of increase in dome overpressure in Table 3b) defined in Table 4 is shown as a dashed line. Note that the probability of dome rupture varies much more slowly with time if the dome overpressure is constant. This is a direct result of the large stress exponent n that determines crack growth rates; the increase in crack growth rate is proportional to the stress increase to the nth power.

For the parameters we use here for the Mount St. Helens dome, the dome would not be expected to rupture for 28 years following the April 1982 eruption if the conduit pressure remained constant. This is comparable to the time required to completely solidify the dome (Dzurisin et al., in press) and the near surface portion of the conduit as well. A linear rate of dome pressurization of 0.52 kPa/day is about the average rate determined from swelling of the dome during the May 1982 eruption, and this is sufficient to reduce the expected rupture time from 28 years (dotted curve in Fig. 4) to between 60 and 120 days (dashed line in Fig. 4).

Therefore the primary factors controlling the time between dome-building eruptions in this model are the rate of pressurization of a hot, ductile conduit

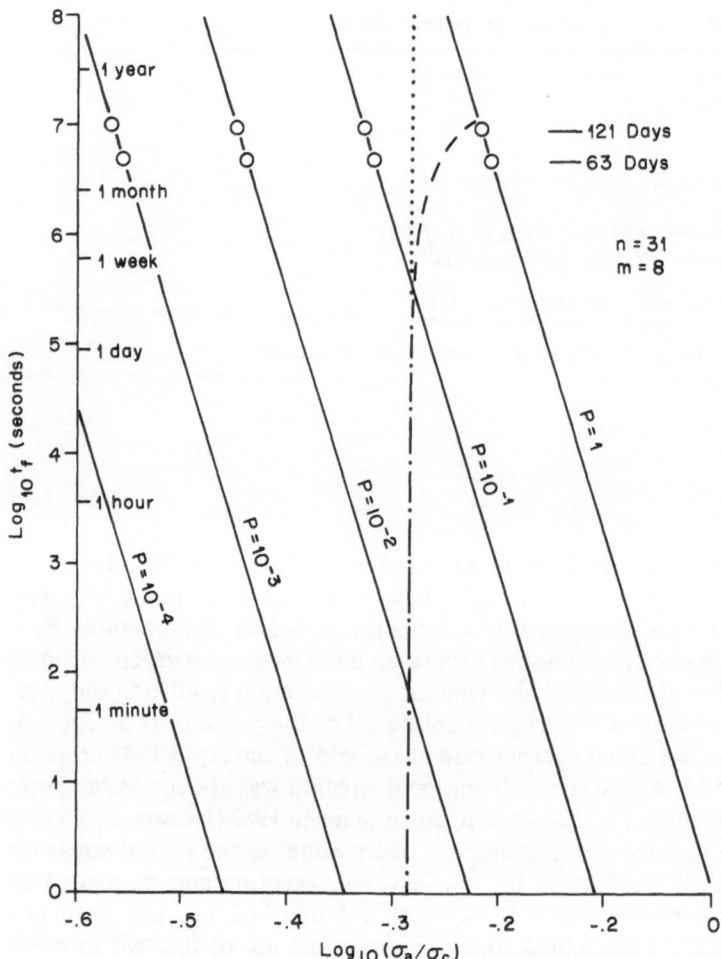

Fig. 4. An estimate of the probability of dome shell rupture versus the duration of loading
(t_f) at a stress (σ_a), where σ_c is the stress at which the shell ruptures immediately. The data
points are average inter-eruptive periods in 1981 (63 days) and 1982 (121 days). The probability of rupture is constant along each *solid line*, which has a slope determined from slow
crack growth in Yugawa andesite (Waza et al. 1980). The position of the *dotted line* is determined from a constant conduit pressure, whereas the *dashed line* is determined from a linear
increase in conduit pressure as defined in Table 4

and dome interior, and the crack growth characteristics of the outer portions
of the dome. The rate of pressurization will reflect the average volumetric rate
of growth, whereas the timing and volume of discrete eruptive episodes reflect
the response of the dome to the pressurization of its conduit.

5 Conclusions

The consistent pattern of accelerating dome deformation observed at Mount St. Helens in 1981 and 1982 is indicative of a creep-rupture or tertiary creep process prior to lava extrusion. The geophysical data (seismic, tilt) for each eruptive episode appear to associate an abrupt drop in conduit pressure with the onset of lava extrusion, supporting the hypothesis that the brittle exterior of the dome is the most significant obstacle to lava extrusion.

By developing this hypothesis, eruptions may be analyzed as the result of the rupture of a brittle shell enclosing a ductile, pressurized interior. Analysis of dome deformation data for the May 14, 1982 extrusion, for which a good data set exists, shows that either slow cracking accompanied swelling of the dome or that the pressure within the conduit increased nonlinearly with time. However, the nearly linear, volume-predictable nature of dome growth over periods of years (Swanson and Holcomb, this Vol.) makes a nonlinear increase in conduit pressure highly unlikely, and accelerating growth of the dome prior to lava extrusion is probably due to slow crack growth. Given this, then the increase in the likelihood of an eruption with increasing time may be calculated based upon the rate of pressure increase within the dome and by the crack growth characteristics of its brittle exterior.

Acknowledgments. I wish to thank Bob Anderson and Dave McTigue for critical reviews of the manuscript. I am also indebted to Don Swanson for lending his insight on the behavior of the dome and for allowing me access to data that were difficult and sometimes hazardous to obtain.

References

Atkinson BK (1982) Subcritical crack propagation in rocks: theory, experimental results and applications. J Struct Geol 4:41–56

Atkinson BK, Meredith PG (1987) The theory of subcritical crack growth with applications to minerals and rocks. In: Atkinson BK (ed) Fracture mechanics of rock. Academic Press, pp 111–166

Cashman KV, Taggart JE (1983) Petrologic monitoring of 1981 and 1982 eruptive products from Mount St. Helens. Science 221 (4618):1385–1387

Chadwick WW, Swanson DA, Iwatsubo EY, Heliker CC, Leighley TA (1983) Deformation monitoring at Mount St. Helens in 1981 and 1982. Science 221 (4618):1378–1380

Chadwick WW, Archuleta RJ, Swanson DA (1988) The mechanics of ground deformation precursory to dome-building extrusions at Mount St. Helens: 1981–1982. J Geophys Res 93,B5:4351–4366

Costin LS (1987) Time dependent deformation and failure. In: Atkinson BK (ed) Fracture mechanics of rock. Academic Press

Denlinger RP (1987) An analysis of dome building eruptions as creep-rupture phenomena, (abstract). Int. Union of Geod Geoph, XIX general assembly, Vancouver, Canada, abstract v. 2, p. 413

Dzurisin D, Denlinger RP, Rosenbaum JG (1989) Cooling rate and thermal structure determined from progressive magnetization of the dacite dome at Mount St. Helens, Washington. J Geophys Res (in press)

Dzurisin D, Westphal JA, Johnson DJ (1983) Eruption prediction aided by electronic tiltmeter data at Mount St. Helens. Science 221 (4618):1381–1382

Goetze C (1971) High temperature rheology of Westerly granite. J Geophys Res 76: 1223–1230

Iverson RM (1989) Lava domes modeled as brittle shells that enclose pressurized magma, with application to Mount St. Helens. IAVCEI Proc Volcanol 2:47–69 (this vol)

Jaeger JC, Cook NGW (1979) Fundamentals of rock mechanics. Chapman and Hall, third ed. London

Kirby SH (1983) Rheology of the lithosphere. Rev Geoph Space Phys 21:1458–1487

Kirsanov IT (1982) Extrusive eruptions on Bezymianny volcano in 1965–1974 and their geologic effect: Akad Nauk USSR, Sibirskoe Otdelenie, Institut Geologii i Geofiziki, Nauka, Moscow, pp 50–69, translated by DB Viatlioano

Malone SD, Boyko C, Weaver CS (1983) Seismic precursors to the Mount St. Helens eruptions in 1981 and 1982. Science 221 (4618):1376–1378

Sneddon IN, Lowengrub M (1969) Crack problems in the classical theory of elasticity. John Wiley and Sons, Inc, New York

Swanson DA, Casadevall TA, Dzurisin D, Malone SD, Newhall CG, Weaver CS (1983) Predicting eruptions at Mount St. Helens, June 1980 through December 1982. Science 221:1369–1376

Swanson DA, Holcomb RT (1989) Regularities in the growth of Mount St. Helens' dome. IAVCEI Proc Volcanol 2:1–24 (this vol)

Swanson PL (1984) Subcritical crack growth and other time and environment-dependent behavior in crustal rocks. J Geophys Res 89:4137–4142

Tokarev PI (1983) Experience in predicting volcanic eruptions in the USSR. In: Tazieff H, Sabroux JC (eds) Forecasting volcanic events. Elsevier, Amsterdam, pp 257–268

Vardar O, Finnie I (1975) An analysis of the Brazilian disk fracture test using Weibull probabilistic treatment of brittle strength. Int J Fracture 11:495–508

Varnes DJ (1983) Time-deformation relations in creep to failure of earth materials: Proceedings of the 7th southeast Asia geotechnical conference, vol. 2, pp 107–130

Waza T, Kurita K, Mizutani H (1980) Tectonophysics 67:25–34

Weibull W (1939) A statistical theory of the strength of materials, Ingvetensk Akad Handl, No 151

Wiederhorn SM, Bolz CH (1970) Stress corrosion and static fatigue of glass. J Am Ceram Soc 53:543–548

Wiederhorn SM, Fuller ER, Mandel J, Evans AG (1976) An error analysis of failure prediction techniques derived from fracture mechanics. J Am Ceram Soc 59:403–411

Wilkins BJS (1980) Slow crack growth and delayed failure of granite. Int J Rock Mech Min Sci Geomech (Abstr) 17:365–369

Appendix 1. Dimensions of variables and constants

D	= equilibrium shape parameters, dimensionless
$\bar{\sigma}$	= tensional breaking strength, M/Lt^2
τ	= shell thickness, L
γ	= unit weight, M/L^2t^2
h	= overpressure, in terms of magma head, L
H	= dome height, L
$N_{\theta\theta}$, $N_{\phi\phi}$	= stress resultants, M/t^2
K_I	= stress intensity factor for tensile loading, $M/L^{0.5}t^2$
E	= energy barrier for crack growth, $1/t^2$
R	= gas constant, $1/t^2 T$
T	= temperature in degrees Kelvin, T
v	= crack velocity, L/t
2c	= crack length, L
Y	= dimensionless geometric factor

Appendix 1 (cont.)

σ_a, σ_c	= stress, M/Lt^2
t_f	= failure time, t
n	= stress intensity exponent for crack velocity
f	= slope of applied stress versus time divided by initial stress, $1/t$
P	= probability values between 0 and 1
σ_0	= product of stress and volume$^{1/m}$, $ML^{3/m-1}/t^2$
t	= time, t
r	= average of horizontal dimensions of dome, L
A	= empirical constant, $L^{0.5n}t^{2n}/M$
m	= Weibull crack distribution parameter, dimensionless
V	= volume, L^3

Subscripts	
o, i	= initial values
c	= critical values at rupture

Appendix 2. An estimate of relaxation time for the dome core. A linear Maxwell model (Jaeger and Cook 1979) may be used to approximate the relaxation of stress within the magma in the dome. For a magma viscosity of 10^{11} poise (Chadwick et al. 1988) and a high frequency shear modulus of 10^3 MPa, the magma will relax in 10 seconds under a sustained static load. The criteria used to estimate shear stresses in the magma is

$$\dot{\varepsilon}\eta \leq G/10 \tag{A1}$$

where G is shear modulus, $\dot{\varepsilon}$ is strain rate and η is viscosity. For $G = 10^3$ MPa, strain rates may be as high as $10^{-2}/s$ before shear stresses induced by the deformation of the dome cause it to deviate from its static equilibrium shape.

Viscoplastic Models of Lava Domes

S. BLAKE

Abstract

It is proposed that lava domes possess a yield strength whose magnitude governs their shape and explosivity. Theoretical and laboratory modelling of the spread of viscoplastic material over a horizontal surface provide means of examining geological data on dome growth in terms of dome rheology. Experiments using kaolin slurries show that a growing dome maintains a state of static equilibrium such that its height, H, radius, R, density, ϱ, and yield strength, τ_0, are always related by $H = 1.76 \, (\tau_0 R/\varrho g)^{1/2}$, a relationship which is independent of the effusion rate. No concentric or radial structures developed on the models' surfaces. Instead, slip plane traces spiralled out from the dome summit in clockwise and anti-clockwise directions. These traces divide the surface into rhombohedral roughness elements which, although their size and roughness is greatest near the dome margins, have an order of magnitude spacing of $\tau_0/\varrho g$. Labelled positions on the surface of a growing dome move radially outward but never reach the dome margin. The experiments predict that the distal surface of a growing low lava dome should experience radial shortening and circumferential stretching.

The relation $H = 5.75 \, R^{1/2}$ describes the growth of the 1979 Soufriere, St Vincent, dome well and implies a yield strength of 2.6×10^5 Pa. Similar yield strengths are calculated for other low lava domes irrespective of composition, suggesting that the strength of chilled magma forming the dome carapace is $10^5 - 10^6$ Pa. In contrast, more common Peleean domes are covered in rock debris and are pyramidal in shape such that $\tan^{-1} (H/R) = \Phi$, the angle of repose of loose talus surrounding these domes ($30° - 45°$). The talus apron rather than magma rheology governs the dome shape directly as these domes are too strong to spread very far under their own weight, indicating yield strengths of at least $10^6 - 10^7$ Pa. Pressures considerably greater than atmospheric (10^5 Pa) must accumulate if Peleean domes are to deform. Explosive disruption is typical of Peleean domes but rare at low lava domes. This difference in hazard potential is accounted for by the larger yield strengths of Peleean domes.

1 Introduction

Lava domes form when magma piles up above and around a vent either because of the flatness of the ground, the restrictive influence of encompassing crater walls or the high viscosity of the magma. Domes are common products of intermediate and silicic volcanism, and Simkin et al. (1981) record 217 dome-forming events in their catalogue of the past 10000 years of volcanism.

Conceptually, domes are the simplest type of lava flow, yet geological observations (Williams 1932; Cotton 1944; Perret 1950; Macdonald 1972; Williams and McBirney 1979; Moriya 1978) show a variety of styles ranging continuously from nearly solid upheaved plugs to viscous coulees and lava streams. Many dome-forming eruptions involve explosive phases which can occur before, during, or after dome emplacement (Newhall and Melson 1983; Heiken and Wohletz 1987). In an attempt to obtain some systematic understanding of the cause of this variability, this chapter presents a study of the quantitative relationships between the rheology, morphology and volcanology of lava domes. The central assumption upon which this work is based is that domes behave as fluids only when they experience a shear stress greater than some yield strength, i.e. they are viscoplastic materials. Yield strengths have been measured in lavas of basaltic to dacitic composition (Shaw et al. 1968; Shaw 1969; Pinkerton and Sparks 1978; McBirney and Murase 1984; Murase et al. 1985) and have been proposed for rhyolitic lava (Fink 1984). It is generally agreed (e.g., McBirney and Murase 1984; Chester et al. 1985) that the yield strength and viscoplastic nature of lava arise because of interactions between bubbles or crystals suspended in a viscous melt. The simplest type of viscoplastic is the Bingham plastic, defined as having a plastic viscosity which is independent of shear rate. This is an oversimplification for magmas as they have plastic viscosities which decrease with increasing shear rate. However, for any viscoplastic which is moving very slowly, such as a growing lava dome, the shear stresses will be just slightly greater than the yield strength. The Bingham model will, therefore, approximate the material's rheological behaviour over this narrow range of shear stresses and rates. The growth of viscoplastic domes can therefore be modelled using the Bingham model.

This chapter treats the geology and rheology of lava domes as follows: In Section 2 four types of lava dome are defined on the basis of their different morphologies, which are in turn attributed to differences in their ability to flow under their own weight. This section provides a geological background against which theoretical models for domes can be seen. Section 3 develops a theoretical description of the spread of a Bingham plastic. Laboratory experiments described in Section 4 are used to examine the theoretical model and provide further insight into the flow of viscoplastics. Geological implications of the modelling are examined in Section 5 and summarized in Section 6.

2 Four Types of Lava Dome

In this section, lava domes are grouped into four types on the basis of those morphological characteristics that are believed to be directly determined by the physical properties of the dome-forming material. This prepares the way for a more quantitative investigation of the link between physical properties and dome behaviour. This approach will use geological data on the free-flow of lava over horizontal surfaces to infer the domes' physical properties using vari-

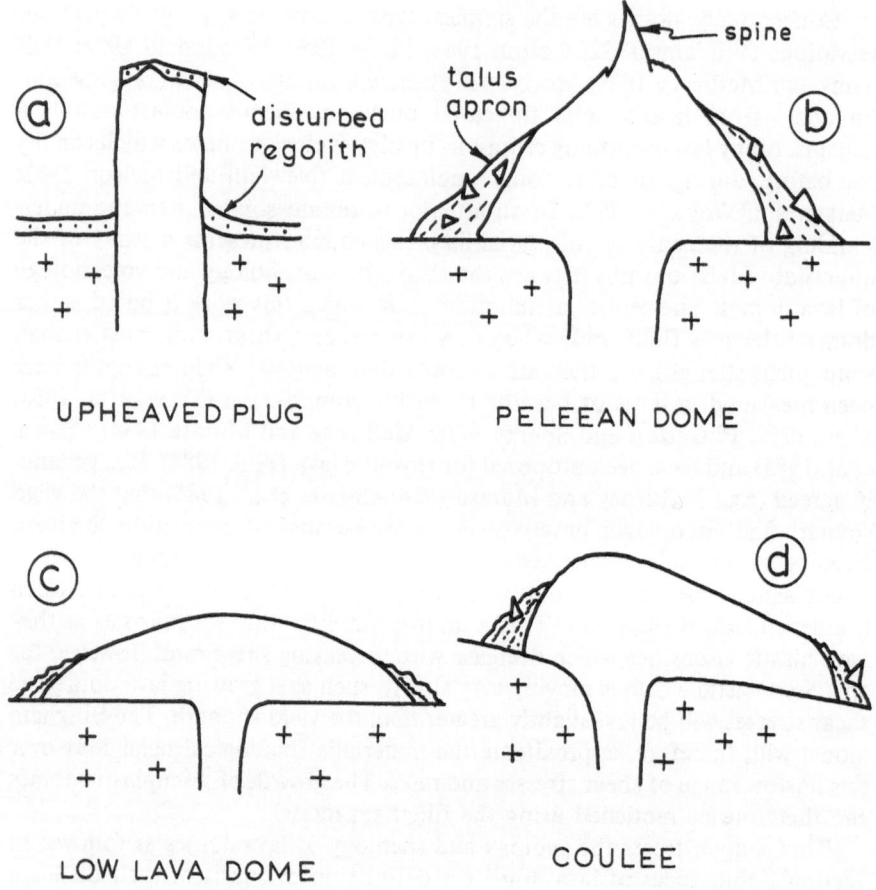

Fig. 1. Sketches showing the distinguishing features, seen in cross-section, of the four types of dome discussed in this chapter

ous fluid dynamic models. Domes that have been contained in deep craters or tephra rings (e.g. Panum dome, California; Sieh and Bursik 1986), and composite domes (e.g. Tauhara, New Zealand; Lewis 1968) cannot reflect the unhindered flow of magma and so are excluded from the following scheme. Similarly cryptodomes (e.g. the 1910 Yosomiyama eruption, Usu, Japan; Moriya 1985) are outside the realm of this study. The various dome types are sketched in Fig. 1, and their height (H) versus radius (R) relationships are plotted in Fig. 2. The most "viscous" type is discussed first, followed by progressively less "viscous" types.

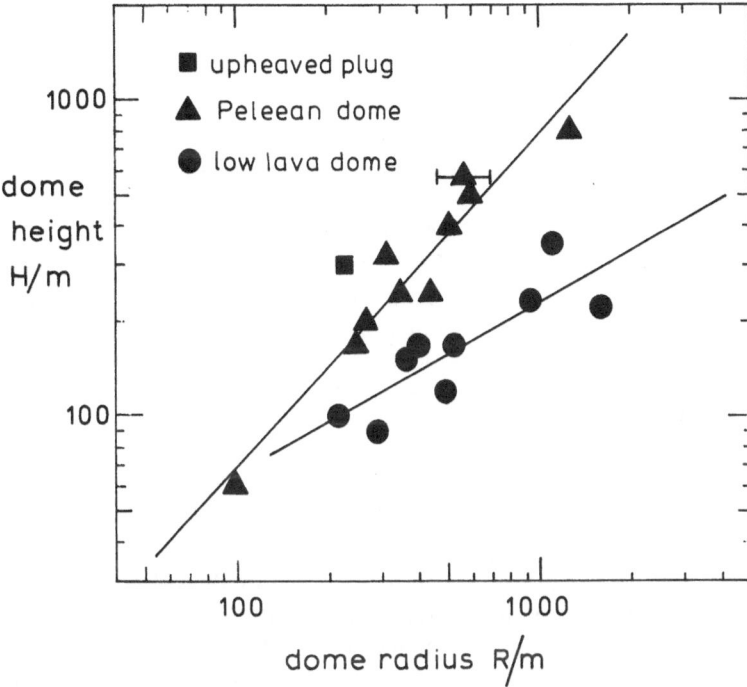

Fig. 2. Height (*H*) versus radius (*R*) plot of selected low lava domes (*circles*), Peleean domes (*triangles*) and an upheaved plug (*square*). A regression line with slope 1.06 passes through the Peleean dome population of (in order of increasing radius) Nevados de Chillan 1981, Sheveluch 1980, Puluweh 1981, Novy 1956, Great Sitkin 1974, Mount St. Helens 1984, Mont Pelee 1903, Lamington 1955, Santiaguito (Santa Maria) 1922 and Lassen Peak. A number of Japanese domes with relatively flat tops and symmetrical outlines have been selected from the compilation by Moriya (1978) to provide a population of low lava domes. They yield a regression line with a slope of 0.55. The single datum point for an upheaved plug gives the dimensions of the O'Usu dome, Japan (Williams 1932). It plots above, and furthest from, the regression line describing the Peleean domes. The height of the O'Usu dome is greater than its radius

2.1 Upheaved Plugs

Magmatic material which is so viscous, or strong, as not to deform on emerging from a conduit will form an upheaved plug (Williams 1932). The plug radius will equal that of the vent and by upward growth the dome can become columnar, such that its height exceeds its radius (Fig. 1 a). The O'Usu dome at the Japanese volcano Usu (Fig. 2) is an example of this type. The dome that protruded through the Roof Mountain cryptodome (Showa-Shinzan) of the same volcano in 1945 (Minakami et al. 1951) and spines thrust up from Peleean domes (see below) are analogous to upheaved plugs. True upheaved plugs are rare, however, probably because they can readily collapse.

2.2 Peleean Domes

The type dome in this category formed during 1902–3 at Mont Pelee, Martinique (e.g. Williams 1932; Macdonald 1972). These domes are topped by crags and spines of rock protruding from a steep collar of debris (Fig. 1 b). The characteristic talus slopes dominate the morphology of these domes, giving them a pyramidal, or conical, shape and ensure that their widths exceed that of the feeder vent (Williams and McBirney 1979).

Peleean domes grow episodically over many months or years (the Novy dome at Bezymianny, Kamchatka, has been growing since 1956; Bogoyavlyen-skaya and Kirsanov 1981) but also suffer major destructive events. During growth periods, towering spines and fins are thrust up from the dome summit, large sectors of the dome can swell outward (e.g. Lamington; Taylor 1958), or relatively fluid magma can be squeezed out of fissures (e.g. Mt. St. Helens; Swanson et al. 1987). The latter process was termed exogenous growth by Williams (1932). Disintegration of the spines and jagged crust leads to the accumulation of the distinctive apron of crumble breccia. More destructive than these rockfalls, however, are explosions that generate devastating nuees ardentes. These explosions can be triggered by the collapse of dome segments exposing hot magma to atmospheric pressure (Perret 1937) or by explosive release of pent-up magmatic or meteoric water (Fisher and Heiken 1982). Repeated sequences of growth and destruction over many years lead to a very jumbled edifice several hundred metres high.

The extensive talus apron determines the geometry of the dome's edifice (Fig. 1 b). It is expected, therefore, that the height/radius ratio of Peleean domes should be similar to the tangent of the angle of repose of cohesionless rubble, Φ. Data from nine historically erupted Peleean domes are plotted in Fig. 2 and show the relationship $H/R = 0.774 \pm 0.160$, which indicates $31° \leq \Phi \leq 43°$ in agreement with Allen (1982) and Borgia et al. (1983) who report $30° \leq \Phi \leq 45°$ for gravel and block-bearing debris. The enormous dome of Lassen Peak, California (Williams 1931) plots on the same trend (Fig. 2). Similarly, a compilation of the dimensions of young Japanese domes by Moriya (1978) shows $H/R < \tan^{-1} 33°$, indicating the influence of Φ in limiting the shape of Peleean domes.

2.3 Low Lava Domes

These domes (e.g. Soufriere 1979 and Galunggung 1918) have smooth, gently rounded profiles (Fig. 1 c) that contrast with the steep conical forms of Peleean domes (Fig. 1 b). Neither of the cited domes supported explosions but steamed profusely whilst cooling. The prominent brecciated slopes and jagged summits of Peleean domes are absent, suggesting that they were formed from less viscous magmas. Inflation of the dome by a continuous magma supply forces the dome to spread under its own weight in the process Williams (1932) identified as endogenous growth.

The heights and radii of several low lava domes from Japan (Moriya 1978 and personal communication) are plotted in Fig. 2. For a given radius these domes are lower than Peleean domes (hence the term low lava dome), which would be consistent with a more fluid behaviour. This, and the insignificance of a rubble collar, implies that the shape of low lava domes may be determined by the rheological properties of the dome rather than by the mechanical properties of a talus collar.

2.4 Coulee

Coulees are stubby lava flows (Fig. 1 d) that are transitional between symmetric low lava domes and long lava flows or streams. The rheological properties of a coulee are such that the lava can flow down a gentle incline rather than build a symmetrical low lava dome. The flow dynamics of a coulee will thus depend on the ground slope, as well as the viscous properties, and may be studied using models presented by Hulme (1974), Wilson and Head (1983), and Pieri and Baloga (1986). The spread of viscoplastic lava on a flat horizontal surface has received much less attention. In an attempt to fill this gap the remainder of this chapter is devoted to developing, testing, and applying a fluid dynamical model of dome growth.

3 Theoretical Models of the Flow of Low Lava Domes

3.1 Basics

Consider a circular lava dome of radius R and height H erupting on a planet with gravitational acceleration g (a glossary of symbols is given in Table 1). The dome is denser than its host environment (commonly air but foreseeably water) by an amount ϱ. The dome is modelled as a Bingham plastic with yield strength τ_0 and plastic viscosity η. By definition, this rheological law relates shear stress (τ) to shear rate ($\dot{\varepsilon}$) as follows

$$\tau = \tau_0 + \eta\dot{\varepsilon} . \tag{1}$$

The dome margin advances at a speed U due to the dome volume, V, growing with time, t. A versatile model of effusion history is given by the relationship (Huppert 1982; Huppert et al. 1982)

$$V = St^\alpha , \tag{2}$$

where S and α are constants. As long as the dome maintains the same geometrical shape then

$$V = c_1 HR^2 , \tag{3}$$

where c_1 is a constant dependent upon the exact shape of the dome; to an order of magnitude we can write

Table 1. Glossary of symbols, their meaning and dimensions (mass M, length L, time T)

A	Surface area of a dome, L^2
A_m	Surface area beyond the radius r_m, L^2
B	Bingham number $= \tau_0/\eta\dot{\varepsilon}$
c_i	i = 1 to 7, dimensionless constant
D	Conduit diameter, L
F_B	Force exerted by the weight of a dome, $M\,L\,T^{-2}$
F_τ	Resistive shear force, $M\,L\,T^{-2}$
g	Gravitational acceleration, $L\,T^{-2}$
H	Dome height, L
h	Vertical coordinate, L
h_m	Vertical coordinate of a marker on the surface of a dome, L
h_0	$\tau_0/\varrho g$, characteristic length scale, L
k	$1.76\,h_0^{1/2}$, $L^{1/2}$
p	V_0/V
R	Dome radius, L
R_f	Vent radius, L
R_0	Radius of a dome with volume V_0, L
R'	Radius of a model dome possessing a talus apron (Fig. 15), L
R^*	Radius of the rocky summit of a model Peleean dome (Fig. 15), L
r	Radial coordinate, L
r_c	Correlation coefficient
r_m	Radial coordinate of a marker on the surface of a dome, L
S	Supply rate coefficient [Eq. (2)], $L^3\,T^{-\alpha}$
T	Transition time-scale [Eq. (8)], T
t	Time, T
U	Speed of the margin of a spreading dome, $L\,T^{-1}$
V	Volume of a dome, L^3
V_0	Volume of a dome when $r_m = 0$, L^3
X	R^*/R'
α	Supply rate exponent [Eq. (2)]
$\dot{\varepsilon}$	Shear rate, T^{-1}
η	Plastic viscosity, $M\,L^{-1}T^{-1}$
θ	Ground slope
ϱ	Density, $M\,L^{-3}$
$\Delta\varrho$	Country rock density minus magma density, $M\,L^{-3}$
τ	Shear stress, $M\,L^{-1}T^{-2}$
τ_0	Yield strength, $M\,L^{-1}T^{-2}$
Φ	Angle of repose

$$V \sim HR^2 \ . \tag{4}$$

The physics that govern the shape of the dome entail a balance between the buoyancy force, F_B, which acts radially outward and is due to the weight of the dome, and the resistive force, F_τ, due to shear stress on the dome base. These forces have order of magnitude values

$$F_B \sim g\varrho H^2 R \ ; \tag{5}$$

$$F_\tau \sim \tau R^2 \ . \tag{6}$$

Implicit in Eq. (6) is the assumption that the area of the vent, R_f^2, is much less than the area covered by the dome, R^2.

The force balance $F_B \sim F_\tau$ describes the equilibrium state of the dome when inertial forces are negligible (as will be the case for a slowly growing viscous lava dome). This force balance can take either of the two general forms discussed below. Firstly, results obtained by Huppert (1982) and Huppert et al. (1982) for the spread of Newtonian fluid ($\tau_0 = 0$, or a Bingham fluid such that $\tau_0 \ll \eta \dot{\varepsilon}$) will be reproduced. Secondly, the shape of a viscoplastic material in the limit $\tau_0 \gg \eta \dot{\varepsilon}$ will be determined along lines originally taken by Orowan (1949) and Nye (1952). The conditions for which the transition between Newtonian and Bingham plastic behaviours occur will also be investigated.

3.2 Spread of a Newtonian Fluid

The following few lines of algebra precis results obtained by Didden and Maxworthy (1982), Huppert (1982) and Huppert et al. (1982). The buoyancy versus viscosity force balance for a slowly spreading Newtonian fluid of dynamic viscosity η is, from Eqs. (1), (5) and (6)

$$g\varrho H^2 R \sim \eta \dot{\varepsilon} R^2 . \tag{7}$$

The shear rate is, to an order of magnitude, $\dot{\varepsilon} \sim U/H$ and similarly $U \sim R/t$. Substituting these scales, together with the conservation equation $V = St^\alpha \sim HR^2$ into Eq. (7) yields

$$R = c_2 (g\varrho S^3/\eta)^{1/8} t^{(3\alpha+1)/8} ; \tag{8}$$

$$H = c_3 \left(\frac{\eta S}{g\varrho}\right)^{1/4} t^{(\alpha-1)/4} ; \tag{9}$$

$$H = c_4 (SR^{2(\alpha-1)} (\eta/g\varrho)^\alpha)^{1/(3\alpha+1)} . \tag{10}$$

The constants c_2, c_3 and c_4 are of order unity. Huppert (1982) has determined c_2 to be a weak function of α; it is expected that c_3 and c_4 will also be dependent on α.

Equation (9) shows that for a constant effusion rate ($\alpha = 1$) the dome height will not change with time. Similarly, for waxing effusion ($\alpha > 1$) H increases with time and for waning effusion ($\alpha < 1$) H decreases with time. Equation (8) shows that R increases indefinitely with time (even when $\alpha = 0$) so that a frozen lava dome may owe its shape to the agencies responsible for halting its motion rather than those that dictated its spreading behaviour.

3.3 Spread of a Viscoplastic, Modelled as a Bingham Plastic

The motion of a Bingham plastic (or in fact any viscoplastic) ceases when the stress falls below the yield strength τ_0. A viscoplastic dome will thus be in equilibrium when, according to Eqs. (5) and (6)

$$g\varrho H^2 R \sim \tau_0 R^2 \; . \tag{11}$$

This, together with the conservation equation $V = St^\alpha \sim HR^2$ leads to

$$R = c_5 \, (S^2/h_0)^{1/5} t^{2\alpha/5} \; ; \tag{12}$$

$$H = c_6 (Sh_0^2)^{1/5} t^{\alpha/5} \; ; \tag{13}$$

$$H = c_7 \, (h_0 R)^{1/2} \; , \tag{14}$$

where c_5, c_6 and c_7 are constants of order unity and the natural length scale h_0 is defined by

$$h_0 = \tau_0/\varrho g \; . \tag{15}$$

Nye (1952) showed that c_7 has an approximate value of $\sqrt{2}$.

Notice that the relationship between H and R at a given moment [Eq. (14)] does not depend on S, α or t. Also, when effusion stops (i.e. $\alpha = 0$) H and R do not continue to change. This is an important contrast with the behaviour of Newtonian fluid and arises because the governing force balance for a Newtonian dome involves the dynamic process of viscous shear whilst for a viscoplastic dome the force balance is attained in a state of static equilibrium. Experiments can show how a viscoplastic dome can grow from one state of static equilibrium to the next.

3.4 Transition Between Newtonian and Bingham Viscoplastic Behaviours

The spread of a viscoplastic can be described by the model of a spreading Newtonian fluid if the viscous stress in the flowing dome is much greater than the yield strength. Thus if the right hand side of Eq. (1) is dominated by the second term ($\eta\dot\varepsilon$) rather than the first term (τ_0), then the fluid behaviour will be approximately Newtonian. To decide if this condition will be met in a given situation it is necessary to calculate the ratio

$$B = \tau_0/\eta\dot\varepsilon \; . \tag{16}$$

When this ratio, known as the Bingham number, is much less than one, then Newtonian behaviour will occur. In this case $\dot\varepsilon \sim U/H$ so that the requirement for Newtonian flow can be written as

$$B = \frac{\tau_0 H}{\eta U} \ll 1 \; . \tag{17}$$

R and H are then given by Eqs. (8) and (9), so that, noting $U \sim R/t$, inequality (17) can be expressed in terms of a time scale, T, such that

$$t^{(\alpha-5)} \gg T^{(\alpha-5)} = \frac{\tau_0^8}{g^3\varrho^3\eta^5 S} \tag{18}$$

is required for Newtonian flow. For $\alpha = 5$, flow will always be Newtonian if the yield strength is so small that

$$\tau_0 \ll (g^3 \varrho^3 \eta^5 S)^{1/8} \ . \tag{19}$$

If this inequality is not satisfied then the dome will always behave as a plastic [Eqs. (12)–(14)]. A more intriguing feature of Eq. (18) is that for $\alpha < 5$ then the flow is at first Newtonian until T, after which the plastic component of the material dominates its spreading dynamics. Conversely, for $\alpha > 5$ the flow starts as a plastic and after time T starts to follow a Newtonian behaviour. These changes in flow regime can be understood by considering, for example, the latter case. Here, a rapidly increasing flow rate ($\alpha > 5$) promotes purely viscous shearing throughout the material, culminating in Newtonian behaviour. Analogous possibilities for changes in flow regime of Newtonian fluids, from viscosity-dominated to inertia-dominated flows, or vice versa, have been recognized by Huppert (1982) and verified experimentally by Maxworthy (1983).

4 Laboratory Models of the Flow of Lava Domes

The flow of viscoplastic material in controlled laboratory experiments provide insights into the behaviour of lavas under complex natural conditions (e.g. Hulme 1974). In order to test the theoretical predictions advanced in the previous section, experiments into dome growth have been conducted with kaolin slurries.

4.1 Experimental Apparatus and Procedure

Experiments were performed using the apparatus sketched in Fig. 3, based on the techniques of Hulme (1974). An approximately 50:50 mix of kaolin and de-ionized water was stored in a 10-l or 20-l polythene jerry can. Water under mains pressure was fed at a monitored rate into a strong balloon sealed into the jerry can. This forced the slurry into a rigid pipe which fed the model lava dome through a 2.8-cm diameter vent. This procedure was intended to maintain a known constant effusion rate (i.e. $\alpha = 1$, $V = St$). The walls of the jerry can became distended during the early part of each experiment, however, so that there was little control on the effusion rate. It will be shown later that this shortcoming was not a hindrance to the quantitative analysis of the experiments.

The model domes were extruded onto a horizontal surface covered with cartridge paper. The paper prevents a lubricating film of water from bleeding from the suspension and so ensures that the dome experiences shear stresses dependent on the dome's rheological properties. A clock, and concentric circles and graduated radial spokes marked on the paper allowed the dome's growth rate to be measured from vertical photographs taken at every centimetre increase in radius. A dome up to 50 cm in diameter was produced in each experiment.

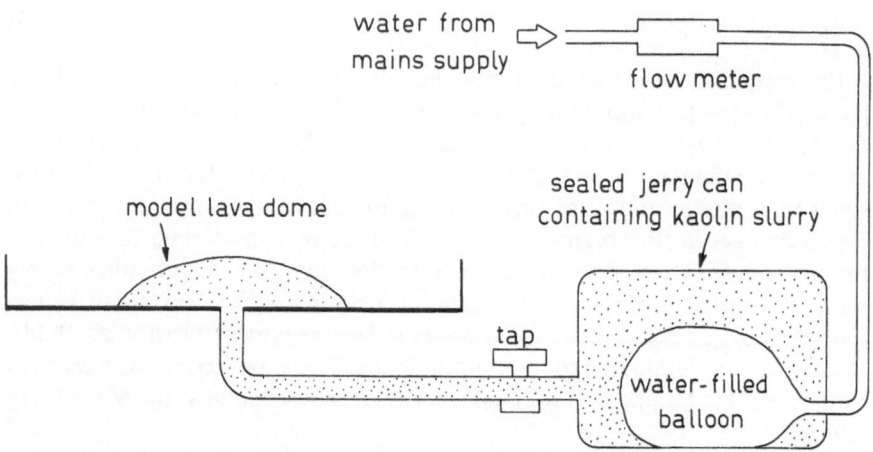

Fig. 3. Schematic of the apparatus used to produce domes of kaolin slurry. See text for details on experimental procedure

Once the supply of slurry had been turned off no further motion occurred. The cross-sectional shape of the dome was measured by traversing across several diameters of the dome with a probe. This allowed the maximum depth (H) of the dome immediately above the vent to be estimated by interpolation.

In two experiments H was measured as a function of time during dome growth with a theodolite. A further technique used in some experiments involved placing markers on the dome surface and observing their motion during dome growth.

The physical properties of the kaolin suspensions were measured using material taken from the domes. The density was determined by weighing 300 ml of slurry. Rheological properties were measured with a Haake Rotovisko viscometer using a shear vane to determine the yield strength and a concentric cylinder apparatus to determine the plastic viscosity. Standard techniques (van Wazer et al. 1963 pp. 55 – 61) were used to invert the raw torque versus rotation rate measurements into shear stress versus shear rate data independently of any presumed rheological model. Typical rheograms are shown in Fig. 4 where, because of non-linearity in the $\tau - \dot{\varepsilon}$ relations, it becomes necessary to define an apparent yield strength τ_0' and two asymptotic plastic viscosities η and η'. The behaviour of the suspensions at low shear rates, as is anticipated for slow dome growth, is limited by the true yield strength, τ_0, and described by the ideal Bingham model investigated above. On the other hand, "Newtonian" flow will be characterized by high shear rates, so the quantities τ_0' and η' are of concern. It is these values which should be used in place of τ_0 and η in Eq. (16) when checking whether the flow conditions can be considered Newtonian. Plastic behaviour is assured if inequality (8) is not satisfied.

The conditions and physical parameters for each experiment are listed in Table 2. The time, T, after which viscoplastic behaviour is anticipated was calculated to be less than 1 s in every experiment. The sole rheological parameter

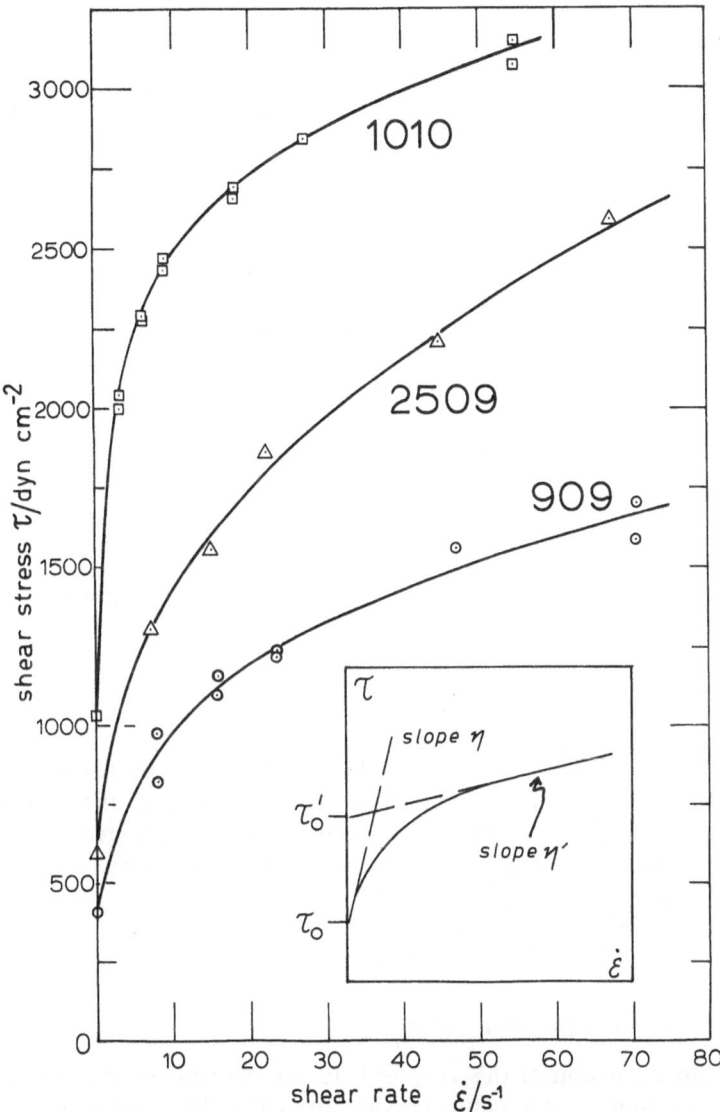

Fig. 4. Rheograms of the kaolin suspensions used in experiments *909, 2509* and *1010* (also *1710*). The *inset* defines various rheological properties of the suspensions. Thus, at low shear rates the material behaves as a Bingham plastic with yield strength τ_0 and plastic viscosity η. At high shear rates, however, the material behaves as a Bingham plastic with plastic viscosity η' and an apparent yield strength τ_0'. Note that τ_0' is a fictitious quantity, and it is the value of τ_0 which limits the material's behaviour as $\dot{\varepsilon} \rightarrow 0$ (10 dyn cm^{-2} = 1 Pa)

Table 2. Experimental conditions and results

Dome number	Properties of kaolin slurry				Recharge rate S/cm^3 $\cdot$s^{-1}	Average radius R/cm	Maximum height H/cm
	ϱ/g $\cdot$cm^{-3}	τ_0/dyn $\cdot$cm^{-2}	η/ poise	h_0/ cm			
909	1.36	420	120	0.315	3.8	22.9	4.75
2509	1.41	580	200	0.419	1.1	30.0	6.4
1010	1.43	1040	320	0.741	1.2	22.5	6.9
1710a*[a]	1.43	1040	320	0.741	*	5.0	3.45
1710b	1.43	1040	320	0.741	*	6.75	4.0
1710c	1.43	1040	320	0.741	*	10.0	4.85
1710d	1.43	1040	320	0.741	*	15.5	6.0
3110	1.43	1200	n.m.[b]	0.855	2.4	18.4	7.0
2002a*[a]	1.42	1270	n.m.	0.912	*	9.75	5.0
2002b	1.42	1270	n.m.	0.912	*	16.75	6.9

[a] * Experiments 1710 and 2002 involved measuring successive additions to a single dome.
[b] n.m. Not measured.

controlling the behaviour of the model domes is thus expected to have been their yield strength τ_0.

4.2 Dome Height as a Function of Yield Strength and Dome Radius

A mound of plastic material will be expected to have a height which is proportional to the square roots of the yield strength and of the dome radius [Nye 1952 and Eq. (14)]. The experimental data (Table 2) agree exceedingly well with this prediction (Fig. 5a), and give the regression lines

$$H/h_0 = 1.746 \, (R/h_0)^{0.503}$$

(with a correlation coefficient $r_c = 0.998$) and

$$H = (1.760 \pm 0.043) \, (R \, h_0)^{1/2} \; . \tag{20}$$

Since one aim of this chapter is to use the heights and radii of growing low lava domes to estimate their yield strengths, the experimental results have also been expressed in the more useful form

$$\tau_0 = (0.323 \pm 0.016) \, (H^2 \varrho g / R) \; . \tag{21}$$

Dome heights measured by theodolite during the growth of domes 1010 and 3110 agree well with Eq. (20), as shown in Fig. 5.

4.3 Dome Cross-Sectional Shape

Nye (1952) has shown that a circular dome with a gently sloping surface will have an equilibrium shape in cylindrical coordinates (r, h) given by

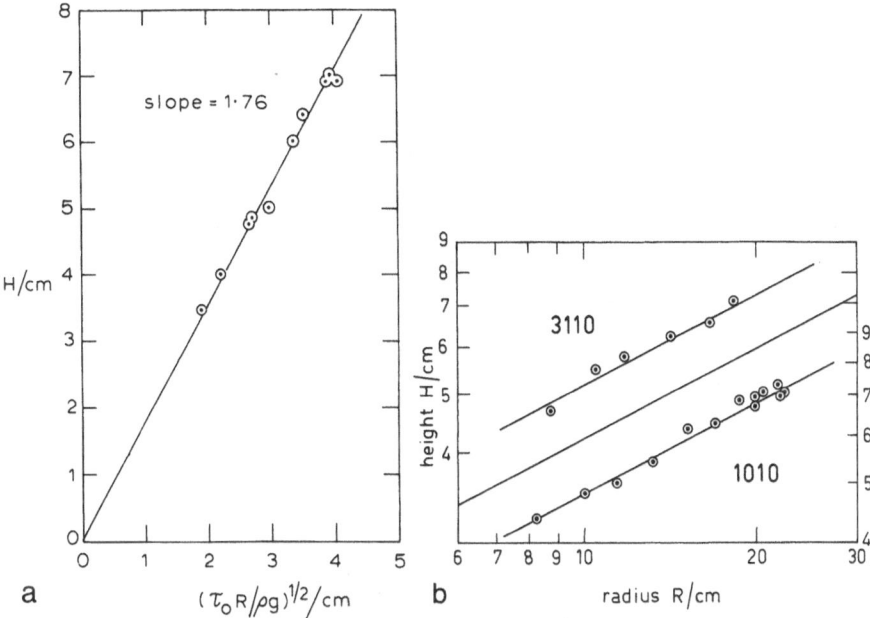

Fig. 5. a The final dimensions of experimental domes shows agreement with the relationship $H = 1.76 \, (\tau_0 R/\varrho g)^{1/2}$ [Eq. (20)]. *b* The heights and radii of domes *1010* and *3110* during their growth are described by the static force balance defined by Eq. (20) and given by the *straight lines* in these plots

$$h = \sqrt{2 \, h_0 \, (R-r)} \ . \tag{22}$$

This parabolic profile is vertical at the snout of the dome (at $r = R$) and has a peak (at $r = 0$) which is not horizontal. It was difficult to decide whether or not the experimental domes showed these two features. Nye's solution must be only approximately correct, however, because his assumption of a gentle slope everywhere on the surface does <u>not</u> hold near the steep dome margins. Thus, whilst Eq. (22) predicts $H = \sqrt{2 R h_0}$, the experiments show that $H = 1.76 \, (R h_0)^{1/2}$ with a precision which precludes the difference between 1.76 and $\sqrt{2}$ (≈ 1.41) being due to experimental uncertainties.

Figure 6 shows the cross-sectional shapes of nine domes. Also shown are the parabolas defined by an empirically modified version of Eq. (22), namely

$$h = 1.76 \, [(R-r)h_0]^{1/2} \ . \tag{23}$$

This forces agreement between the model and observation at the peak and at the edge (defined by the mean radius) of each dome but also provides a service-able description of the dome's shape. The influence of the finite-sized feeder beneath the domes, uncertainties in measurements of r and h, and departures from perfect circular symmetry contribute to ambiguities in assessing the dif-ference between Eq. (23) and the observations plotted in Fig. 6. The largest and most symmetrical domes will be expected to give the most meaningful compar-

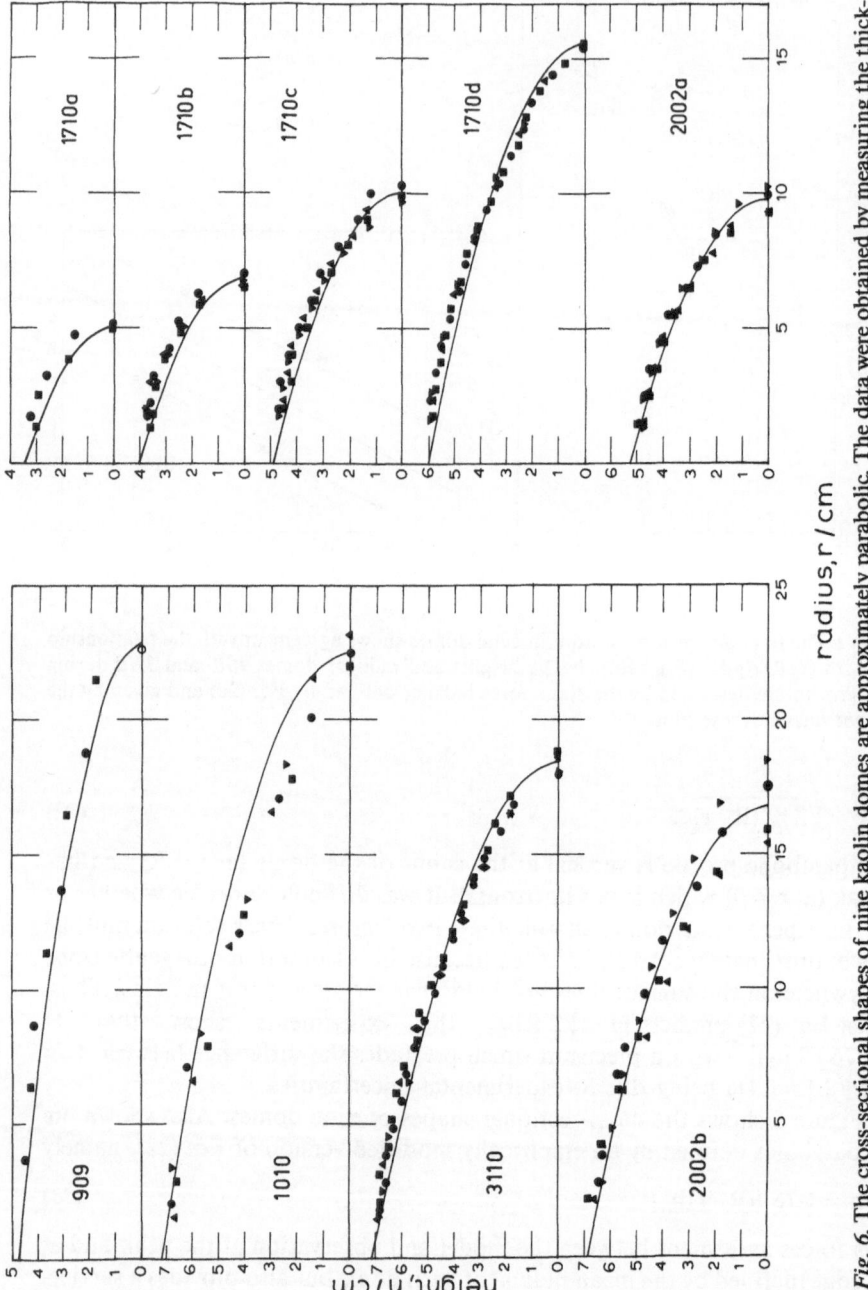

radius, r / cm

Fig. 6. The cross-sectional shapes of nine kaolin domes are approximately parabolic. The data were obtained by measuring the thickness of each dome across one or two diameters. The peak height (H in Table 2) was interpolated from the plotted data. The *circles* and *squares*, and the *upright* and *inverted triangles* pertain to profiles measured on the same diametral traverses. The *curve* drawn alongside each set of data is given by $h = 1.76\ (h_0\ (R-r))^{1/2}$ where R is the average dome radius

ison between the modelled and observed profiles. These qualities of size and symmetry were best met in experiments 1710c, 1710d, 3110 and 2002a. The results from these experiments (Fig. 6) indicate that Eq. (23) tends to underestimate the height close to the summit and overestimate height close to the margin. The summit region is also slightly flatter than given by Eq. (23). Overall, however, the simple parabolic curve is a satisfactory model of the dome shape and can be integrated to describe a number of geometrical properties. Thus, it is found that

average height, $\bar{H} = 8\,H/15$;   (24)

volume, $V = 8\,\pi\,1.76\,(h_0\,R^5)^{1/2}/15$;   (25a)

or [cf. Eq. (3)] $V = 8\,\pi HR^2/15$.   (25b)

The surface area, A, is less simple but, defining $k = 1.76\,h_0^{1/2}$,

$$A = (\pi/32)\,\{2\,[R\,(4\,R + k^2)]^{1/2}(8\,R - k^2)$$
$$+ k^2\,(16\,R + k^2)\,\ln\,[(2\,R^{1/2} + (4\,R + k^2)^{1/2}/k]\}\ .$$ (26)

4.4 Surface Textures

It has been shown in Fig. 5 that the heights and radii of both stationary and expanding domes are related by the same equation. This common relationship [Eq. (20)] indicates a state of static equilibrium such that the force imposed by the dome's weight is balanced by a basal shear force limited by the yield strength. It is interesting to consider how a growing dome can deform in a way which allows it to remain statically stable whilst undergoing the dynamic process of growth. The experiments suggest a way in which this is achieved.

The surface of every dome was seen to be divided into a network of rhombohedral segments defined by two intersecting sets of spiral grooves (Fig. 7). These markings are interpreted to be the traces of slip planes that cut through the dome such that one set of slip plane traces spirals out from the dome summit in a clockwise direction whilst the other spirals out anti-clockwise. The viscoplastic material is thought to deform along the complex array of slip planes within the dome, enabling it to continuously re-adjust its shape to remain in the stable configuration described by Eq. (20).

The characteristic spacing of the slip plane traces are typified by observation of the 15.5 cm radius dome 1710d. From the peak of this dome out to about 2 cm radius, the surface texture was indistinct. Between 2 and 4 cm radius, the rhombohedral texture became noticeable and between 4 and 7 cm the distance between successive spiral traces was less than 1 mm. This spacing gradually became larger with greater radial position, being about 2 mm at 11 to 13 cm and about 4 mm beyond 13 cm. The magnitude of the surface relief caused by these spiral markings also coarsened towards the dome margins, and was most pronounced in domes having large yield strengths. This suggests that, at a given radial position, the spacing between slip plane traces would be pro-

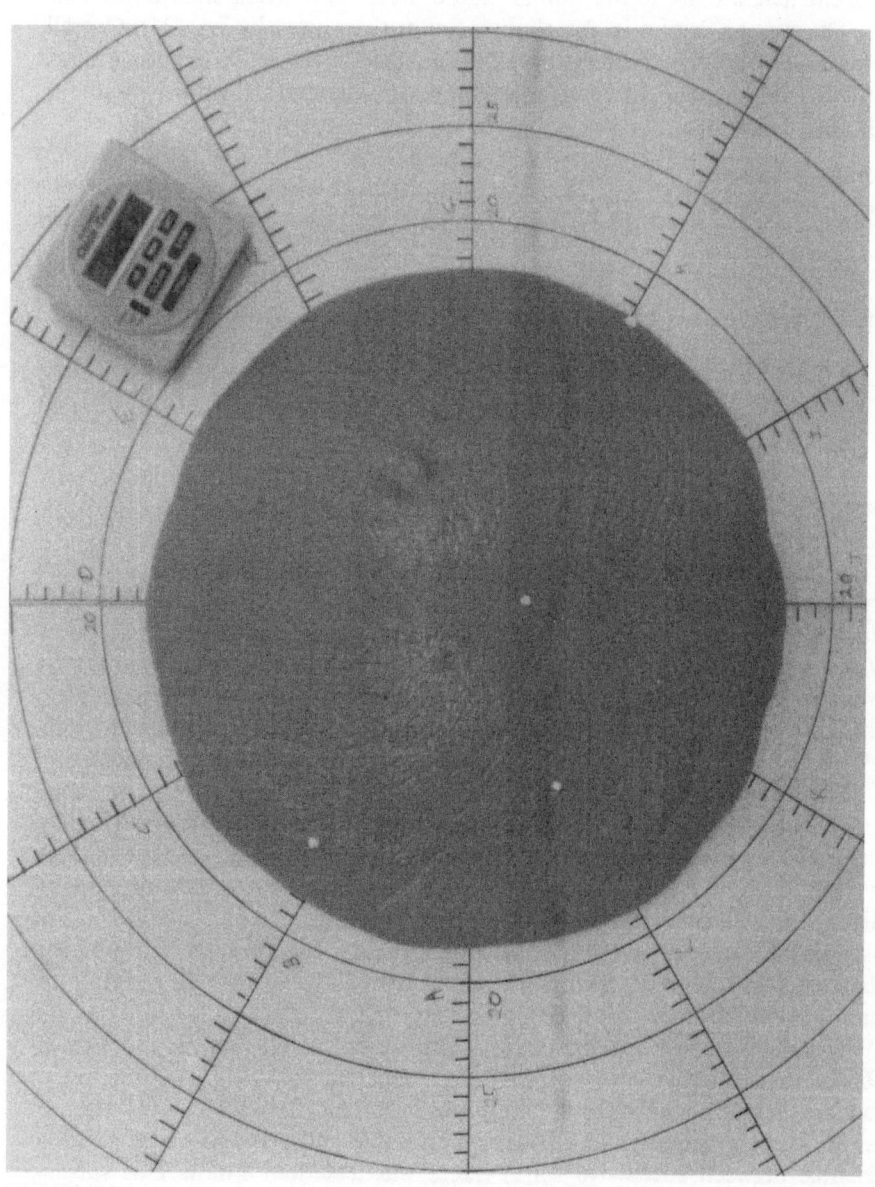

Fig. 7. Vertical photograph taken during the growth of dome 3110 after 26 min and 13 s (mean radius R = 17.6 cm). The radial scale is marked in centimetres, with concentric circles every 5 cm. Note the two sets of spiral slip plane traces which fan out from the dome's summit. Four white polystyrene marker beads are also visible. Their positions reflect

portional to the characteristic length scale of the material h_0 (7.41 mm for this dome).

Other surface textures or structures were not observed in the experiments. Even in experiments 1710 and 2002, which involved phases of effusion separated by periods of no effusion, the dome surfaces remained smoothly curved and were traversed by even patterns of spiral slip plane traces. Changes in effusion rate to a higher or lower value produced neither concentric nor radial structures.

4.5 Dome Radius and Height as Functions of Time

The fact that the equilibrium relation for a viscoplastic dome [Eq. (20)] was maintained during the growth of domes 1010 and 3110 (Fig. 5b) allows us to take an important step towards describing the expansion rate of growing domes. Specifically, the volume of the dome at any time can be determined from its dimensions using Eq. (25b) and equated to its value defined by Eq. (2). Manipulation of these equations allows Eqs. (12) and (13) to be solved for c_5 and c_6,

$$R = 0.65 \, (S^2/h_0)^{1/5} \, t^{2\alpha/5} \; ; \tag{27}$$

$$H = 1.4 \, (Sh_0^2)^{1/5} \, t^{\alpha/5} \; . \tag{28}$$

Data from the two experiments (909 and 3110) in which the kaolin slurry was stored in the smaller reservoir are plotted in Fig. 8 and are seen to asymptote to the trends predicted by Eqs. (27) and (28) with $\alpha = 1$. At early times

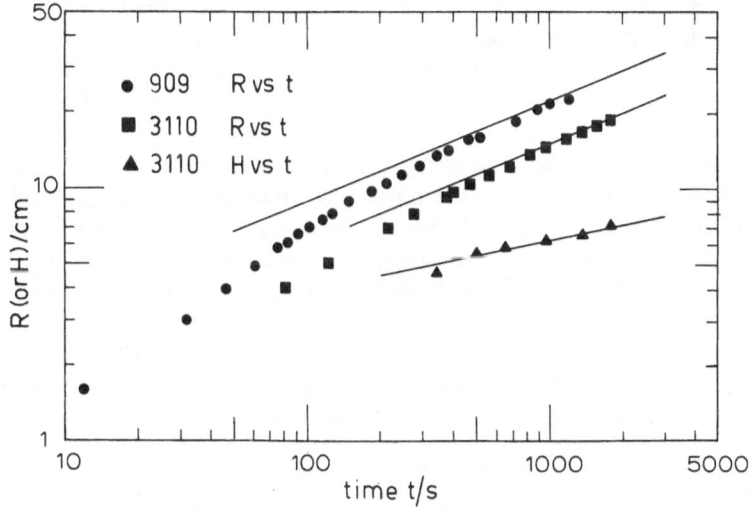

Fig. 8. Radii of domes *909* and *3110*, and height of dome 3110 plotted against time are found to agree with the behaviour anticipated by Eqs. (27) and (28)

the data plot below the theoretical predictions. This discrepancy between the observed R–t and H–t trends and the model relationships $R \propto t^{0.4}$ and $H \propto t^{0.2}$ at early times can be ascribed to a combination of (1) changes in α and S when the reservoir was expánding, and (2) the influence of the finite-sized outlet ($R_f = 1.4$ cm). Figure 5b indicates that the latter aspect was unimportant by the time $R \simeq 3 R_f$. The effects of reservoir expansion seem to have taken longer to be overcome. Only in two experiments 909 and 3110 did the volume and pressure in the reservoir apparently become stabilized (presumably due to the small capacity of the reservoir used in these particular experiments) such that a constant effusion rate ($\alpha = 1$) was reached. In these experiments the mean effusion rate (total dome volume ÷ experiment duration) was less than 10% lower than the reservoir recharge rate. This encourages the view that these experiments reached a steady state in which the efflux rate equalled the monitored recharge rate so as to give the internally consistent match portrayed in Fig. 8 between the data and the model [Eqs. (27), (28)].

4.6 Advance of Surface Markers

During experiments 1010, 3110 and 2002, small (ca. 5 mm diameter) beads of expanded polystyrene were placed onto the summit area of the growing domes (e.g. Fig. 7). The motion of each marker bead was followed using the timed photographs of the experiments. Each bead was carried outward on a purely radial path. The radial speed of a given bead decreased as it approached the dome margin, however, so that no marker ever reached the extreme front of the dome. Thus, unlike low viscosity basalts that advance by over-running their flow margins in the manner of a tractor tread (Macdonald 1972, pp. 76) the surficial material of these viscoplastic models did not descend to the flow front. Instead, stretching of the surface must have occurred to permit radially moving markers to "pile-up" close to but not at the flow margin.

A model of this behaviour which can be tested quantitatively using data from the experiments is illustrated schematically in Fig. 9. The diagram shows two stages in the growth of a dome. At the earlier stage (Fig. 9a) a position on the surface is marked at (r_{ma}, h_{ma}) when the dome margin is at (R_a, 0). At the later stage (Fig. 9b) the marker has moved to (r_{mb}, h_{mb}) and the margin to (R_b, 0). It is proposed that the surface area on the dome's curved surface below elevation h_{ma} on the earlier dome equals that below h_{mb} on the later dome. In other words the shaded regions of the domes in Figs. 9a and 9b are of equal area.

This idea is tested by calculating areas from measured values of R and r_m and assuming a dome shape given by Eq. (23). The surface area below an elevation h_m is denoted by A_m and is calculated to be

$$A_m = (\pi/32\,k^4)\,\{2\,h_m\,(4\,h_m^2 + k^4)^{1/2}\,(16\,R\,k^2 - 8\,h_m^2 - k^4)$$
$$+ k^6\,(16\,R + k^2)\,\ln\,[(2\,h_m + (4\,h_m^2 + k^4)^{1/2})/k^2]\}\,, \qquad (29)$$

where $k = 1.76\,h_0^{1/2}$ and $h_m = k\,(R - r_m)^{1/2}$.

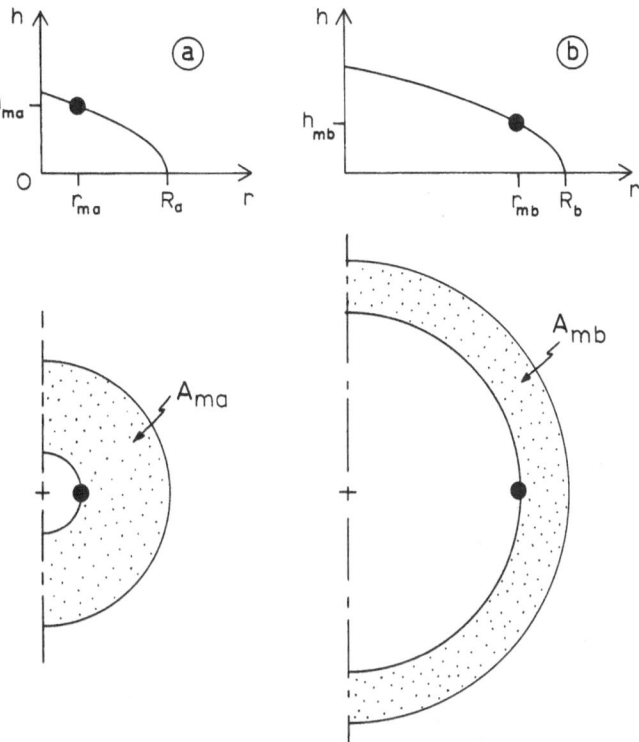

Fig. 9. A marker (*filled circle*) on the surface of a growing dome moves from the position (r_{ma}, h_{ma}) to (r_{mb}, h_{mb}) as the dome spreads from radius R_a to R_b. In *a* and *b* the *upper diagram* shows a section through half of the domes, and the *lower diagram* shows a plan view of each half dome. In order to explain the experimental observation that a given marker never reaches the dome margin (i.e. $r_{mb} < R_b$) it is proposed that the *stippled regions* A_{ma} and A_{mb} on the dome surface between the marker and the dome margin are equal in area

It is possible to calculate A_m from the physical properties which determine the characteristic length scale and the measured distances r_m and R. Equation (29) reduces to Eq. (26) when $h_m = H = kR^{1/2}$.

Data from experiments 1010 and 3110 are plotted in the form A_m versus dimensionless marker position, r_m/R, in Fig. 10. There are three trends in these plots. At positions close to the dome summit ($r_m/R \lesssim 0.3-0.5$) the surface area beyond a marker is calculated to have increased as the marker moved outward. Uncertainties of ± 2 mm in both r_m and R (due to errors in making measurements from the photographs and variability in the radius of slightly asymmetric or uneven domes) produce uncertainties in A_m which are less than the changes in area calculated for successive positions. This implies that fresh material breaches the surface of the dome within this summit region. Between $0.3 \lesssim r_m/R \lesssim 0.8-0.9$, the area beyond a marker is calculated to remain constant within limits which are much less than the anticipated uncertainties (Fig. 10). This implies that the surface of a dome experiences radial shortening and

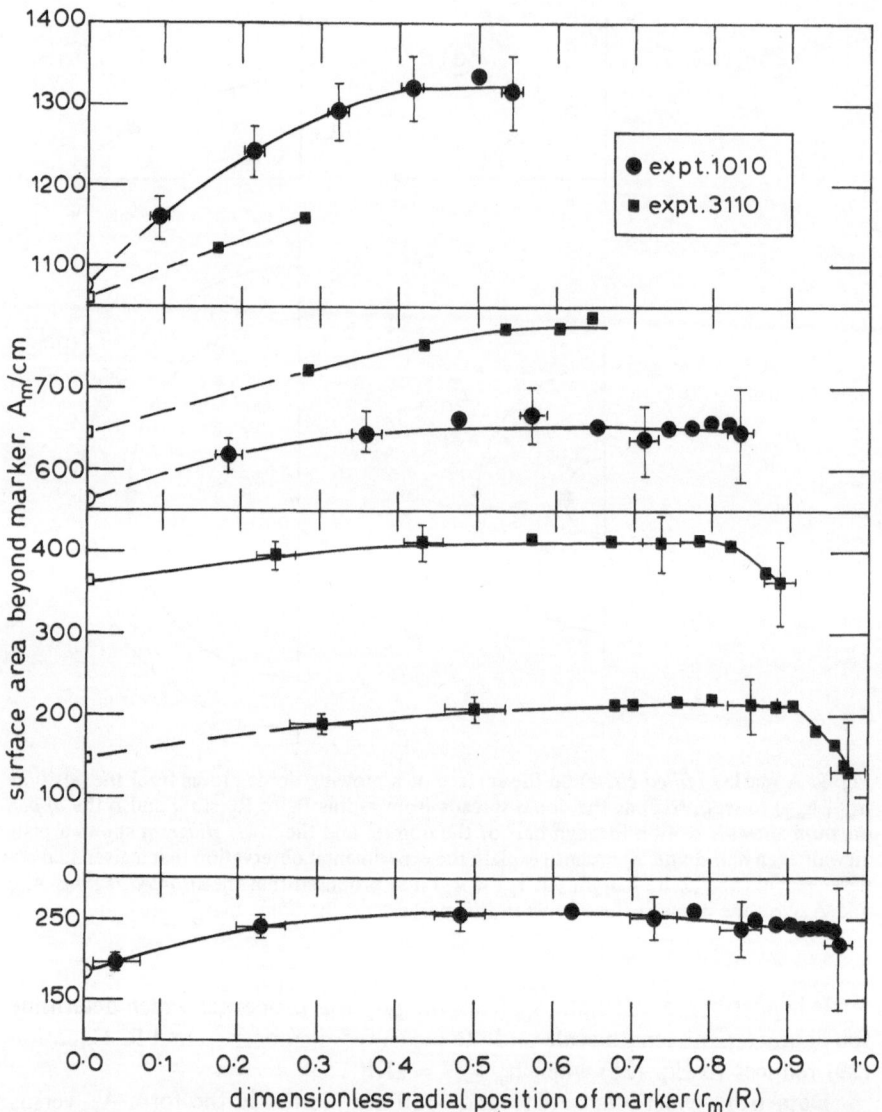

Fig. 10. The surface area between a given marker and the dome's edge, A_m, is calculated from Eq. (29) and plotted against the marker's position r_m/R. Results obtained from the paths of three markers in experiment 1010 (*circles*) and four markers in experiment 3110 (*squares*) show that A_m remains constant (within limits calculated for ± 2 mm uncertainties in the measurements) beyond $r_m/R \simeq 0.3$ to 0.5. The *dashed lines* are back-extrapolated trends giving the total dome surface area (*open symbols*) when the given marker would have been on the peak of the dome

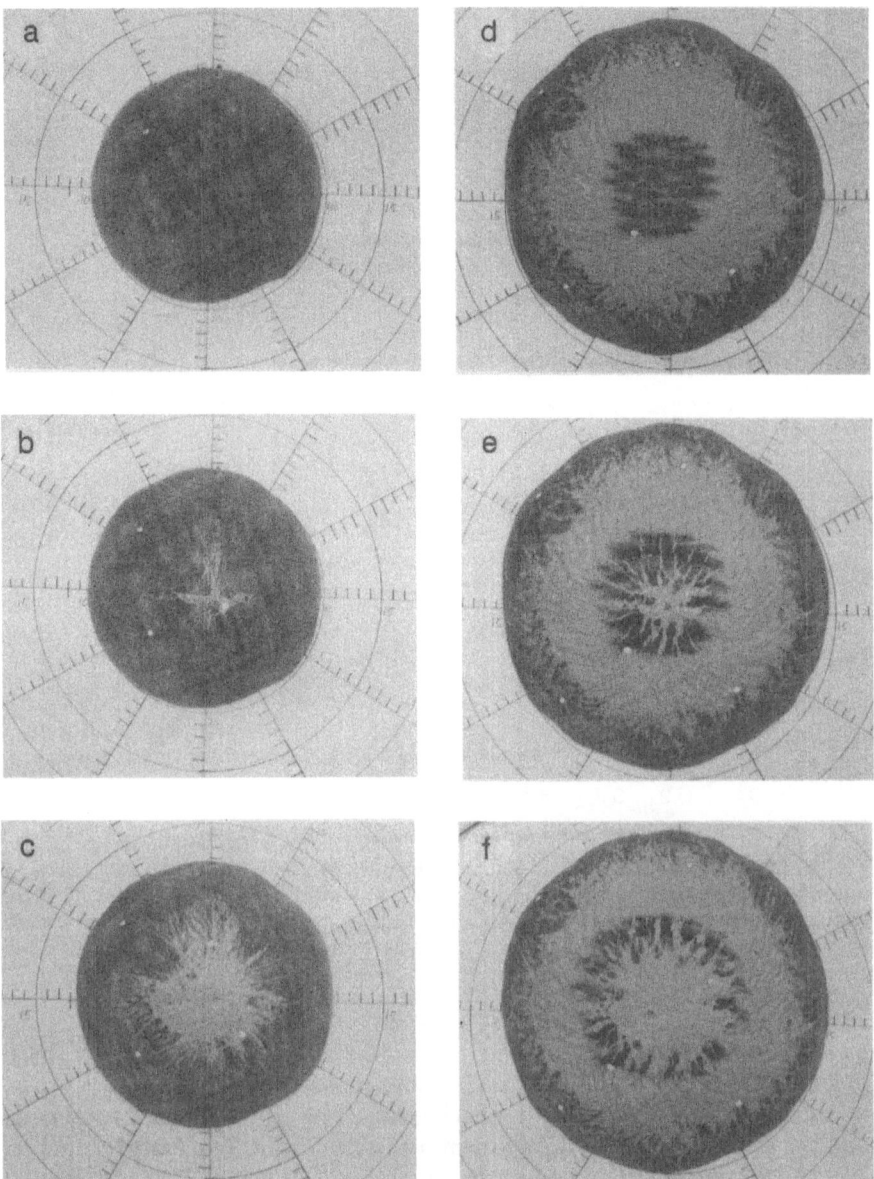

Fig. 11 a−f. Stages in the growth of dome 2002b (scale on base as in Fig. 8). *a* Kaolin effusion has been halted and the entire surface of the dome is covered with a sprinkling of dark sand. *b* Growth of the dome has been restarted, causing the summit to be torn apart so that light-coloured kaolin slurry is being exposed along short radial openings in the old sandy surface. *c* The dome surface displays two concentric regions of different age. The dark, sand-covered, outer part has not been disturbed by fresh material. The light summit region has appeared since the time shown in *a* (cf. Fig. 9). *d* Dome effusion has been stopped again and a capping of sand sprinkled over the dome's summit. Note that the peripheral sand-covered band has become narrower since the previous frame. *e* The second coating of sand is breaking up because of dome expansion. Note that the spiral slip plane traces do not appear to be related to the "tears" in the sand. *f* The second covering of sand has been stretched aside at the dome peak, revealing fresh kaolin. All of the sand which initially coated the entire dome surface (*a*) is now concentrated around the rim of the dome in a narrow band (cf. Fig. 9)

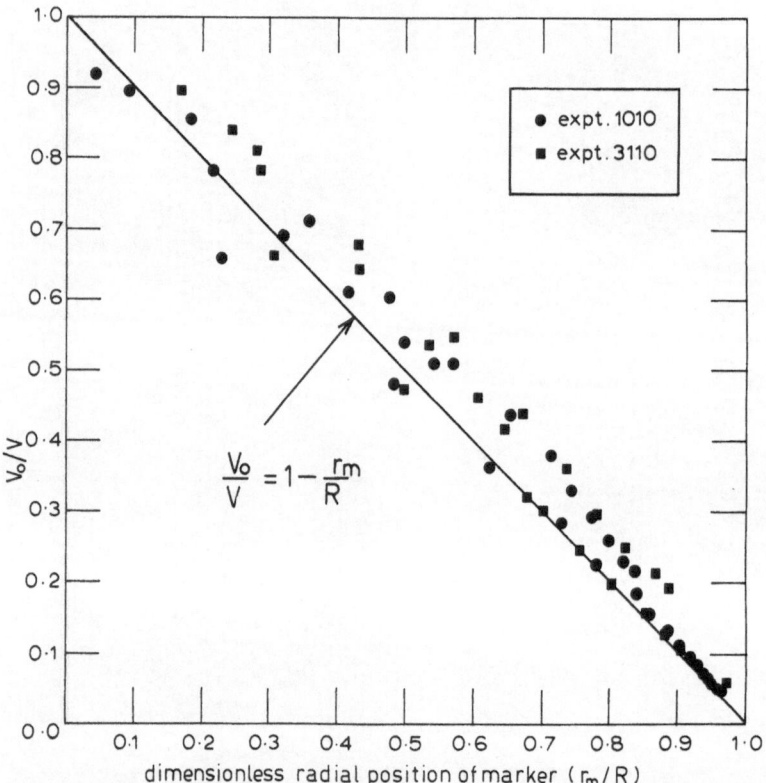

Fig. 12. Material which has just appeared at the peak of a dome of volume V_0 will move out to a radial position r_m whilst the dome grows to a new volume V and radius R. These variables have been determined using data from experiments 1010 (*circles*) and 3110 (*squares*) and show, to a first approximation, agreement with the line $V_0/V = 1 - r_m/R$

circumferential stretching. Beyond $r_m/R \simeq 0.8-0.9$, the plots show a sharp downturn, but there are a number of reasons for suspecting this to be spurious. Variation in the dome radius approaches the size of the difference between R and r_m, so that Eq. (29), which assumes perfect circular symmetry, cannot be expected to cope with an uneven dome outline when $r_m \simeq R$. It can also be seen in Fig. 6 that the very edge of the dome is the location where the parabolic profile [Eq. (23)] can show large departures from the measured profiles. The consequence of these problems are made plain by uncertainties in r_m and R of ± 2 mm yielding uncertainties in A_m of over 70% (Fig. 10).

The favoured interpretation of Fig. 10 is that it shows the dome surface to be divisible into two regions or domains. In the inner region ($r_m/R \lesssim 0.3-0.5$) new material can appear at the surface. Beyond this region no new material appears and no old material disappears. Further verification of this division comes from a further experiment in which the entire surface of dome 2002a was covered with sand (Fig. 11 a). On resuming the growth of this dome, fresh material became visible only near the summit (Fig. 11b, c, e, f). No evidence was found for the surface coating of sand being consumed into the dome or of it being over-ridden at the dome front.

The data on marker position and movement can be applied to the analysis of the disposition of parts of the dome erupted at different times. This is of concern when attempting to estimate the volumetric proportions of chemically zoned domes, or those containing an uneven distribution of inclusions (e.g. Koyaguchi 1986). Consider a growing dome which has reached a volume V_0 and radius R_0. The material on the peak of the dome ($r = 0$, $h = H_0$) will move outward to a new position at $r = r_m$ as the dome grows to volume V and radius R. We wish to determine how the labelled position r_m relates to the original volume V_0. This can be done for domes which are close to circular in outline using the experimental results. The dome radius, R_0, at the moment when $r_m = 0$ was obtained by back extrapolating a plot of r_m versus R. V_0 is then calculated from R_0 using Eq. (25a), and V is similarly calculated from R. A useful relationship between these quantities is sought by plotting V_0/V versus r_m/R in Fig. 12. This reveals, to a first approximation, that

$$V_0/V = 1 - r_m/R . \tag{30}$$

Thus, a sample which first reached the surface at the dome peak after a proportion p ($= V_0/V$) of the final symmetrical dome has been erupted will be found at a radial position $(1-p)R$. The converse of this result is not necessarily true, however, since material does not always reach the dome surface at the peak (Figs. 10, 11).

5 Analysis of Lava Domes as Viscoplastics

We now turn to the consideration of natural lava domes. The first part of this section shows that Huppert et al.'s (1982) data on the Soufriere dome can be interpreted satisfactorily in terms of the viscoplastic model developed above. The yield strengths of this and other low lava domes [such as the June 1980 Mount St. Helens dome which Murase et al. (1985) studied using Eq. (22)] are calculated and discussed. The relationships between yield strength and coulee formation on a gentle slope, and between yield strength and the behaviour of Peleean domes are also investigated.

5.1 1979 Soufriere Dome

The 1979 eruption of La Soufriere on the West Indian island of St. Vincent has been described by Shepherd et al. (1979) and Westercamp and Tomblin (1979). Starting on April 13th a series of explosive eruptions destroyed a pre-existing lava dome and crater lake, sending ash to heights of 18 km and generating nuees ardentes and mudflows. The magma was a basaltic andesite with

45% crystals. The last explosion was on April 26th and by May 3rd, a steaming lava extrusion had appeared on the dry crater floor. The subsequent behaviour of this extrusion has been documented by Shepherd et al. (1979) and Huppert et al. (1982) in remarkable detail, sufficient to test fluid dynamic models of spreading lava. The dome increased in size until mid-October when it had reached a height of 133 m and radius of 434 m (volume $47.3 \times 10^6 \, m^3$). Observations of the dome indicated that the period of most significant growth lasted from May 6th to August 4th (i.e., 90 days). After that date the volume of the dome did not change appreciably, although the dome shape changed slightly, possibly as a result of movement and avalanching of rubble on the dome flanks und surface. Huppert et al. (1982) found the first 90 days of growth to be described by

$$R = 0.051 \, t^{0.580} \quad (r_c = 0.98) \; , \tag{31}$$

where R is in metres and t in seconds. Their data also yields

$$H = 4.95 \, R^{0.524} \quad (r_c = 0.99) \quad \text{and} \tag{32}$$

$$H = 0.927 \, t^{0.312} \quad (r_c = 0.99) \; . \tag{33}$$

These three equations utilise the three measured properties R, H and t directly. Huppert et al. (1982) estimated the dome volume from H and R and then obtained the regression line for this derived quantity

$$V = 0.0248 \, t^{1.36} \quad (r_c = 0.99) \; , \tag{34}$$

which, with reference to Eq. (2) shows the parameters $S = 0.0248 \, m^3 \, s^{-1.36}$ and $\alpha = 1.36$. Equations (31) to (34) constitute the basic information with which the fluid dynamic models can be compared.

Huppert et al. (1982) proceeded to show that for $\alpha = 1.36$ then Eq. (8) anticipates $R \propto t^{0.63}$, which is in reasonable agreement with observational data [Eq. (31)]. This seemed to justify the assumption of an effectively Newtonian rheology and, comparing Eqs. (31) and (8), implied a viscosity of 2×10^{11} Pa-s. A problem with this interpretation, however, is that if $\alpha = 1.36$, then Eq. (9) predicts $H \propto t^{0.090}$ whilst Huppert et al's (1982) observations give $H \propto t^{0.312}$ [Eq. (32)]. Furthermore, after re-arranging Eq. (8) in the more appropriate form, analogous to Eq. (21),

$$\eta = \left(\frac{c_2}{R}\right)^8 \varrho g S^3 t^{(3\alpha+1)} \; , \tag{35}$$

it is found that viscosities calculated from this theory for constant-viscosity lava increase by over two orders of magnitude over the 90 day period of interest. These inconsistencies could be explained if either the growth model $V = St^\alpha$ did not apply, the time origin selected by Huppert et al. (1982) was inappropriate, or the Newtonian rheological model was invalid. Verifying either of the first two possibilities is problematic because of uncertainties in the manipulation of the measured quantities H, R and t needed to calculate V and choose a time origin.

In order to apply the viscoplastic model of lava dome growth it is not essential to know the volume or time. The basic relationship which must be satisfied if the dome behaved as a viscoplastic with a constant bulk yield strength is that $H \propto R^{1/2}$ [Eqs. (14) or (20)]. Clearly, the observations of the Soufriere dome ($H \propto R^{0.524}$) are in good agreement with the viscoplastic model. Furthermore, taking $\alpha = 1.36$, then Eqs. (27) and (28) predict $R \propto t^{0.544}$ and $H \propto t^{0.272}$ which are closer to the observed trends $R \propto t^{0.580}$ and $H \propto t^{0.312}$ than are the analogous relationships implied by the Newtonian model. For the first 90 days of the Soufriere dome's growth

$$H/R^{1/2} = 5.75 \pm 0.47$$

so that Eq. (21) estimates the yield strength as $(2.6 \pm 0.5) \times 10^5$ Pa for $\varrho = 2500$ kg m^{-3} and g = 9.81 m s^{-2}. The yield strengths calculated from each pair of height and radius measurements do not show a substantial variation with time.

The viscoplastic model is in closer agreement with the field data than is the Newtonian fluid model. To confirm that the viscoplastic model is justified physically, the Bingham number must be large, or equivalently the time-scale T at which Newtonian behaviour fails to be applicable must be much shorter than the eruption duration. In fact T should be less than the time of the earliest measurement thought to reflect viscoplastic behaviour. Knowledge of both τ_0 and η are required to calculate T [Eq. (18)] however. Whilst τ_0 is estimated from the above model it is not possible to estimate the corresponding value of η. Similarly, Huppert et al's (1982) Newtonian model gives an estimate of η but cannot quantify τ_0. Using Eq. (18), however, it is calculated that for T < 1 day requires $\eta \lesssim 5 \times 10^9$ Pa-s if in fact $\tau_0 = 2.6 \times 10^5$ Pa, or $\tau_0 \lesssim 2 \times 10^6$ Pa if in fact $\eta = 2 \times 10^{11}$ Pa-s. Both of these constraints on the unknown quantities η and τ_0 would be anticipated from current knowledge of magma rheology [e.g. Huppert et al. (1982) anticipated an apparent viscosity of $10^6 - 10^9$ Pa-s and a yield strength of $500 - 10^4$ Pa for the Soufriere magma] so that it is not possible to make a confident decision as to which model is more correct. I favour the viscoplastic model because of its closer agreement with the field data in comparison with the Newtonian model. It is concluded that the Soufriere dome maintained a shape during its growth which is indicative of its having a yield strength of about 2.6×10^5 Pa.

5.2 Calculated Yield Strengths of Selected Low Lava Domes

The analysis of the heights and radii of the growing 1979 Soufriere dome illustrates how the model presented earlier provides information about dome rheology. Similar data is unavailable for other low lava domes, so that estimates of their rheological properties must be made on a less rigorous basis, relying on the assumption that the final dimensions of a dome reflect its yield strength at the moment when effusion ceased. Data from rhyolite dome fields at La Primavera (Mexico), Coso (California), and Maroa (New Zealand) are

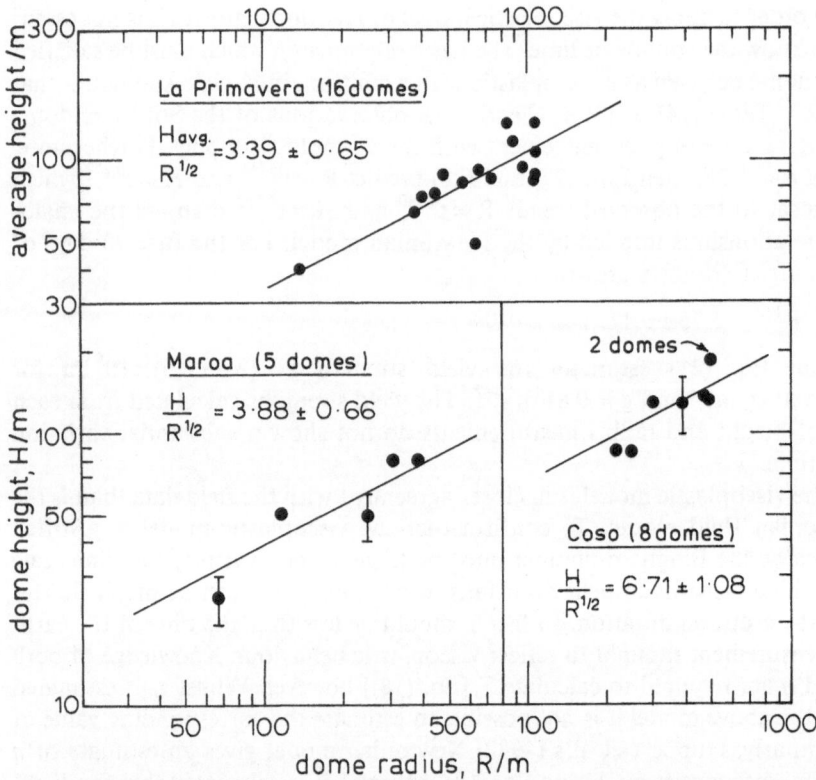

Fig. 13. The heights and radii of equidimensional silicic lava domes from three volcanoes are described by the relations $H/R^{1/2}$ = a constant ± 1 SD. Data for La Primavera comendites is from Clough et al. (1982) where the height is the average thickness of the dome. The basal areas and maximum heights of the rhyolite domes at Maroa and Coso were determined from maps by E. F. Lloyd (unpublished) and Duffield and Bacon (1981). The radius of each dome was calculated according to R = (basal area/π)$^{1/2}$. Domes which are banked against an encompassing tephra ring, are partially buried, asymmetric or composite in nature were excluded from these compilations. The heights of domes for which *error bars* are shown could not be determined very precisely because of uncertainties in interpreting the relationships between contour lines and dome margins

plotted in Fig. 13 and produce the trend $H \propto R^{1/2}$ anticipated for domes with equal yield strengths. This agreement is such that yield strengths calculated for each dome at a given volcano cover a range of less than one order of magnitude (Table 3). This substantiates the interpretation that a dome's dimensions reflects its yield strength rather than some other physical control. In other words Eq. (21) will have physical validity and is not a meaningless algebraic manipulation of a dome's dimensions.

Yield strengths calculated with Eq. (21) for individual lava domes at other volcanoes are also listed in Table 3. The June 1980 dome of Mt. St. Helens had dimensions indicative of a yield strength of 1.3×10^5 Pa which is equal to the value measured by Murase et al. (1985) at the near-eruptive temperature of

Table 3. Dimensions and calculated yield strengths of low lava domes of different compositions

Volcano	Dome	SiO_2 wt%	R(m)	H(m)	τ_0(Pa)[a]	Data source
Primavera	Field	74–78	138–1093	68–243[c]	$(3.3\pm1.2)\times10^{5}$[b]	Clough et al. (1982)
Maroa	Field	Rhyolite	67–389	25–80	$(1.2\pm0.4)\times10^{5}$[b]	Lloyd (unpublished maps)
Coso	Field	74–77	213–500	85–183	$(3.6\pm1.2)\times10^{5}$[b]	Duffield and Bacon (1981)
Soufriere	1979 First	55				Huppert et al. (1982)
	90 days		30–410	30–123	$(2.6\pm0.5)\times10^{5}$[b]	
	Final shape		434	133	3.2×10^{5}	
Mt. St. Helens	June 1980	62.5	183	45	1.3×10^{5}	Moore et al. (1981)
	Oct. 1980	61	113	37	9.6×10^{4}	
Galunggung	1918	54.7	250	85	2.3×10^{5}	Allard (1983)
Novarupta	1912	76.9[d]	190	65	1.7×10^{5}	Hildreth (1983)
Barcena	1952	61	30	8	1.7×10^{4}	Richards (1959)
Ukinrek	E. Maar	48	60	40	2.1×10^{5}	Kienle et al. (1980)
Abu	"10"	52.7	240	30	2.9×10^{4}	Koyaguchi (1986), Moriya (1978)
Santorini	Liatsikas	63.4	20	9–13	$(3.2-6.7)\times10^{4}$	Georgalas (1962)

[a] τ_0 is calculated from Eq. (21) with $\varrho = 2500$ kg m^{-3}, $g = 9.81$ m s^{-2}.
[b] Mean (and 1 SD) value.
[c] Maximum heights, calculated from the reported average heights assuming $H = 15\,\hat{H}/8$ [Eq. (24)].
[d] Mixed magma comprising rhyolite lava with andesite inclusions.

about 900 °C. Similarly, the low lava domes of Abu, Barcena and Liatsikas have calculated yield strengths (Table 3) which are broadly appropriate to hot magmas of their respective compositions (cf. McBirney and Murase 1984). In other cases, however, the yield strengths given in Table 3 are of the order of 10^5 Pa irrespective of SiO_2 content. For example the basaltic dome at Ukinrek, Alaska (Kienle et al. 1980), has the dimensions of a material with a yield strength of 2.1×10^5 Pa, yet basaltic magma would not be expected to have a yield strength greater than 10^3 or 10^4 Pa (Shaw 1969; McBirney and Murase 1984). The Ukinrek dome formed by effusion of alkali basalt magma onto the floor of a maar, where a torrent of water from the disturbed water table quenched the exterior of the dome, leaving the interior hot and incandescent (Kienle et al. 1980). The quenched carapace of the dome is assumed to have given it the very high yield strength implied by its shape.

Similarly, the yield strength of the 1979 Soufriere dome is an order of magnitude greater than yield strengths expected for basaltic andesites at magmatic temperatures (McBirney and Murase 1984). It is likely that the high yield strength reflects the strength of the dome's carapace rather than that of hot interior magma. [Huppert et al. (1982) invoked similar reasoning to explain their very high viscosity estimate.] This interpretation is encouraged by noting that magma which is less dense than country rock by an amount $\Delta \varrho$ will rise through a circular conduit of diameter D only if

$$\tau_0 < g \Delta \varrho D / 4 \tag{36}$$

(Wilson and Head 1981). If the minimum observed dome diameter (60 m) is equated with the maximum possible conduit diameter, and assuming $\Delta \varrho = 300$ kg m^{-3}, then Eq. (36) implies a maximum yield strength for the magma in the conduit of 4.4×10^4 Pa. This and lesser values of yield strength are appropriate to basaltic andesite magma (McBirney and Murase 1984). The much higher strength of the dome thus implies that the dome's rheology was different from that of the magma which fed it.

If the bulk yield strength of many low lava domes is controlled by the strength of a cool surface layer rather than that of the hot interior magma, then this questions the use of the constant-property models discussed in Section 3. A scaling analysis of the growth of a flat dome beneath a viscoplastic cover still yields Eq. (11), however, so that there should still be a square root relationship between H and $h_0 R$ as long as the carapace is at least of order h_0 thick. This is an essential requirement if the carapace is to be able to deform along the shear planes which have a spacing of ca. h_0. When this is not the case, the gravitational stresses set up by the weight of the dome may not deform the shell and excess pressures will build up in the dome. Iverson (this Vol.) has developed a mathematical model to treat such cases and finds it describes one of the Mt. St. Helens (Peleean) domes well. The data from low lava domes (Fig. 13 and Table 3) agree with the viscoplastic models, however, implying that in these cases the surface layers do not act as a brittle shell.

Further application of the laboratory observations to low lava domes with deformable surfaces can be made. The viscoplastic models showed evidence for circumferential stretching and radial shortening. Structural studies by Fink (1983) on the rhyolitic Little Glass Mountain flow revealed patterns of folding and fracturing which resulted from similarly oriented extensional and compressive forces. In contrast, the common occurrence of surface folding on lava domes and coulees was not a feature of the models. This implies that a viscous skin is essential for folding to develop (cf. Fink and Fletcher 1978; Fink 1980). The stability of a deformable skin is beyond the scope of the present investigation, but it seems that some can remain undeformed and encase low lava domes whilst others become folded into ogives; the role of downslope flow may be critical but this must await proper investigation.

Two other important aspects which can be considered together are the observations of flow banding in real lava domes and of slip plane traces in the laboratory domes. If a lava dome has a yield strength of 10^5 Pa, slip plane traces would be expected to cause a regular surface texture at a scale of $h_0 = 10^5/2500/10 = 4$ m. This is the same order of magnitude as blocks on many dome surfaces but I do not know of any domes which show a recognizable surface pattern (such as shown in Fig. 7) which could be unequivocally related to slip plane traces. In the experiments, slip plane traces were obscured by a coating of sand particles (Fig. 11) so it may be optimistic to expect to see them on real lava domes which are covered either in rubble or a brittle shell.

Inside lava domes, flow banding is typical, and it is tempting to interpret this texture in terms of shearing in a Newtonian fluid. This would appear to contrast with the kaolin domes which show, at a given time, shearing on localized slip planes. However, the slip planes must continuously relocate themselves according to the dictates of the changing stress field in the growing dome. Consequently, slip planes will migrate through the dome so that all of the dome interior experiences shear at some time. Thus, on a time-averaged basis, the entire dome will have been sheared. The presence of flow banding need not, then, be incompatible with a viscoplastic rheology. A detailed structural analysis of flow bands and crystallite orientation may be able to discriminate between shear histories caused by different rheologies (either of the continuous silicate liquid or the bulk lava).

5.3 The Transition from Low Lava Dome to Coulee

Magma effusing onto a gentle slope will tend to flow downhill. Short thick lava flows which form in this way are termed coulees (e.g., Cotton 1944). Coulees reflect the departure from the formation of a symmetrical low lava dome to the development of a long lava flow (Fig. 1 d). We now examine the conditions which allow a coulee rather than a low lava dome to form.

Magma with a yield strength extruded onto slightly sloping ground will first pile up around the vent to form a symmetrical dome with $H \sim (h_0 R)^{1/2}$. After some growth, however, the dome will start to flow downhill and a short lava flow or coulee will form. The motion of the coulee will be determined by a force balance which is different from that which controls dome growth. In

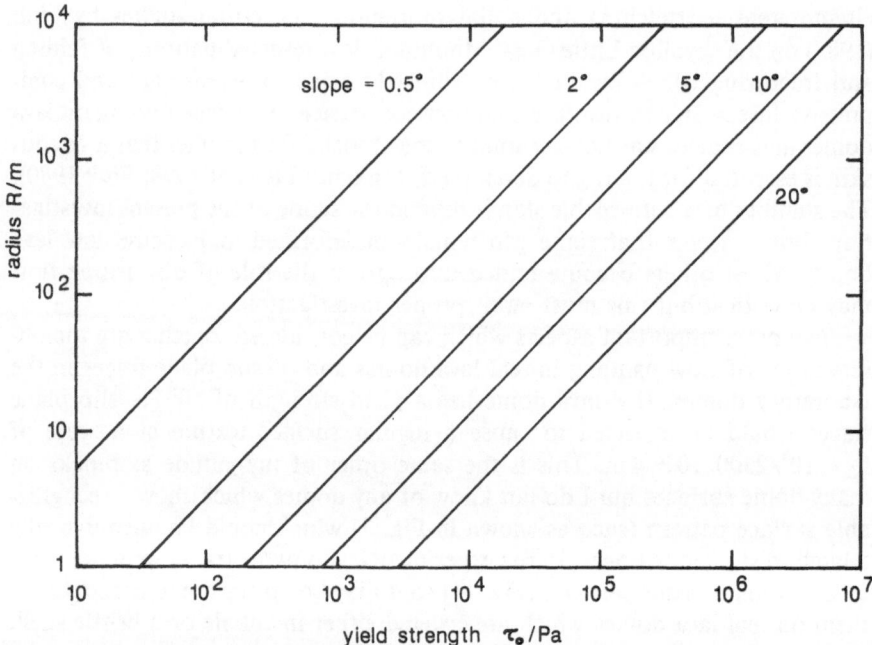

Fig. 14. When magma effuses onto a sloping surface, a low lava dome will form initially. After exceeding a critical radius, dependent on the yield strength and ground slope [Eq. (37)] as graphed in this diagram, it will start to flow downhill, developing into a coulee, and then a lava flow

particular, for a coulee on a slope of θ degrees then the pressure on the base $\varrho g H \sin \theta$ cannot be less than the yield strength, τ_0, so that the minimum thickness is

$$H = \tau_0/\varrho g \sin \theta = h_0/\sin \theta$$

(Robson 1967; Hulme 1974). A lava extrusion will develop into a coulee if its thickness can be reduced by so doing. A coulee will therefore develop once

$$h_0/\sin \theta < (h_0 R)^{1/2} , \quad \text{i.e.}$$

$$R > h_0/\sin^2 \theta . \tag{37}$$

Hence, a gentle slope will allow a dome to spread to a greater distance before moving off as a coulee than would a steep slope. Similarly, for a given slope, the larger the yield strength the less likely a coulee is to form.

Equation (37) has been graphed in Fig. 14. This shows that lava with $\tau_0 = 10^5$ Pa will not form a coulee on a 2° slope until it is about 4 km in diameter. This distance is considerably larger than most lava domes (Fig. 2). On the other hand, lava with $\tau_0 = 100$ Pa (Hawaiian basalt) will form a flow on a 2° slope after spreading to just 4 m across. For typical eruptive volumes, domes will form from lava with moderately high yield strengths ($\gtrsim 10^4$ Pa) whilst coulees and flows develop from lavas with low yield strengths ($\lesssim 10^4$ Pa).

5.4 Peleean Domes

So far, only domes that carry small amounts of rubble or talus have been modelled. In this section, the model of viscoplastic lava domes is modified to include the geometrical influence of a talus collar on dome shape. Application of this model allows the minimum strengths of Peleean domes to be estimated and it is shown that the calculated values are consistent with other volcanological properties of Peleean domes.

The model illustrated in Fig. 15 shows a mound of viscoplastic of height H and radius R whose margins are obscured by rubble having an angle of repose Φ. The talus extends the margin of the dome edifice to R'. In this model, rubble accumulates where the surface of the core is steeper than the angle of repose. This division in the dome's surface occurs at a radius R*, where

$$dh/dr = -\tan\Phi .$$ (38)

To a first approximation the core of the dome is assumed to be governed by a force balance similar to that of Eq. (11). This is because the cohesionless rubble has a smaller density and height than the core and so will not be an appreciable barrier to the spread of the lava. For the sake of argument it is assumed that the dome's core can be described by Eq. (23). The talus apron presents a profile described by

$$h = (R'-r)\tan\Phi .$$ (39)

Together with Eqs. (22) and (37), this allows the position R* to be related to the external dimensions H and R' and the physical properties Φ and h_0 as follows

$$R* = R' - \frac{1.76^2 h_0}{2\tan^2\Phi} \quad \text{and}$$ (40)

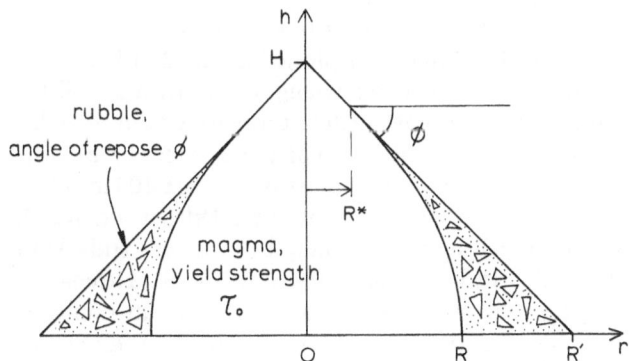

Fig. 15. This model of a dome has a core of viscoplastic magma which is surrounded by an apron of talus extending out to a distance R'. The magma/talus contact outcrops at a radius R*, where the slope of the magmatic core equals the angle of repose of the talus

$$H = 1.76\,h_0^{1/2} \left(R' - \frac{1.76^2\,h_0}{4\tan^2\Phi} \right)^{1/2} . \tag{41}$$

These two equations model the dimensions of a dome of viscoplastic encircled by a rubble collar. Before examining the properties of this model it is profitable to consider the conditions leading to domes with either small or large rubble collars. It follows from Eq. (41) that when

$$\frac{1.76^2\,h_0}{4\tan^2\Phi} \ll R' \tag{42}$$

the amount of rubble is negligible and Eq. (23) is a good approximation to the shape of the entire dome. On the other hand, Eq. (40) indicates that a dome will be completely covered in rubble when

$$1.76^2\,h_0/2\tan\Phi \le R' . \tag{43}$$

In this case the dimensions of the dome will be such that $H/R' = \tan\Phi$.

A further step is to consider the ratio of the radial extent of rubble-free dome, R^*, to the total dome radius R'. Let this be denoted by $X = R^*/R'$. Manipulating Eqs. (40) and (41) gives

$$H = \frac{1.76^2\,h_0}{2\tan\Phi} \left(\frac{1+X}{1-X} \right)^{1/2} . \tag{44}$$

For Peleean domes X is small, whilst for low lava domes X is closer to unity (cf. Fig. 1 b, c). The difference between the two dome types might be identified on morphological grounds by defining domes with $X \le 1/2$ to $1/3$ as Peleean. Equation (44) can then be used to describe the transitional conditions in terms of dome height and the physical properties Φ and τ_0. This has been done graphically in Fig. 16 for realistic ranges of X and Φ. The minimum yield strengths of Peleean domes can be estimated from this diagram. For example, a Peleean dome with a height of 100 m must have $\tau_0 \gtrsim 10^6$ Pa according to this model. Similarly, $H = 1000$ m implies $\tau_0 \gtrsim 10^7$ Pa. With reference to the heights of Peleean domes plotted in Fig. 2, it follows that the exceedingly large Lassen Peak dome had a strength of at least 5×10^6 Pa, whilst the strengths of large historical domes such as those of Mount St Helens, Novy and Santa Maria may exceed 2×10^6 Pa. For these values to be other than rough estimates the carpace of cool rock must be at least 100 m (i.e., order h_0) thick and this seems unlikely. The cores of these Peleean domes should, therefore, be described much more accurately by Iverson's (this Vol.) model, and indeed he shows this to be so for a dome at Mt. St. Helens.

The present empirical model of Peleean domes is still useful for understanding the transition between low lava dome and Peleean dome morphologies. Figure 17a describes the shapes of two growing domes with yield strengths of 10^5 Pa and 5×10^6 Pa erupted from a vent of radius 5 m. At the start of their extrusion the hypothetical domes form small upheaved plugs,

Fig. 16. Peleean domes can be
distinguished from low lava
domes by the former's abundant talus apron. The model
of domes presented in Fig. 15
and discussed in the text allow
this difference to be expressed
in a plot of yield strength
against dome height. The
transition between the dome
types (*stippled area*) is defined by Eq. (44) with
$1/3 \leq X \leq 1/2$ and
$30° < \varPhi < 45°$. Other physical
constants are $\varrho = 2500$ kg m^{-3}
and $g = 9.81$ m s^{-2}

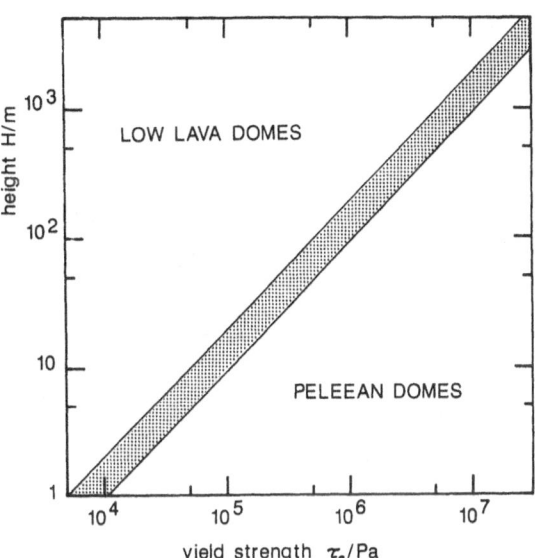

with radius equal to that of their vent, so that their dimensions initially plot
as the dashed near-vertical trend in Fig. 17a. This behaviour was seen at the
start of each kaolin dome experiment. Eventually the plug-dome topples or
crumbles to form a rubble-covered mound which will be modelled by Eq. (41)
[subject to the constraint of Eq. (43)].

Continued growth of the high strength ($\tau_0 = 5 \times 10^6$ Pa) dome results in the
pyramidal rubble-covered mound growing in height whilst retaining its extensive talus collar, and this dome follows a growth path which is described by
$H = R' \tan \varPhi$, giving a trend of unit slope in Fig. 17a. On the other hand, the
weaker magma ($\tau_0 = 10^5$ Pa) spreads more readily and, once $H \simeq 10$ to 20 m
(cf. Fig. 16) forms a low lava dome surrounded by a narrow rubble collar (i.e.
$R' \simeq R$). The resultant low lava dome follows a trend described by $H = 1.76$
$(h_0 R)^{1/2}$. The $H-R'$ path sketched by the model of a low-strength dome in
Fig. 17a can be seen in the behaviour of the Mount St. Helens dome of October 18–27th., 1980 (Moore et al. 1981) plotted in Fig. 17b. This correspondence implies that Eq. (41) provides a satisfactory method of empirically describing the transition between the end-member trends first identified in Fig.
2 as

$$H = R' \tan \varPhi \qquad \text{(Peleean domes)} \qquad \text{and}$$

$$H = 1.76 (h_0 R)^{1/2} \qquad \text{(low lava domes)}.$$

5.5 Consequences of the High Yield Strengths of Peleean Domes

It has been argued that the shapes and dimensions of Peleean domes are consistent with their having a bulk yield strength of at least 10^6 to 10^7 Pa (Fig. 16).

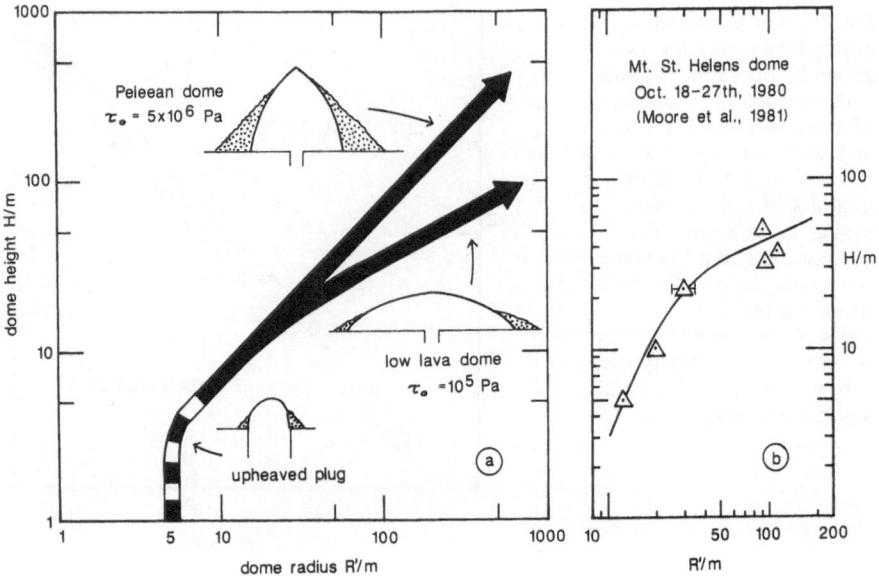

Fig. 17. a The calculated height versus radius relations followed by two growing domes of yield strength 5×10^6 and 10^5 Pa which emerge from a vent of radius 5 m. At early stages both magmas form upheaved plugs but eventually the weaker magma spreads out to form a low lava dome $[H \propto (R')^{1/2}]$ whilst the stronger magma maintains the pyramidal form of a Peleean dome $(H \propto R')$. *b* The growth of the October 1980 dome at Mt. St. Helens (Moore et al. 1981) followed an $H - R'$ path which is similar to that described by the model of low lava dome growth depicted in *a*. All but the last data points refer to rapid growth over the first day, when the dome was steep-sided and covered in brecciated slabs and spines. The final shape developed slowly as the dome sagged and spread under its own weight. The final dimensions are consistent with a bulk yield strength of 9.6×10^4 Pa (Table 3)

Apart from their distinctive morphologies, Peleean domes are also characterized by the occurrence of explosions during their growth. Reflection on the requirements for bubble growth in a viscoplastic provides a way of understanding the cause of this destructive behaviour and its restriction to the strongest types of dome.

Vapour bubbles within a viscoplastic must attain internal overpressures of order τ_0 if they are to force their host material aside in order to grow. The yield strengths calculated for Peleean domes thus necessitate overpressures of at least 10 to 100 times atmospheric pressure (10^5 Pa) before unhindered bubble expansion can occur. These values are appropriate to the generation of high velocity blasts (Wilson 1980) which can accompany nuees ardentes during the lifetimes of Peleean domes (Macdonald and Alcaraz 1956). In contrast, gases exsolving from weaker domes or lavas such as low lava domes with $\tau_0 \lesssim 5 \times 10^5$ Pa can expand without generating significant overpressures. It is notable that blasts do not occur during the growth of low lava domes (e.g., Shepherd et al. 1979; Huppert et al. 1982). Explosions at these domes always seem to be related to the decompression of newly arrived gas-rich magma at the surface (e.g. the 1980 domes of Mount St. Helens; Moore et al. 1981). The yield strengths inferred for various lava domes thus correlate with the occurrence of internally generated explosions at these domes, supporting the modelling of these bodies as viscoplastics.

6 Summary

The geometry and volcanological behaviour of lava domes gives information about the physical properties of the dome-forming material. Viscoplastic domes have been modelled as Bingham plastics in order to determine the rheological properties associated with various styles of eruptive behaviour. The results of this combined theroretical, experimental and geological approach can be summarized as follows:

1. Peleean domes have height, H, to radius, R, ratios of about 0.6 to 0.9. These values are similar to the tangent of the angles of repose ($30°-40°$) of the apron of crumble breccia which covers the flanks of these domes.
2. Certain other domes (e.g. 1918 Galunggung, 1979 Soufriere) have lower heights for a given radius, less jagged surfaces and smaller talus aprons. These domes are termed low lava domes and are postulated to possess shapes and dimensions which directly reflect their yield strength.
3. Laboratory experiments, using kaolin suspensions, have shown that a slowly growing dome of viscoplastic material maintains a state of static equilibrium such that $H = (1.76 \pm 0.04) \, (\tau_0 R/\varrho g)^{1/2}$. This result qualifies the often quoted work of Nye (1952) who calculated, to a first approximation, that the multiplicative constant is $\sqrt{2}$. The shape of the dome is approximately parabolic (Fig. 6) and its surface is traversed by two sets of spiral slip plane traces (Fig. 7). These traces are believed to be the surface expression of an array of planes within the dome along which movement takes place so as to allow the dome to remain statically stable whilst it grows in size. The experiments verify a theoretical model for dome growth [Eqs. (20), (27), (28)] based on a balance between dome weight and yield strength [Eq. (11)].
   Further experimentation showed that the surface of the dome did not become overridden in "caterpillar-track" fashion at the dome front. It has also been shown that beyond a near-summit region, surface material is neither removed nor replenished but becomes radially compressed and circumferentially stretched as the dome grows.
4. The dimensions of the growing 1970 Soufriere low lava dome indicate the spread of a material with a bulk yield strength of 2.6×10^5 Pa. This and similar values for other low lava domes apparently reflect the strength of cool material in the domes' carapace. Only material with a lower yield strength can rise through feeder conduits of realistic dimensions.
5. Peleean domes are inferred to be stronger than low lava domes, probably reflecting some combination of higher crystal content, longer cooling his-

tories and thicker carapaces. Minimum strengths of $10^6 - 10^7$ Pa are implied for these domes. Such strengths are compatible with the explosive release of pressurized magma which characterizes Peleean eruptions but which is absent during low lava dome emplacement.

6. For magma erupted onto gentle slopes, lava flows and coulees develop if $\tau_0 \lesssim 10^3 - 10^4$ Pa, low lava domes and coulees develop if 10^4 Pa $\lesssim \tau_0 \lesssim 10^6$ Pa and Peleean domes and upheaved plugs develop if $10^5 - 10^6$ Pa $\lesssim \tau_0$.

7. The best test of the model presented here is its application to measurements of low lava dome growth and it is hoped that such applications will be made as suitable data [of the type presented by Huppert et al. (1982) for the Soufriere dome] become available. A log-log plot of H versus R is the best means by which to start analyzing dome-growth data. A trend with unit slope will be expected to describe a Peleean dome, which can then be analyzed in detail using the model of Iverson (this Vol.). A low lava dome will give a trend with a slope of $1/2$ if the dome dimensions are determined by a constant yield strength. The yield strength is calculated using Eq. (21). A dome which behaves as a Newtonian fluid is likely to give a different slope. The apparent viscosity can be calculated [e.g., Eq. (35)] if H and R have been measured as functions of time and if the volume (preferably calculated independently of H and R, say from digitized topographic maps) increased such that $V = St^\alpha$.

Acknowledgements. K. Johnston is thanked for his help with the experiments, which were supported by the University of Auckland Research Fund. I am grateful to E. F. Lloyd, I. A. Nairn, I. Moriya, C. R. Bacon and S. Bronto for sharing their knowledge of lava domes and to G. A. Mahood for useful review comments. B. Tricklebank and T. Gray kindly arranged access to their viscometer in the School of Engineering, University of Auckland. Thanks also go to R. Bunker, J. Williams and C. Whale for typing the manuscript and R. M. Harris for preparing the diagrams.

References

Allard P (1983) Facing hazards from the reawakening of dormant explosive volcanoes: the examples of Mt. St. Helens, El Chichon, and Galunggung in 1980–1982. In: Tazieff, H, Sabroux, J-C (eds) Forecasting Volcanic Events. Elsevier, Amsterdam, pp 561–584

Allen JRL (1982) Sedimentary structures. Their character and physical basis. Vol. 1. Elsevier, Amsterdam, pp 1–593

Bogoyavlyenskaya GY, Kirsanov IT (1981) Twenty five years of activity at Bezymianny. Volc. Seism. (See Volcano News 13 & 14, 1983)

Borgia A, Linneman S, Spencer D, Morales LD, Andre JB (1983) Dynamics of lava flow fronts, Arenal volcano, Costa Rica. J Volcanol Geotherm Res 19:303–329

Chester DK, Duncan AM, Guest JE, Kilburn CRJ (1985) Mount Etna, anatomy of a volcano. Chapman Hall, London, pp 1–404

Clough BJ, Wright JV, Walker GPL (1982) Morphology and dimensions of the young comendite lavas of La Primavera volcano, Mexico. Geol Mag 119:477–485

Cotton CA (1944) Volcanoes as landscape forms. Whitcombe and Tombs, Christchurch, New Zealand, pp 1–416

Didden N, Maxworthy T (1982) The viscous spreading of plane and axisymmetric gravity currents. J Fluid Mech 121:27–42

Duffield WA, Bacon CR (1981) Geologic map of the Coso volcanic field and adjacent areas, Inyo County, California, scale 1 : 50,000. Misc. Geol. Inv. Map I-1200 US Geol Surv, Menlo Park, California

Fink JH (1980) Surface folding and viscosity of rhyolite flows. Geology 8:250–254

Fink JH (1983) Structure and emplacement of a rhyolitic obsidian flow. Little Glass Mountain, Medicine Lake Highland, northern California. Geol Soc Am Bull 94:362–380

Fink JH (1984) Structural geologic constraints on the rheology of rhyolitic obsidian. J Noncryst Solids 67:135–146

Fink JH, Fletcher RC (1978) Ropy pahoehoe: surface folding of a viscous fluid. J Volcanol Geotherm Res 4:151–170

Fisher RV, Heiken G (1982) Mt Pelee, Martinique: May 8 and 20, 1902 pyroclastic flows and surges. J Volcanol Geotherm Res 13:339–371

Georgalas GC (1962) Catalogue of the active volcanoes of the world, vol XII Greece. IAVCEI, Rome, pp 1–40

Heiken G, Wohletz K (1987) Tephra deposits associated with silicic domes and lava flows. Geol Soc Am Spec Pap 212:55–76

Hildreth W (1983) The compositionally zoned eruption of 1912 in the Valley of Ten Thousand Smokes, Katmai National Park, Alaska. J Volcanol Geotherm Res 18:1–56

Hulme G (1974) The interpretation of lava flow morphology. Geophys. J Roy Astron Soc 39:361–383

Huppert HE (1982) The propagation of two-dimensional and axisymmetric viscous gravity currents over a rigid horizontal surface. J Fluid Mech 121:43–58

Huppert HE, Shepherd JB, Sigurdsson H, Sparks RSJ (1982) On lava dome growth, with application to the 1979 lava extrusion of the Soufriere of St Vincent. J Volcanol Geotherm Res 14:199–222

Iverson RM (1989) Lava domes modeled as brittle shells that enclose pressurized magma, with application to Mount St. Helens. IAVCEI Proc Volcanol 2:47–69 (this vol)

Kienle J, Kyle PR, Self S, Motyka RJ, Lorenz V (1980) Ukinrek Maars, Alaska I. April 1977 eruption sequence, petrology and tectonic setting. J Volcanol Geotherm Res 7:11–37

Koyaguchi T (1986) Textural and compositional evidence for magma mixing and its mechanism, Abu volcano group, Southwestern Japan. Contrib Mineral Petrol 93:33–45

Lewis JF (1968) Tauhara volcano, Taupo Zone. Part 1 – Geology and structure. N Z J Geol Geophys 11:212–224

Macdonald GA (1972) Volcanoes. Prentice Hall, New Jersey, pp 1–510

Macdonald GA, Alcaraz A (1956) Nuees ardentes of the 1948–1953 eruption of Hibok-Hibok. Bull Volcanol 18:169–178

Maxworthy T (1983) Gravity currents with variable inflow. J Fluid Mech 128:247–257

McBirney AR, Murase T (1984) Rheological properties of magmas. Annu Rev Earth Planet Sci 12:337–357

Minakami T, Ishikawa T, Yagi K (1951) The 1944 eruption of Volcano Usu in Hokkaido, Japan. Bull Volcanol ser 2 11:45–157

Moore JG, Lipman PW, Swanson DA, Alpha TR (1981) Growth of lava domes in the crater, June 1980–January 1981. US Geol Surv Prof Pap 1250, 541–547

Moriya I (1978) Topography of lava domes. Bull Dep Geogr Komazawa Univ 14:55–69 (in Japanese)

Moriya I (1985) Volcanoes in Hokkaido, Japan. Geogr Rep Kanazawa Univ 2:1–42

Murase T, McBirney AR, Melson WG (1985) Viscosity of the dome of Mount St Helens. J Volcanol Geotherm Res 24:193–204

Newhall CG, Melson WG (1983) Explosive activity associated with the growth of volcanic domes. J Volcanol Geotherm Res 17:111–131

Nye JF (1952) The mechanics of glacier flow. J Glaciol 2:82–93

Orowan E (1949) in report of a Joint meeting of the British Glaciological Society, the British Rheologists' Club and the Institute of Metal. J Glaciol 1:231–236

Perret FA (1937) The eruption of Mt Pelee 1929–1932. Carnegie Inst Wash Publ 458:1–126

Perret FA (1950) Volcanological observations. Carnegie Inst Wash Publ 549:1–162

Pieri DC, Baloga SM (1986) Eruption rate, area, and length relationships for some Hawaiian lava flows. J Volcanol Geotherm Res 30:29–45

Pinkerton H, Sparks RSJ (1978) Field measurements of the rheology of lava. Nature 276:383–385

Richards AF (1959) Geology of the Islas Revillagigedo, Mexico. 1. Birth and development of Volcan Barcena, Isla San Benedicto (1). Bull Volcanol Ser II 22:73–123

Robson GR (1967) Thickness of Etnean lavas. Nature 216:251–252

Shaw HR (1969) Rheology of basalt in the melting range. J Petrol 10:510–535

Shaw HR, Peck DL, Wright TL, Okamura R (1968) The viscosity of basaltic magma: an analysis of field measurements in Makaopuhi lava lake, Hawaii. Am J Sci 266:225–264

Shepherd JB, Aspinall WP, Rowley KC, Pereira J, Sigurdsson H, Fiske RS, Tomblin JF (1979) The eruption of Soufriere Volcano, St Vincent, April-June 1979. Nature 282:24–28

Sieh K, Bursik M (1986) Most recent eruption of the Mono Craters, eastern central California. J Geophys Res 91:12539–12571

Simkin T, Siebert L, McClelland L, Bridge D, Newhall C, Latter JH (1981) Volcanoes of the world. Smithsonian Institution, Hutchinson Ross Stroudsberg, pp 1–232

Swanson DA, Dzurisin D, Holcomb RT, Iwatsubo EY, Chadwick Jr WW, Casadevall TJ, Ewert JW, Heliker CC (1987) Growth of the lava dome at Mount St. Helens, Washington (USA), 1981–1983. Geol Soc Am Spec. Pap 212:1–16

Taylor GAM (1958) The 1951 eruption of Mount Lamington, Papua. Bur Miner Resour Geol Geophys Bull 38 (2nd ed), pp 1–129

Van Wazer JR, Lyons JW, Kim KY, Colwell RE (1963) Viscosity and flow measurement. A laboratory handbook of rheology. Wiley Interscience, New York, pp 1–406

Westercamp D, Tomblin J-F (1979) Le volcanisme récent et les éruptions historiques dans la partie centrale de l'arc insulaire des Petites Antilles. Bull BRGM Ser 2 sect IV N 3/4:293–319

Wiliams H (1931) The dacites of Lassen Peak and vicinity, California, and their basic inclusions. Am J Sci 22:385–403

Williams H (1932) The history and character of volcanic domes. Bull Dep Geol Sci, Univ Calif Publ 21:51–146

Williams H, McBirney AR (1979) Volcanology. Freeman & Cooper, San Francisco, pp 1–391

Wilson L (1980) Relationships between pressure, volatile content and ejecta velocity in three types of volcanic explosions. J Volcanol Geotherm Res 8:297–313

Wilson L, Head JW (1981) Ascent and eruption of basaltic magma on the earth and moon. J Geophys Res 86:2971–3001

Wilson L, Head JW (1983) A comparison of volcanic eruption processes on Earth, Moon Mars, Io and Venus. Nature 302:663–669

Mafic Lava Flow

Surfaces of Aa Flow-Fields on Mount Etna, Sicily: Morphology, Rheology, Crystallization and Scaling Phenomena

C. Kilburn

Abstract

Most of Etna's historical lavas have produced aa flow-fields. Despite the classification, they support a variety of surfaces in both the pahoehoe and aa categories. From observations of surface features at metre to submillimetre scales, two morphological series have been recognized: a sequence from pahoehoe to aa along a single flow, and an evolutionary trend among pahoehoe surfaces near boccas down a flow-field. The first series is more prominent and is superimposed on the second. Both series are associated with increasing crystallinity. The pahoehoe-aa sequence, however, is characterized by relatively higher crystallization rates and shear rates, indicating greater undercoolings and imposed stresses during emplacement. The identification of crystallization rate as a controlling factor demonstrates the influence of crystallization kinetics on lava development. The pahoehoe-aa transition also shows evidence of self-similar patterns at scales from at least millimetres to tens of centimetres, and may be accompanied by a self-feeding mechanism which couples large- and small-scale changes in lava rheology and shear rate.

1 Introduction

Only the outer parts of lavas are normally accessible for monitoring their evolution. Understanding the growth of lava crusts is therefore important for fully interpreting observations during effusions; it is also important for better reconstruction of the emplacement conditions of old lava flows. The surface textures of a lava reflect the state of the crust immediately before solidification. This state is related to conditions in the mobile lava beneath the crust, and both crust and mobile interior influence the subsequent dynamical evolution of the flow. Information on the internal condition and dynamics of a flow may thus be provided by studies of surface features.

Such a connection between surface expression and flow conditions is particularly convenient since subaerial lavas, silicate (e.g. Macdonald 1953) and non-silicate (Watanabe 1940; Dawson 1962), are traditionally classified according, first, to the relative continuity of their surfaces and, second, to the nature of their surface textures. Surface structure depends on the rheological resistance to motion of the cooling crust and the deforming stresses to which it is subjected. An unbroken surface is formed when the crust is able to deform continuously until its strength exceeds the imposed stress; a fragmental surface is formed when the imposed stresses exceed crustal resistance by an amount large enough to cause failure.

The governing conditions, however, are complicated by the number of processes which affect stress and rheological behaviour at large and small scales. Thus, while lava rheology is sensitive to bulk composition, crystallization and vesiculation, factors including gravity, thermal gradients and crystal and vesicle motion all influence the stress distribution in the crust.

Consequently, a large number of ideas have been proposed to explain differences in the surface morphology of lavas, the controlling factors most commonly suggested or alluded to being:

1. Differences in lava crystallinity (Scrope 1856; Dutton 1884; Wood 1917; Washington 1923; Emerson 1926; Jaggar 1930; Perret 1950; Macdonald 1953; Walker 1967; Sparks and Pinkerton 1978; Kilburn 1981; Rowland and Walker 1987).
2. Vesiculation and escaping gases cooling and disrupting the surface (Dana 1890; Geikie 1903; Mercalli 1907; Daly 1911, 1914; Day and Shepherd 1913; Jaggar 1917; Washington 1923; Perret 1950; Rittmann 1962; Kieffer 1979).
3. Lava discharge rate (Pinkerton and Sparks 1976; Sparks and Pinkerton 1978; Rowland and Walker in press).
4. Agitation, or rate of deformation, of the lava (Emerson 1926; Jaggar 1930; Macdonald 1953; Peterson and Tilling 1980; Kilburn 1981).

Attention is here focussed on the surface morphology of historical lavas of Mount Etna, Sicily, because: (1) data have also been collected from active Etnean lavas; and (2) most previous detailed studies have concentrated on Hawaiian lava surfaces (e.g. Macdonald 1953, 1967, 1972; Wentworth and Macdonald 1953; Peterson and Tilling 1980; Rowland and Walker 1987; but see also Jones 1943). The Etnean descriptions thus provide a basis not only for interpreting flow evolution, but also for highlighting similarities and differences with comparable lavas from Hawaii.

2 General Features of Etnean Lavas

Most subaerial basaltic lavas can be classified as *pahoehoe* or *aa*. Condensing the original definitions of Dutton (1884), Macdonald (1953) described pahoehoe as "the type of lava that in solidified form is characterized by a smooth, billowy, or ropy surface" and aa as lava with "a rough, jagged, spinose, and generally clinkery surface".

Virtually all Etna's historical lavas are of the aa variety. Flow-fields are usually composed of several distinct flows. The flows may emanate from a common source or they may be interconnected to form a system of distributaries, one flow budding from the periphery of another (Kilburn and Lopes 1988; Figs. 1 and 2). Individual flows may cover areas from a few square metres to more than a square kilometre. Generally, however, their areas lie between $10^2 - 10^5 \, m^2$, with lengths ranging from tens of metres to several kilometres. They are mainly hawaiitic in composition (Chester et al. 1985).

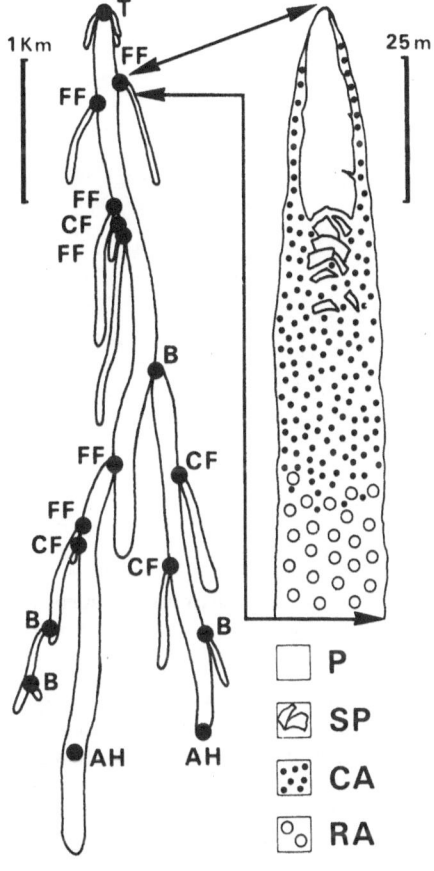

Fig. 1. Left: Hypothetical distribution of different near-bocca pahoehoe surfaces down a flow-field. *Filled circles:* locate boccas. *T,* Threaded pahoehoe; *FF,* finely filamented pahoehoe; *CF,* coarsely filamented pahoehoe; *B,* bladed pahoehoe; *AH,* arrow-head pahoehoe. The widths of the flows are exaggerated about ten times. *Right:* Distribution of pahoehoe and aa along a single *flow. P,* Pahoehoe; *SP,* slab pahoehoe; *CA,* cauliflower aa; *RA,* rubbly aa. The surfaces (except rubbly aa) are described in text

Even when classified as aa, individual flows commonly contain subordinate stretches of pahoehoe (Figs. 1 and 2). Both these main types of surface can be subdivided, and along any *flow* after emplacement a common downstream sequence is (Fig. 1): pahoehoe, slab pahoehoe, cauliflower (initial) aa and rubbly (mature) aa. Only the first three are considered in this chapter.

The link between large- and small-scale processes controlling flow evolution is here examined using descriptions of the gross and detailed features of pahoehoe and aa surfaces, supplemented by observations of their formation on active lavas. The key changes recognized (summarized in Table 1, Section 6) are used to identify: (1) two major morphological trends among Etnean lava surfaces; (2) crystallinity, crystallization rate and shear rate as governing factors; (3) the similarity of aa features at length scales differing by several orders of magnitude; and (4) the possible role of self-feeding processes in flow development.

Fig. 2. Distal secondary boccas (outlets; with *arrows*) in Etna's main 1981 north-flank flow-field. Surfaces downstream (to *top left*) from each bocca have cauliflower aa margins alongside pahoehoe, which changes to slab pahoehoe with tilted slabs (example *circled*), and to cauliflower aa (see also Fig. 1). Crescentic, concave-downstream tears are just visible (as darker traces) across the pahoehoe surface by the nearest bocca (*A*). Note colour contrast with surrounding, paler rubbly aa of parent flow. Textures at the vent for parent flow resemble those near secondary boccas shown here

3 Pahoehoe Surfaces

3.1 Field Aspect

Pahoehoe surfaces commonly occur at the head of a flow (where it is rarely more than a few metres wide) and continue downstream for metres or tens of metres (Figs. 1 and 2). They may be flanked by *cauliflower aa* (the initial type of aa described in Sect. 4; Figs. 1 and 2) or may extend to the flow margins, either forming a smooth, continuous edge or breaking into contorted fragments as much as tens of centimetres across. Occasionally the surface appears as the amalgamated crusts of adjacent currents within a flow, their junctions often being defined by incipient aa.

   Five typical large-scale aspects have been recognised for pahoehoe surfaces:
1. Flat, even surfaces.
2. Cleaved surfaces, cut across their widths by crescentic, concave-downflow, planes of failure (Fig. 2), whose gapes generally dip downstream at 25° or more, steeper dips often occurring upflow. These surfaces are commonly found close to feeding outlets, or boccas.

3. Humped surfaces, also common near boccas, have individual humps that can be as wide as the whole pahoehoe surface and commonly have crescentic outlines, again concave downstream. Humps may overlap, those upflow lying on top, and have total variations of relief of tens of centimetres or less. They are not inflated and their form suggests emplacement in stages, such as during the last pulses of effusion or during the progressive extension downstream of a hood over an active bocca.
4. Bonded surfaces of torn pahoehoe crust bound together by undisturbed pahoehoe from beneath. The torn crusts are generally raised by a few centimetres and frequently have arcuate outlines, concave downflow. As on cleaved surfaces, the curved tearing paths may cut across the middle of a surface and resemble those of glacial crevasses formed during accelerating laminar shear (Nye 1952).
5. Ropy surfaces, the ropes of which first appear about the centreline of a crust, but may stretch across the whole surface only a few metres downstream. They may be tightly or loosely packed, are often curved with convex margins downflow, and shallow outwards from central thicknesses that may reach tens of centimetres. When ropy crusts tear, they tend to do so along the joins between ropes and thus along paths curved in the direction opposite from that in smoother pahoehoe. Such behaviour illustrates the influence of local morphology on surface disruption.

At its downstream end, the continuous pahoehoe gradually gives way to widening marginal zones of broken pahoehoe slabs and cauliflower aa (Figs. 1 and 2). Eventually the whole crust is disrupted to slab pahoehoe (Wentworth and Macdonald 1953; Peterson and Tilling 1980; this form appears to be equivalent to the slabby aa of Lipman and Banks 1987), a collection of irregular tablets (Fig. 2) which are typically 5–20 cm thick but have widths and lengths as much as 2–3 m (Figs. 2 and 6).

When new flows bud from the margins of existing flows, stretches of pahoehoe may be found near secondary boccas throughout an aa flow-field (Figs. 1, 2 and 5). Most of these stretches share the large-scale characteristics described above. Along some short flows, however, especially towards the periphery of a flow-field and where secondary boccas have formed through the roofs of flows, a pastier pahoehoe may be extruded which does not grade into cauliflower aa.

Such pahoehoe often has *ripply* or *grooved* features. Ripply pahoehoe surfaces (Fig. 3) have rugged longitudinal cross-sections. Steep slopes, 1–3 cm high, are backed by gentler slopes extending upstream for about 10–20 cm. Both slopes commonly dip upstream (Fig. 4). The traces where adjacent slopes meet appear as angular ripples across the surface and have wavelengths reaching several tens of centimetres. These features may have been caused by irregular tearing of the exposed surface, or by the lava pulling away from and then rubbing against the roof of its feeding conduit (Fig. 4).

Grooved surfaces are characterized by grooves running parallel with the direction of flow (Fig. 5). Generally 10–20 cm wide and a few centimetres

Fig. 3. Ripply pahoehoe. Flow direction is to *bottom left.* Helmet gives scale

deep, such grooves have been attributed to lava being issued from a jagged-edged orifice (Wentworth and Macdonald 1953). Apparently similar surfaces have been mentioned by Nichols (1938), Bullard (1947), Krauskopf (1948), Einarsson (1949), Macdonald (1953, 1967, 1972) and Rowland and Walker (1987) and have acquired the general name of *toothpaste* lava. However, grooving is the only common feature and the associated small-scale surface textures may differ. For example, while the small spines on the grooved surfaces in Fig. 5 (Sect. 3.2) point downstream, Rowland and Walker (1987) have documented spines pointing upstream on Hawaiian toothpaste lava, and Macdonald (1967) has reported grooved lava with a granular small-scale texture.

The preservation of grooves and tear-structures demonstrates the low fluidity of grooved and ripply pahoehoe. This is emphasized in extreme cases when such lava is extruded as arcuate ridges or tongues (Fig. 5). The ridges are seldom more than a few metres long. Whether upright or arched, they usually stand less than a metre above the surrounding surface and taper towards their rims, which may be only 10–30 cm thick. The tongues are usually 1–2 m wide, less than 10 m long, and as much as 50 cm thick by their boccas. The crusts normally contain only small vesicles, less than 2–3 mm across, which are sometimes sufficiently abundant to give their interiors a honeycomb texture.

Ridges typically develop through the surface of a flow, while tongues may occur both at the surface and around a flow's perimeter, particularly in the

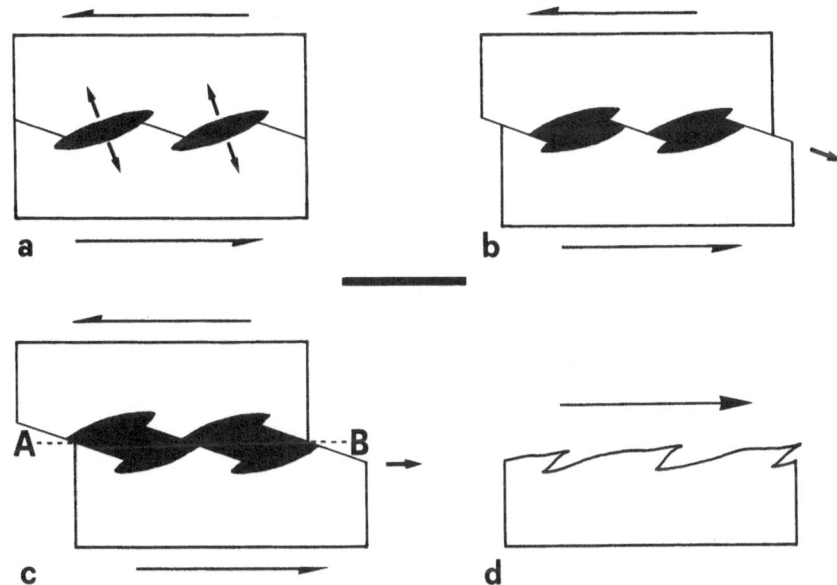

Fig. 4a–d. Ripples and arrow-head spines appear to form by a similar process, but at different scales. The central *scale bar* represents about 1 cm for arrow-head spines and about 10 cm for ripples. Upstream side is to the *left* of each diagram. Shearing in nearly solidified lava initiates inclined tension gashes. Upstream-dipping gashes widen first (*a*). Slippage occurs on complementary gashes (*b*) as lava downflow leaves bocca. After separation along *A–B* (*c*), upper level (static) becomes roof of conduit and lower level becomes mobile lava surface. The escaping lava scrapes against roof, accentuating small-scale spines (or large-scale ripples) pointing downstream. (*d*) Rugged surface of emerging lava. *a* and *b* after Gamond (1983)

depressions between adjacent lobes (Fig. 5). Both appear to be thin relatives of the wedge-shaped squeeze-ups documented by Nichols (1938), who attributed their downward thickening to widening of the surface outlet during extrusion.

3.2 Small-Scale Features

Pahoehoe surfaces show a wide variation in small-scale texture and only the more common or distinctive types are described here. These form part of a gradational series and intermediate types do occur.

The most widespread forms of pahoehoe belong to the *filamented* category of Wentworth and Macdonald (1953). These normally have glassy surfaces (confirmed by SEM studies) and are characterized by some form of raised lineament, or filament, with subordinate rounded knobs of submillimetre size. In order of decreasing abundance on Etna, the three commonest types are:

1. *Finely Filamented Pahoehoe.* Fine filaments are narrow ridges about a millimetre in height and basal width and rarely longer than 3–4 cm. They

Fig. 5. Distal, curved tongues of grooved pahoehoe. Source is at junction of tongues, near base of adjacent lobe of rubbly aa. Helmet gives scale

may stand alone, a few millimetres apart, or group together to form wider, multiridged features. The filaments are usually subparallel, straight or sinuous but, on occasion, may show a braided pattern for a few centimetres. The average alignment of the filaments often follows the principal direction of flow (Wentworth and Macdonald 1953), although their preferred orientations can sometimes change in the downstream direction through 90° within 5 – 10 cm. Such changes in orientation reflect changes in surface tearing, perhaps as a result of eddies in the lava.

Finely filamented surfaces may appear even or irregular on a centimetre scale. Variations in relief can reach from 3 – 4 to over 5 – 10 cm. The raised portions sometimes become detached wholesale from below, pockmarking the surface and littering it with fragments.

2. *Coarsely Filamented Pahoehoe.* Coarsely filamented surfaces might be analogous to the 'drawn surfaces' of Foster and Mason (1955). Their general appearance resembles the coarse brushstrokes of an oil painting. In detail, their filaments commonly are either wider versions of fine filaments or have crudely T-shaped cross-sections. The latter are usually a few millimetres high and as much as 5 – 6 mm across their tops. Typically a few millimetres apart, coarse filaments show only a poor mutual alignment over distances of a few centimetres.

Below the glassy veneer, vesicles associated with coarsely filamented pahoehoe appear to be larger and vertically flatter than those below finely filamented surfaces. The T-sections of some coarse filaments might thus result from their being parts of torn vesicles, the surface skin under tension preferring to fail where it thins over a shallow vesicle.

3. *Bladed Pahoehoe.* Bladed pahoehoe is characterized by an irregular array of drawn-out filaments and flared blades which, typically a few millimetres high, may widen to 5−6 mm.

Filamented pahoehoe can occur near boccas in most parts of a flow-field (Fig. 1). Also widely distributed, but much less abundant, is a form of *spiny pahoehoe* (Peterson and Tilling 1980), characterized by tiny projections which resemble the stubby spines common to a category of aa (described below).

Other pahoehoe surfaces, however, have a more selective distribution. In the upper reaches of a flow-field, especially at the main vent, surfaces may occasionally be laced by delicate glassy threads, typically thinner than fine filaments. The tracery of this *threaded pahoehoe* is only millimetres deep, easily detachable and crunchy underfoot. In contrast, *arrow-head pahoehoe* is more common towards the lower reaches of a flow-field. Its dull surface has a small-scale roughness caused by rows of millimetre-sized spines running irregularly across the direction of flow. The spines are squat and broad-based and, looking like tiny uneven arrowheads, usually have an apex pointing downstream. They are commonly associated with the pasty ripply and grooved surfaces described in Section 3.1 and may be the result of lava tearing away from and scraping against its conduit (Fig. 5).

Despite the dominance of filamented textures, the preferred locations of each variety of pahoehoe texture suggests the following ideal trend down a flow-field (Fig. 1): from threaded, through filamented and spiny, to arrow-head pahoehoe. This trend corresponds to an increasing pastiness of small-scale textures and so, presumably, to an increase in crystallinity and rheological resistance towards the distal periphery of a flow field. More prominent, however, are the contemporaneous changes from pahoehoe to aa morphologies.

4 Cauliflower Aa

4.1 Field Aspect

An initial aa crust resembles a field of maltreated cauliflowers: a relatively even surface punctuated by irregular protrusions with heights and distances from each other in the range of centimetres to tens of centimetres. Smaller protrusions tend to be equidimensional, while larger ones usually widen upwards, showing lumpy tops with weak branching on a centimetre scale. In fine detail, all have extremely irregular surfaces, composed of millimetre-sized spines in chaotic arrangement. Because of their general appearance, the protrusions are here called *aa cauliflowers* and surfaces on which they are the predominant component (whether as fixed protrusions or detached fragments) are classified as *cauliflower aa*.

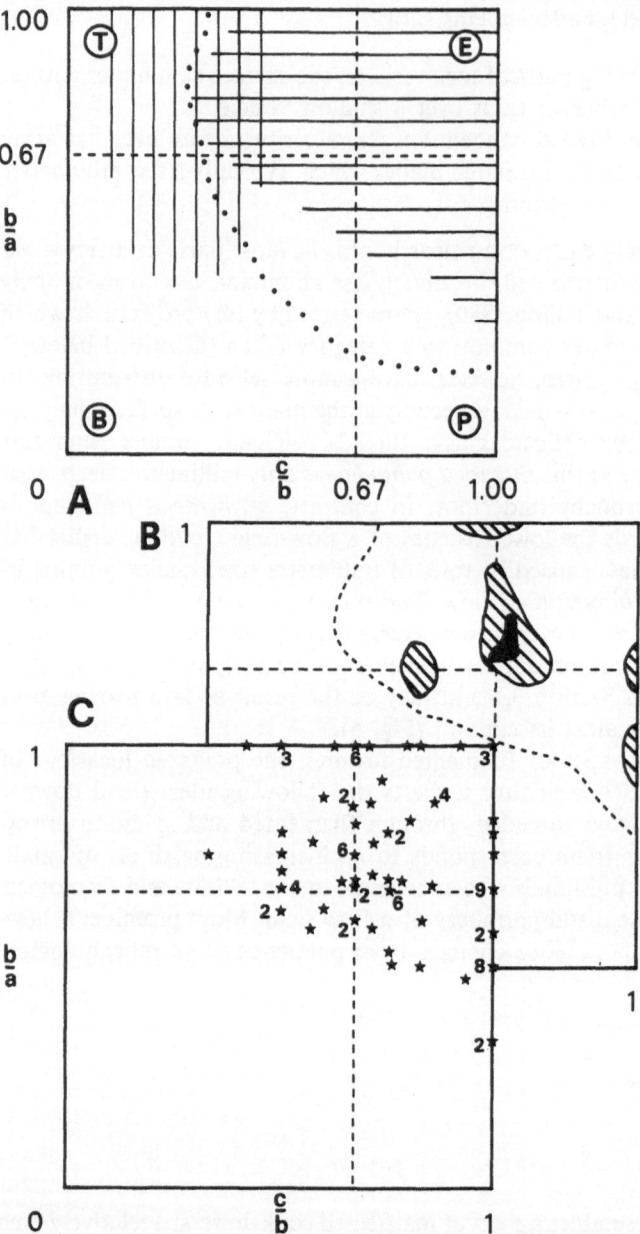

Fig. 6A–C. Shapes of surface fragments on Zingg (1935) diagram. Axial ratios of fragments max : intermediate : min = $a:b:c$. *A Horizontal lines*, aa cauliflowers; *vertical lines*, pahoehoe slabs; *dots*, enclose field (to *right*) of smaller pahoehoe fragments. Aa field based on *6C* and other data. Pahoehoe fields based on general aspect and random measurements; they are general guides only. Shapes separated by *dashed lines* are: *T*, tabular; *E*, equant; *B*, bladed; *P*, prolate. (Note that "bladed" as used here does *not* have the same meaning as the "bladed" for detailed pahoehoe surface textures.) *B, C* Shapes of 94 aa cauliflowers from one of Etna's 1983 flows, about 10 m from the bocca. *Numbers* against stars *(6C)* give number of cauliflowers with same shape. The data points are concentrated just inside equant field *(6B)*. Concentration at a point is number of points within a 0.05 radius. Concentrations: *black*, 11 or more; *diagonal lines*, 5–11; *blank*, less than 5. *Dashes* show extent of distribution. Most with an axial ratio of 1 have lengths of 2 cm or less

As emphasized by Jaggar (1930), protruding cauliflowers are primary features and grade uninterruptedly into the massive lava beneath. Downstream, they tend to become bigger or to maintain the same size distribution; however, they do not show consistent size variations across different flow surfaces, presumably because the centre of a final crust is younger than its edges and may have developed under a wide range of different conditions.

Cauliflower aa is normally first found in quantity downstream from slab pahoehoe or at the feeding outlet itself. In typically less than a few metres, the surface has developed a loose veneer of broken aa cauliflowers and, if downflow from slab pahoehoe, subordinate pahoehoe fragments. Pahoehoe debris and broken aa cauliflowers can usually be distinguished by their small-scale surface textures, if not by their gross form. Textural changes are gradational, however, and it is not uncommon to find pahoehoe filaments and aa spines on a single fragment.

Cauliflowers generally have dense interiors and break into equant fragments (Fig. 6) which can sometimes be more than 30 cm across. The loose cover is rarely thicker than a few tens of centimetres, i.e. rarely more than two or three large fragments deep. The underlying surface of continuous lava retains the form of a field of attached cauliflowers. The fragmental deposit is poorly packed and contains a large proportion of voids. It shows a net reverse grading, in part because pieces broken from existing fragments tend to settle through the voids during lava movement (Macdonald 1972) and in part because, as new cauliflowers form and become detached, the smaller remain around the base of the larger. Consequently, the size distribution of cauliflowers across the upper surface of a loose deposit is not representative of the whole deposit, but is biased in favour of the larger fragments. Indeed, preliminary size-frequency studies suggest that while cauliflowers smaller than 10 cm may contribute only 10% − 15% (by number) of upper surface fragments, they represent as much as 90% of the total cauliflower population.

4.2 Small-Scale Features

The surfaces of aa cauliflowers are uneven at scales from less than a millimetre to more than a centimetre. A characteristic feature in this size range is the presence of spines, with two basic types being recognized: *stiletto spines* and *stubby spines*.

Stiletto spines are the slender-bladed daggers, millimetres long, common on the surfaces of glassy black cauliflowers. The spines are associated with contorted and broken filaments and appear to have been formed by pulling-apart of the congealing surface. Often superimposed are submillimetre, rounded and glassy knobs, similar to those on some pahoehoe surfaces.

Stubby spines are the more widespread type on Etnean cauliflowers. Roughly equidimensional in cross-section, they are rarely longer than a few millimetres and tend to have dull surfaces, often tinted ochre or magenta-brown by oxidation.

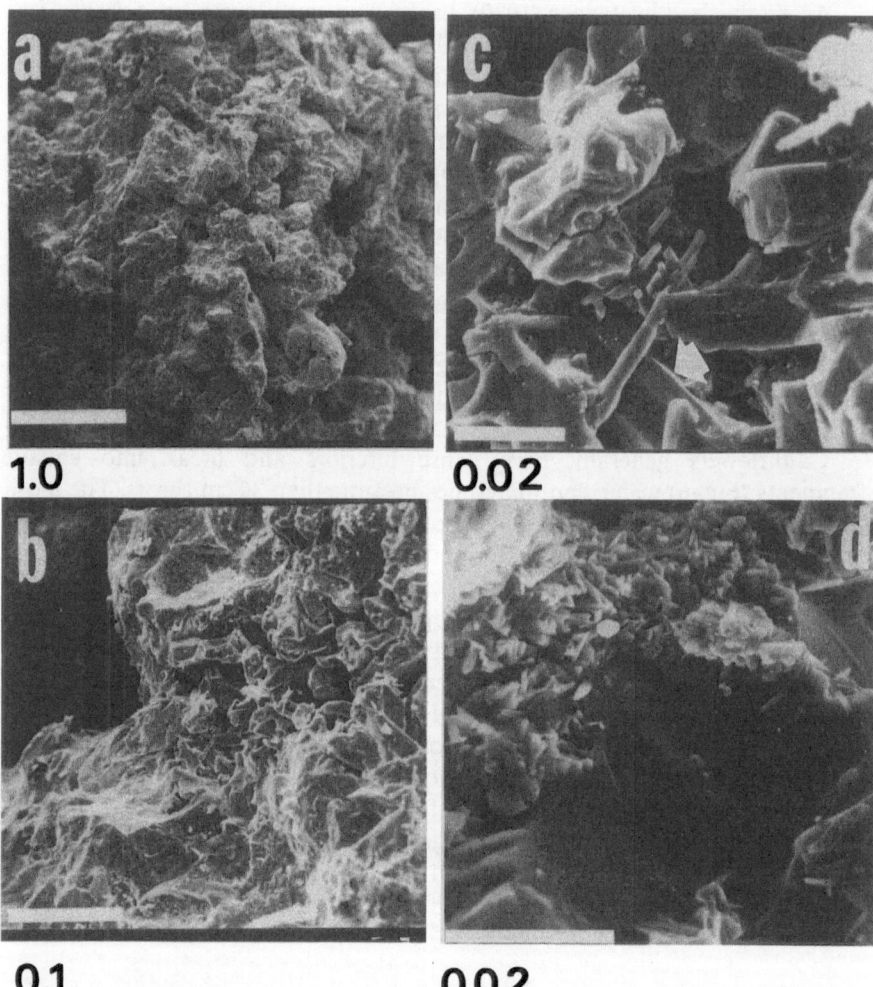

Fig. 7a–d. SEM photographs of stubby-spined aa cauliflowers from Etna. *a* to *c* 1979 lava; all are of the surface at the same orientation. *d* 1980 lava from Northeast Crater. Length scales are in mm. Most crystals in *c* are 0.04–0.05 mm long and euhedral; glassy veneers give them serrated edges. Note also the twinning direction along large crystal, *centre bottom* (*arrow, c*). Apparently thicker glassy veneers give hackly fracture surfaces in *b*. Crystals in *d* have different sizes (0.003–0.004 mm long), although the surface at lower magnification is similar to that in *a*

The change in spine morphology from stiletto to stubby is gradational. Considering a sequence of cauliflowers observed to chill at greater distances from the feeding bocca, the stubby spines first tend to appear in localized matt clusters (≤ 1 cm^2 in area) within a glassier surface that has mainly stiletto spines. Possibly evolving from the glassy rounded knobs, the stubby spines increase in proportion until they become the dominant, if not only, morphological type.

SEM images (Fig. 7) suggest that stubby spines are associated with clusters of unbroken groundmass crystals partially veneered with glass. Within each cluster the crystals tend to have preferred shapes and sizes (Fig. 7 c, d). Crystal appearance, however, can change among clusters. Such changes (Fig. 7 c, d) do not appear to be an orientation effect and are attributed to differences in cluster age, mineralogy, or cooling history.

The fact that different crystal clusters produce similar stubby spines suggests that local crystal content is a more important controlling factor than crystal habit and, hence, that stubby spines are related to inhomogeneous crystallization at millimetre or submillimetre scales. Two possible links between the growth of stubby spines and crystallization are:

1. Crystals nucleate and grow in local clusters around imperfections at the lava surface. The imperfections may be existing crystals, fragments of torn skin, or locally cooler zones where the lava surface is slightly raised (on a millimetre scale). The growing crystal clusters present a greater resistance to motion than the liquid between them. Deformation is concentrated in the liquid zones, where the surface becomes depressed and leaves the crystal clusters, with a liquid veneer, projecting as stubby spines.
2. Crystal clusters form near the lava surface as outlined above. Because crystallization is rapid, the host liquid is subject to a sudden decrease in volume and voids form just ahead of an advancing crystal face. Within an intimate network of nucleating and growing crystals, some of the voids, singly or after coalescence, intersect the lava surface. If at the point of intersection the interstitial liquid is sufficiently abundant and fluid, the lava surface may move apart by flowage; otherwise it deforms by tearing. In either case, parts of the crystal cluster will be left in relief as stubby spines. A similar process, operating at much smaller scales than in lava, has been proposed to explain the failure of some types of steel (Steele and Lentz 1978).

The change from stiletto to stubby spines is also associated with an increase in surface cracking at scales upward from about a millimetre (i.e. as seen in a hand specimen). The inferred nature of the association starts with stubby spines being surrounded by hairline declivities (cooling cracks, crystallization tears, or the result of vesicles breaking the surface). As stubby spinosity increases, the declivities become more widespread and intersect to form "troughs" enclosing areas of as much as a few square centimetres; the surface thus develops a tesselated pattern of uneven polygonal clusters of stubby spines (Fig. 8). Further deformation concentrates along the troughs, which widen and deepen into "gullies" (Fig. 8) and separate adjacent clusters.

The result is a series of irregularities, the smaller superimposed on the larger (Fig. 8) and all related to the process which forms stubby spines; such a superposition of irregularities showing broadly similar shapes at each scale suggests that aa cauliflowers may be characterized by a fractal (scale-invariant) geometry (Mandelbrot 1977). The troughs and gullies also typically have knobby and stubby features. Conditions may therefore favour a self-feeding mecha-

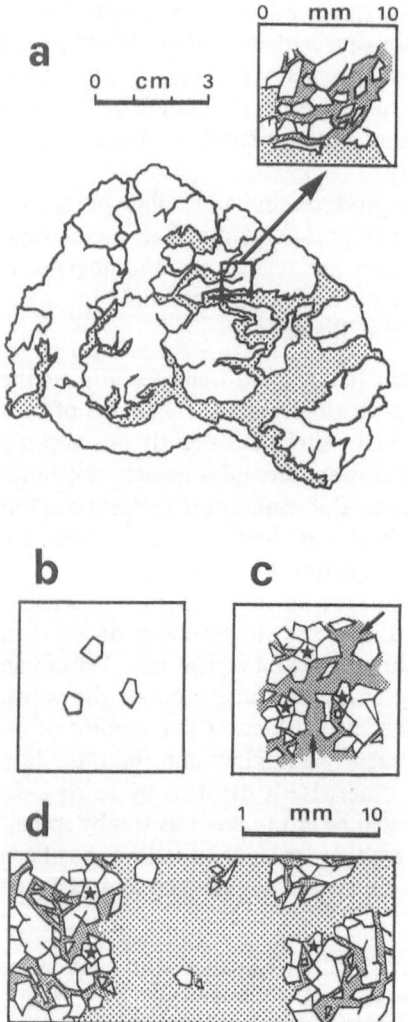

Fig. 8a–d. The basic features of a stubby-spined aa cauliflower. Raised areas (*white*) are separated by irregular cracks (or "troughs") and larger "gullies" (*light stipple*). *a, Inset*: In detail, the raised areas and troughs in *a* can be resolved into collections of stubby spines (*white*) and finer cracks (*heavy stipple*); the *light stipple* corresponds to the margin of a gully in *a*. Note how, in spite of differences in scale, the relations between spines and cracks resemble those between the raised areas, troughs and gullies in the main diagram. The inferred development of the irregularities follows the growth of isolated stubby spines (*b*) into clusters of spines (*c*). The clusters are separated by fine cracks (*heavy stipple* in *c*), along which deformation is concentrated. Some of the cracks (*arrowed* in *c*) eventually open into gullies (*light stipple* in *d*). The *starred* spines in *c* and *d* refer to the original three in *b*

nism, the surfaces of extending depressions themselves being cut by new depressions until limited by chilling. Other aspects of self-feeding processes and scale-independent features will be discussed in Section 6.4.

Underneath, a cauliflower's massive interior continues to the surface and, in hand specimen, does not appear to show systematic changes in vesicularity. The vesicles tend to be smaller and more contorted than in associated pahoehoe crusts. Macdonald (1953) attributed the contortions to twisting of vesicles during flow. Another possibility, suggested by petrographic relations, is that vesicles growing in crystal-rich lava prefer to extend themselves into zones of lower resistance (e.g. between phenocrysts, clusters of microlites or, possibly, liquid inhomogeneities caused by polymerization before crystal nucleation), whether or not the lava is moving.

5 Observations on Active Lavas

The preceding descriptions of surface features suggest that a lava's crystallinity and styles of crystallization contribute to whether it has a pahoehoe or aa crust. This is not surprising since crystallization changes a lava's rheological properties and hence the way it deforms. However, the mode of deformation also depends on the imposed stresses, for which qualitative data have been obtained from active lavas.

5.1 The Pahoehoe-Aa Transition

Repeated observations of the crustal transition from pahoehoe to aa along channelled lava were made during Etna's 1983 and 1984 effusions. The channels had active widths from about 1.0 to 1.5 m. Velocities along surface centrelines varied from about 10 m h^{-1} to 3 m in 10 s (ca. 0.003 to 0.3 ms^{-1} on average), the lower velocities occurring along narrower channels. Velocities were approximately constant across the central one-third to one-half of a channel, decreasing to zero towards its edges. Surface deformation was therefore concentrated within marginal zones, across which the average shear rate (rate of deformation) is estimated to have differed by more than an order of magnitude among the flows (Appendix). Yet, in spite of the large differences in shear rate, the nature of the crustal transformation remained the same.

The transition process is illustrated in Fig. 9. When still an incandescent orange-yellow, the lava surface appeared similar to lumpy porridge (or Italian *polenta*), consisting of domes and hollows with maximum diameters approaching 10 cm but with total variations in relief of only a few centimetres. Incipient darkening occurred preferentially over the tops of domes, presumably cooler than the hollows, with thin, spiderlike threads spreading from their apices towards the depressions (Fig. 9a).

The chilled threads may have been following local surface ridges. They rapidly thickened and coalesced to form distinct clots (Fig. 9b), as previously described by Jaggar (1930) and Peterson and Tilling (1980). Across the middle of the surface, the region of lowest applied shear stress and shear rate, the clots merged to form a continuous, glassy skin, typically of filamented pahoehoe. Gas escape was very low and most tended to occur along channel margins. In these instances, therefore, the filamented textures were more likely the result of flow-induced tearing, on scales of millimetres to centimetres, rather than the product of gas escape (Wentworth and Macdonald 1953, p. 35).

Closer to the channel margins, shear rates were too high for the growing clots to merge as a continuous crust and they remained as isolated domes, rotating about their axes with the shear gradient (Fig. 9b). They also began to twist upwards, creating aa cauliflowers and accentuating the bumpiness of the surface. With further twisting and rolling forward, the cauliflowers eventually detached themselves from below, other cauliflowers starting to develop in the

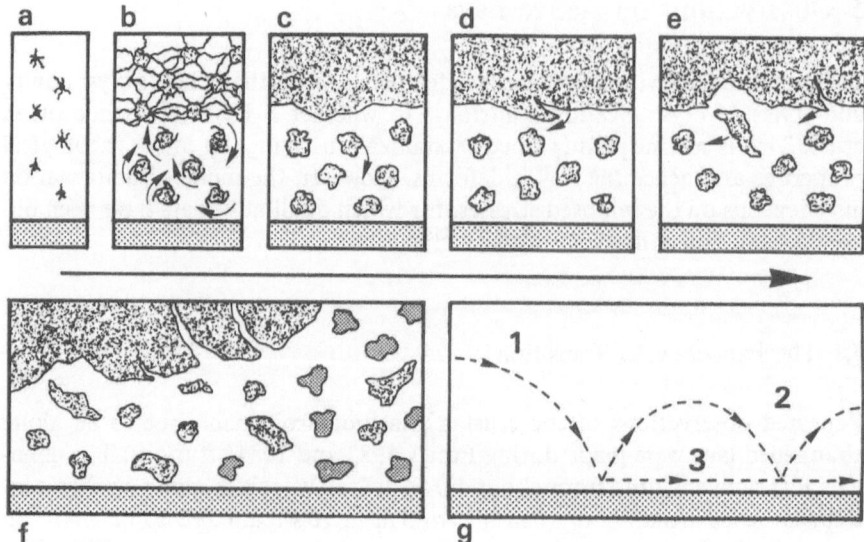

Fig. 9a–g. Schematic illustration of the pahoehoe-aa transition. *Light stipple:* channel margin. *Irregular stipple:* pahoehoe and stiletto-spined aa. *Heavy stipple:* stubby spined aa. *Top* of each diagram coincides with centreline of channel surface. *Central arrow* shows flow direction. Surface chilling spreads radially (*a*) from discrete nuclei. The skin grows and coalesces to pahoehoe across centreline of channel surface (*b, c*), the zone of low shear rate. Shear rates are higher nearer the margin and the skin is twisted into discrete stiletto-spined cauliflowers (*b, c*). When detached, the cauliflowers migrate to the margin (*c*). Pieces tear away from the edge of the central pahoehoe sheet (*d, e*) and also migrate to the margin. The pahoehoe surface eventually narrows (*f*), giving way laterally to stiletto-spined aa cauliflowers and pahoehoe slabs and, finally, to stubby-spined aa cauliflowers. Debris migrating to channel margins (*g*) may repeatedly be pushed back into the stream and again to the margin (*2*) or dragged along the margin (*3*)

newly exposed surface. A similar process of cauliflower growth may have been observed by Finch (reported by Macdonald 1953).

Together with fragments torn from the edges of the central pahoehoe crust, detached cauliflowers tended to migrate towards the channel margins (Fig. 9c–e). Few pieces, however, attached themselves to the inner sides of the levees bounding the channel: they were either dragged downflow along the margin or pushed back into the stream, after which they returned to the margin for the cycle to be repeated (Fig. 9g). Apparently the skins of the fragments were too cold, or the marginal shear rates too large, for significant accretion against the levees. The growth of accretionary levees (Sparks et al. 1976) must therefore indicate specific combinations of marginal shear rates and surface temperatures of migrating fragments.

Across channel centrelines, the continuous pahoehoe crust initially approached a steady width, the pieces tearing away from its edges being rapidly replaced by the surface exposed underneath. Further downflow, however, the continuous skin gradually narrowed to give way to a reemerging lumpy surface

(Fig. 9f), which frequently resembled the glowing coals described by Jaggar (1930) and Perret (1950). Sometimes the change was hastened by wholesale disruption of the continuous crust into slab pahoehoe, as the lava was subjected locally to rapid accelerations or decelerations. Although the new lumpy surface usually developed as cauliflower aa, it was occasionally able to form another pahoehoe central crust. The sequence of events would then be repeated until the crust became wholly and irreversibly aa, even in the zones of low shear stress and shear rate across the middle of the channel.

Observing times ranged from 30 to 120 min, during which discharge rates appeared to be steady within an estimated 20% observational error. Nevertheless, the state of the surface emerging from the bocca often varied during a single observing period. The initial amount of darkened crust could fluctuate from ca. 20 to 80 areal % and, on one occasion, the centreline crust at a bocca changed from ropy pahoehoe to cauliflower aa and back again. Such behaviour must be due to changes in lava rheology or imposed stress, which themselves ought to have changed flow velocities, channel shape or lava level. The changes, therefore, must have occurred within the range of observational error, suggesting that, in the transitional state between pahoehoe and aa, only small differences in effusion conditions (as measurable in the field) may cause significant changes in crustal morphology.

5.2 Styles of Chilling

Across filamented pahoehoe and early cauliflower surfaces, the change from incandescence to blackness occurred smoothly within about 20–30 s. The resulting surfaces were largely glassy. Later aa cauliflowers with stubby spinosity chilled more slowly and less evenly, passing through a speckled phase of orange and black points before darkening completely. On occasion, the two styles of chilling appeared to occur together across a channel, smoothly darkening pahoehoe being flanked by speckled aa cauliflowers.

The change from smooth to speckled chilling is unlikely to be only a morphological effect since all the surfaces have small-scale irregularities. Evenness of chilling reflects thermal homogeneity, while greater speed of chilling reflects less excess heat to be lost. Liquid surfaces chill evenly and quickly to glass because they are structurally (and thermally) homogeneous and do not have to lose latent heat of crystallization; highly crystalline surfaces would also chill evenly, but at a different rate. Surfaces of intermediate crystallinity (stubby-spined cauliflower aa) chill unevenly because they are heterogeneous and crystallizing, and more slowly because they have to lose the latent heat generated at local pockets of crystallization.

6 Interpretations and Discussion

The surfaces of Etna's aa lavas show two morphological trends, from pahoehoe to aa along individual flows, and from threaded to arrow-head textures on the

Table 1. Features of surface morphological trends on Etnean lavas

---

A Pahoehoe-Aa Trend

1. A gradation in small-scale surface textures from dominantly filaments (threads to blades) and knobs (*filamented pahoehoe*), through stiletto-spines, filaments and knobs (*stiletto-spined cauliflower aa*), to stubby spines (*stubby-spined cauliflower aa*).

2. An increase in groundmass crystallinity, from filamented pahoehoe and stiletto-spined cauliflower aa to stubby-spined cauliflower aa.

3. Slower and more uneven chilling of surfaces from filamented pahoehoe and stiletto-spined cauliflower aa to stubby-spined cauliflower aa.

4. An increase in the amount of autobrecciation from pahoehoe to aa.

5. Changes in the style of autobrecciation, pahoehoe crusts producing slabs and aa crusts producing cauliflowers.

6. When crusts of pahoehoe and aa (normally stiletto-spined cauliflowers) form simultaneously across a channel's surface, the pahoehoe occurs down the middle (zone of low stress and shear rate) and the aa near the margins (zones of high stress and shear rate) of the channel.

7. Stubby-spined cauliflower aa may first develop near the margins of a channel but eventually covers the whole surface of a flow (i.e. it extends into zones of lower stress and shear rate with distance downstream). When confined to channel margins, the aa flanks pahoehoe or slab pahoehoe crusts.

B Near-Bocca Pahoehoe Trend

8. A decrease in surface glassiness from threaded to arrow-head pahoehoe, the change overlapping the decrease associated with the pahoehoe-aa transition.

9. Stubby spines occur only rarely in the pahoehoe textural sequence.

---

pahoehoe near boccas down a flow-field (Fig. 1). Both result from a hierarchy of processes acting at different scales, with the first trend being more prominent and superimposed on the second.

The changes recognised for each trend are summarized in Table 1. They can all be interpreted in terms of crystallinity, crystallization rate and shear rate (deformation rate). A clear parallel exists with the studies of Macdonald (1953) and Peterson and Tilling (1980), who identified apparent viscosity and deformation rate as controls on the pahoehoe-aa transition. For a closer comparison with these earlier studies, the influence of bulk crystallinity (related to bulk apparent viscosity) and large-scale shear rate will be considered first.

6.1 Large-Scale Textural Development

No ranges of crystallinity or shear rate are clearly exclusive to either pahoehoe or aa. Stiletto-spine cauliflower aa and filamented pahoehoe crusts form at the same time across a channel and have similar crystallinities (feature 2 in Table 1), but the aa occurs in zones of higher shear rate (feature 6). On the other hand, while the change from filamented pahoehoe to stubby-spined cauliflower aa is characterized by an increase in surface crystallinity (features 2, 3),

Fig. 10a–c. Conditions inferred for pahoehoe and aa surfaces. Large scale (*a*): relations between pahoehoe (*P*) and aa (*A*) in terms of bulk crystallinity (ϕ) of crust and shear rate ($\dot{\varepsilon}$). Small scale (*b* to *c*): relations between detailed surface textures in terms of *local* crystallinity (ϕ_L), crystallization rate ($\dot{\phi}$), local deformation rate [$\dot{\phi}/(1-\phi_L)$] and $\dot{\varepsilon}$. *Numbers* specify fields of different textures; where not separated by *broken* or *solid* lines, fields may overlap. *1* Glassy threads, filaments, knobs and stiletto spines; *2* filaments; *3* arrow-head spines; *4* stubby spines. ϕ_g is the minimum crystallinity at which stubby spines can form. The *shaded* transition in *b* indicates uncertainty if boundary is vertical or has a negative slope. Fields *2* and *3* are of textures more common to pahoehoe. Field *4* is more common to aa. Compare *a* with Fig. 11d

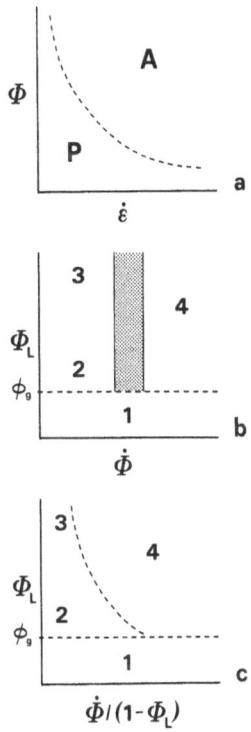

both types of crust can form in zones of low shear rate (features 6, 7). In addition, the pahoehoe-aa and near-bocca pahoehoe trends both involve increases in surface crystallinity (features 2, 3, 8). The relations inferred between crystallinity (ϕ), shear rate ($\dot{\varepsilon}$) and surface features are therefore as shown in Fig. 10a.

The rheological implications of Fig. 10a can be explored by taking crystallinity as an indirect measure of rheological resistance, both increasing together when the shear rate is constant. Rheological behaviour is conventionally illustrated in terms of variation of shear rate with applied shear stress (τ) for laminar flow. Figures 11a and b show typical τ-$\dot{\varepsilon}$ diagrams for Bingham and pseudoplastic materials, the two types of non-Newtonian fluid that have been used to examine the field characteristics of lava flows (Robson 1967; Shaw et al. 1968; Shaw 1969; Hulme 1974; Pinkerton and Sparks 1978; Peterson and Tilling 1980; Kilburn 1981; Fink 1984; Chester et al. 1985; Dragoni et al. 1986; Kilburn and Lopes 1988; Pinkerton and Wilson, in press; Blake, this Vol.). After simple rearrangement, the relations in the τ-$\dot{\varepsilon}$ diagrams (Figs. 11a,b) can be represented in terms of apparent viscosity (η_a; see caption to Fig. 11) and $\dot{\varepsilon}$ (Fig. 11c) and of ϕ and $\dot{\varepsilon}$ (Fig. 11d).

Comparison of the two ϕ-$\dot{\varepsilon}$ diagrams (Figs. 10a, 11d) suggests that the pahoehoe-aa boundary in Fig. 10a is analogous with the failure envelope in Fig. 11d. Such an analogy is consistent with field observation in general (i.e.

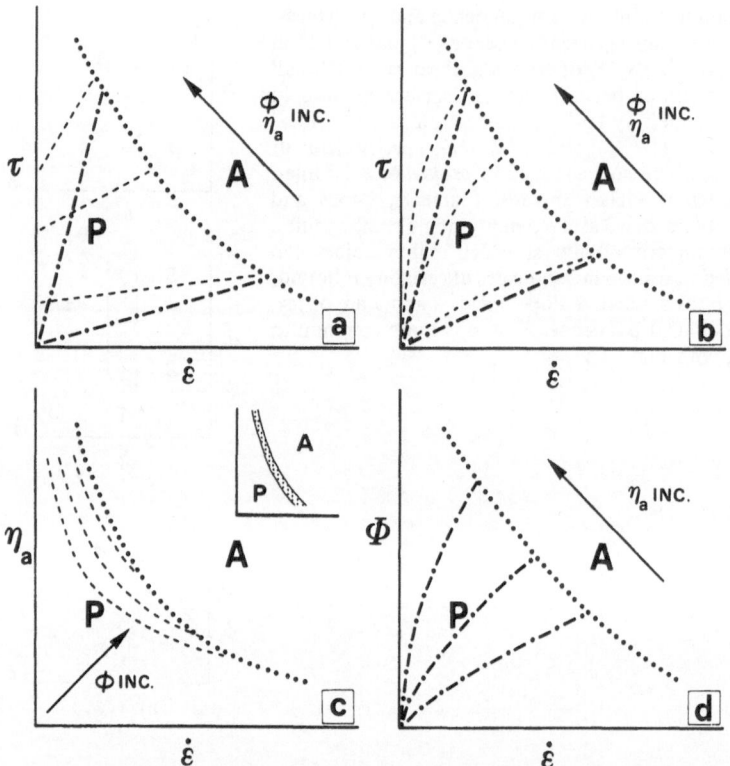

Fig. 11a–d. Rheological interpretation of changes in large-scale surface texture. Shear stress
(τ) – shear rate ($\dot{\varepsilon}$) diagrams for Bingham (a) and pseudoplastic (b) fluids. *Dashed curves:*
rheological flow curves, describing a fluid's resistance to flow; greater resistances are indicated by steeper slopes. *Dash-dot curves:* lines of constant apparent viscosity ($\eta_a = \tau/\dot{\varepsilon}$).
Each shows the conditions for rheologically different lavas to *appear* to have a particular
viscosity to the naked eye; they also show that a lava of constant rheology (following a single
rheological flow curve) will appear to become less viscous at higher shear rates. *Dots:* failure
envelope, above and to the right of which is the field of autobrecciation. *A* aa; *P* pahoehoe.
For increasing bulk crystallinity (ϕ), autobrecciation requires larger stresses (a, b). c shows
relations in a and b in terms of η_a and $\dot{\varepsilon}$. For pseudoplastics, the η_a-lines in c may level to
constant η_a at low $\dot{\varepsilon}$, the limiting η_a increasing with ϕ. *Inset* of c shows shape of pahoehoe-
aa transition zone proposed by Peterson and Tilling (1980). d shows the relations in a and
b in terms of ϕ and $\dot{\varepsilon}$. Compare d with Fig. 10a

the typical continuity of pahoehoe surfaces and discontinuity of aa surfaces)
and with the model of Peterson and Tilling (1980) in particular. From studies
of active Hawaiian lavas, Peterson and Tilling (1980) deduced the conditions
for pahoehoe and aa formation in terms of apparent viscosity and shear rate
(Fig. 11 c inset). The pahoehoe-aa boundary which they inferred has a form
similar to both the rheological flow curves and the failure envelope in Fig. 11 c.
In an earlier analysis, Kilburn (1981) associated the trend of the Peterson-Tilling boundary with that for rheological flow curves, so linking the pahoehoe-aa
transition to a change in lava rheology. This interpretation, however, appears

to be insufficient, since it does not account for the fact that among lavas of the same bulk composition, both pahoehoe and aa can form at comparable crystallinities and, by implication, under comparable rheological conditions.

The re-interpretation of the Peterson-Tilling boundary as a failure envelope allows for both the unidirectional nature of the pahoehoe-aa transition and the overlapping crystallinity ranges of pahoehoe and aa. The rheological diagrams of Fig. 11 are valid only for continuous deformation. Thus, while deforming continuously (at observational scales of centimetres to tens of centimetres), the behaviour of a piece of lava surface will be described by the flow curve appropriate to its rheology. Such a surface may be pahoehoe or aa *with attached cauliflowers*. The behaviour of a surface of loose aa, however, can no longer be represented on the rheological diagrams (although the behaviour of the hidden continuous surface might be). Within itself, of course, a piece of detached crust may still undergo continuous deformation. This is irrelevant, however, since the scale of such deformation is smaller than that of the observed autobrecciation on which the pahoehoe-aa distinction is based.

6.2 Small-Scale Textural Development

The development of detailed surface textures may be examined in terms of local crystallinity (ϕ_L) and crystallization rate ($\dot{\phi}$). ϕ_L is used as a measure of local rheological resistance, while $\dot{\phi}$ has been recognised empirically as an important factor because: variations in $\dot{\phi}$ affect rates of heat loss from the surface and so its mode of chilling; $\dot{\phi}$ has been inferred to influence the shape of stubby spines; and differences in $\dot{\phi}$ could explain why stubby spines are common to aa, but rare on surfaces of similar inferred crystallinities on the near-bocca pahoehoe trend. In addition, the ratio $\dot{\phi}/(1-\phi_L)$ can be used to describe deformation rates in the liquid caused by crystallization. The interpretation follows if $\dot{\phi}$ is considered not as rate of crystallization but as rate of shrinkage of host liquid. Since $(1-\phi_L)$ is the volume fraction of liquid, $\dot{\phi}/(1-\phi_L)$ is seen to be a measure of the relative rate of liquid shrinkage, i.e. the (volumetric) rate of liquid deformation.

The proposed relations between detailed surface textures, ϕ_L, $\dot{\phi}$, and $\dot{\phi}/(1-\phi_L)$ are shown in Fig. 10b and c. Among glassy surfaces, with local crystallinities less than a critical value ϕ_g, the textures of both pahoehoe and aa are restricted to combinations of threads, filaments, knobs and stiletto-spines (Zone 1, Fig. 10b). When lava is more crystallized, the greater number of crystals and probably higher effective viscosity of the host liquid (which is expected usually to become more siliceous) favour pastier and coarser surface textures (Zones 2, 3, Fig. 10b). Stubby spines also form at local crystallinities greater than ϕ_g. However, they are not ubiquitous among surfaces of such crystallinities, but tend to occur mainly on aa crusts. It is postulated that stubby spines can form only at higher crystallization rates because (1) larger crystallization stresses are induced and these can locally rupture the surface, and (2) the more rapidly the surface is deformed, the less time it has to relax to a smoother shape before chilling. In Fig. 10b, therefore, stubby spines are shown to be associated with local crystallinities greater than ϕ_g and relatively high crystallization rates (Zone 4).

The comparable crystallinity ranges of the near-bocca pahoehoe and the pahoehoe-aa trends, but the preference of stubby spines to form on aa surfaces, can be attributed to lavas with different undercooling histories being exposed after different periods of time. According to standard crystallization theory (Chalmers 1964; Kirkpatrick 1976; Dowty 1980), crystallization rates are low at large and small undercoolings and faster at intermediate undercoolings. Because of the prevailing temperature gradient (Pinkerton and Sparks 1976; Fitton et al. 1983; Armienti et al. 1987), lava in a flow becomes increasingly undercooled from centre to surface. The lava may also tend to become less undercooled with time (before quenching after exposure to the atmosphere) owing to crystallization (Brandeis and Jaupart 1987). Near-bocca surfaces are thus interpreted as windows onto lava that has spent most of its time at relatively low undercoolings in the body of one or more flows. Stubby spines are rare because of low crystallization rates; when they do occur, a form of spiny pahoehoe is produced. In contrast, the pahoehoe-aa sequence represents the evolution of a flow's upper levels which, immediately before exposure and quenching, are crystallizing at relatively higher undercoolings. Fast crystallization rates are maintained when local crystallinities exceed ϕ_g and so stubby spines are common textural features.

6.3 Links Between Different Scale Processes

The ϕ_L-$\dot{\phi}$ and ϕ-$\dot{\varepsilon}$ relations (Figs. 10b and 11d) are both needed for a more complete interpretation of changing surface features. Although the two relations refer to processes operating at different scales, they are linked by crystallinity. Changes in ϕ_L affect local deformation; they also change ϕ (hence the bulk rheology of the crust) and so the way the whole surface deforms under larger-scale differences in stress. The overlap of detailed textures between aa and pahoehoe demonstrates that the larger-scale autobrecciation and detailed textural transitions do not coincide exactly; neither, however, is the overlap very large, otherwise the associations of different-scale features would not be so prominent. A narrow overlap is suggested particularly clearly by the almost simultaneous onset of stubby-spine growth and wholesale surface autobrecciation to aa. Indeed, it may be this association which Macdonald (1953, p. 190) described as the near coincidence of the "granulation" and "fragmentation" points during aa development in basalts.

A physical connection is therefore implied between crystallization and large-scale shearing in the transition from pahoehoe to stubby-spined aa. Two possible connections are as follows:

1. An increase in stress (and shear rate) fragments the surface to reveal a level of more crystalline and faster crystallizing lava. Such a mechanism requires

that, in spite of increasing temperatures, the upper layers of lava increase in crystallinity downwards, crystallization being controlled by kinetic processes rather than by thermal gradients alone. If these conditions do occur, they must be accounted for when interpreting observations and physical measurements on active lava surfaces.

2. An increase in shear rate (and stress) accelerates crystallization in the undercooled surface (Emerson 1926; Gibb 1974; Corrigan 1982; Kouchi et al. 1986), greater relative movement in the lava favouring the chances of atoms in the liquid coming to the arrangement necessary to form a nucleus (Mullin and Raven 1962; Gibb 1974); any breaking of existing crystals might also provide a means of accelerating crystallization, by providing more surfaces for heterogeneous nucleation (Mullin and Raven 1962). With the stress sustained, the increase in crystallinity leads to conditions favouring autobrecciation (Fig. 11 a, b). If this is the process which occurs, it has interesting practical implications for mitigating lava hazard (Kilburn 1983; Chester et al. 1985).

For both interpretations, an increase in stress is needed before autobrecciation. The stress increase must be large-scale, since the whole width of a flow's surface is affected, and it must be caused by a common process, since the autobrecciation itself is common. General downstream acceleration is unlikely, because flow fronts decelerate during cooling even when discharge rate increases at a bocca. Most terrain, though, is uneven and acceleration over topographic irregularities may be the most likely source.

6.4 Self-Regulating Behaviour

If crystallinity and bulk shear rate are related according to the second interpretation above, a self-governing process across different scales can be envisaged for which: increasing $\dot{\varepsilon}$ accelerates crystallization; increased $\dot{\phi}$ increases local shear rates and ϕ_L; increased ϕ_L increases ϕ; and increased ϕ affects $\dot{\varepsilon}$, since the lava's rheological resistance has changed. Possibly, therefore, large changes in crustal condition may result from only small initial changes in $\dot{\varepsilon}$, caused externally, which are then amplified by the effects on crystallinity and crystallization rate.

Crystallization behaviour as a link between different-scale processes may also be responsible for resemblances in the mode of formation of stubby spines on aa cauliflowers (Fig. 8) and of the cauliflowers themselves (Fig. 9). As shown in Fig. 8, small clusters of groundmass crystals produce mm-sized stubby spines; clusters of spines (crystal macroclusters) produce cm-sized lumps; and clusters of lumps (crystal superclusters) produce 10-cm-sized cauliflowers. Extrapolating this process leads to clusters of cauliflowers (crystal hypoclusters) producing an aa surface (m-scale). In other words, by creating rheological inhomogeneities in the surface, crystal hypoclusters (related to the initial clusters of groundmass crystals to produce spines) concentrate deformation in the more liquid regions between them and accentuate the growth of cauliflowers across a congealing aa surface.

6.5 Porridge Textures and Aa Slabs

Two other features related to stress patterns are the lumpy, porridgelike textures observed on incandescent surface, and the scarcity of crustal slabs with aa texture. On porridgelike surfaces, the bigger lumps are either smoothed-out and chill to form pahoehoe, or they become accentuated to form stiletto-spined aa cauliflowers. Although the difference in evolutions can be explained in terms of the stress regime in which the lumps happen to find themselves (Fig. 9), the reason why there should be lumps at all remains equivocal. Possibly they are the products of heterogeneous groundmass crystallization (which, since they often chill to largely glassy surfaces with low groundmass crystallinities, implies an extreme sensitivity to crystal distribution), or they may be connected with failure directions that cut across a flow's surface because of non-Newtonian lava rheology (Hulme 1974; A. J. Bond, personal communication, 1987; Blake, this Vol.).

The rarity of crustal slabs with aa texture is emphasized by the commonness of slabs of pahoehoe. Chilling quickly at low shear rates, pahoehoe crusts appear to develop large lateral extents with respect to their thicknesses. They can remain thin, however, because the initially rapid chilling is slowed by higher crystallization rates (and release of latent heat) a few centimetres below the surface. Hence pahoehoe slabs may represent only an outer chill zone on the flow, the crust defined by crystallinity extending deeper. The greater surface disruption of aa (and also, for stubby-spined aa, faster crystallization rates) would prevent a comparably extensive chill zone from developing. The formation of slabs with aa texture would thus require the thicker, more crystallized crust to be disrupted. Compared with pahoehoe slabs, therefore, slabs with aa surfaces can only form under much greater imposed stresses and so are less commonly produced.

7 Concluding Comments

The evolution of lava surfaces has been related to the interplay between imposed stress and rheological resistance. Measures of this interplay are crystallinity, crystallization rate and shear rate. Of these, crystallization rate has not previously been identified. Its inclusion indicates the need to account for the kinetics of crystallization, not only to interpret surface morphology and motion but, since kinetic processes influence cooling rate (Brandeis and Jaupart 1987), also to model the thermal evolution of a crust, which has an important effect on the development of flows (Pinkerton and Sparks 1976; Pieri and Baloga 1986; Guest et al. 1987; Kilburn and Lopes 1988; Pinkerton and Wilson, in press) and flow-fields (Kilburn and Lopes 1988).

The close association between changes in large- and small-scale surface features during the pahoehoe-aa transition therefore suggests that the important differences in ϕ are determined by changes in ϕ_L, and are only secondarily related to initial differences in phenocryst content. This inference is consistent with previous arguments by Sparks and Pinkerton (1978).

Changes in surface texture have been described in terms of two trends, *near-bocca pahoehoe* and *pahoehoe-aa*. The trends are related to different undercooling histories for different batches of lava and require that the interior of a flow crystallizes more slowly than its exterior. Consequently, mechanisms for accelerating crystallization throughout a flow, such as degassing (Sparks and Pinkerton 1978; Kilburn 1983; Lipman et al. 1985; Lipman and Banks 1987), are not considered to be primary causes of the two trends, although they may reduce the time a lava takes to cross a particular textural transition.

The relations proposed here are qualitative and need to be verified quantitatively. Especially important will be studies of the quantities and size-frequency distributions of groundmass crystals in different surfaces (to test the relation with surface texture) and with depth in a flow (to examine variations in undercooling; Kirkpatrick 1976; Dowty 1980; Brandeis and Jaupart 1987). To avoid the obscuring effects of post-emplacement crystallization, quenched samples from active lavas would ideally be used. Controlled cooling experiments could also examine the influence of ϕ_L and $\dot{\phi}$ on surface textures.

By experimentally quantifying the conditions associated with different surface textures, it will be possible to monitor changes in crystallization conditions (e.g. undercooling, crystal growth rate) during lava emplacement. Such data would be useful for refining predictive models of flow-field development and for testing general models of magmatic crystallization.

Acknowledgements. The author is grateful to Don Peterson and Bob Tilling, whose reviews substantially improved the original version of this chapter. Thanks are also due to Dave Rooks and John Bell for photographic expertise, and to Pete Woods for manipulating the SEM (all at University College London). Parts of this study were funded by the U.K. Natural Environment Research Council and by the Royal Society, London.

Appendix: Surface Shear Rates Across Observed Lava Channels

Consider steady flow down a lava channel symmetrical about its middle. Let w be the distance (measured perpendicular to flow direction) across the surface from the *surface centreline*, and let u be the local surface velocity.

The width of the channel is W. The central, weakly deforming portion of the surface has width w_c. The weak deformation is considered negligible and the central portion of the crust assumed to have a constant velocity u_c (also the maximum surface velocity). At the channel margins, $u = 0$. Surface shearing is thus concentrated within two marginal zones alongside the undeforming crust. The width of each zone, w_s, is $(W-w_c)/2$.

For the same rheological model, the velocity profiles across the marginal zones have the same shapes. By definition, the average shear rate, $\dot{\varepsilon}_{av}$ is:

$$\dot{\varepsilon}_{av} = u_c/w_s = 2 \cdot u_c/(W - w_c) = 2 \cdot u_c/W \left[1 - (w_c/W)\right] . \tag{A1}$$

For a given change in u_c, the minimum change in $\dot{\varepsilon}_{av}$ occurs when the change in $W[1 - (w_c/W)]$, or in w_s, is greatest. After differentiation, the change in $w_s(\Delta w_s)$ may be expressed as:

$$\Delta w_s = [1 - (w_c/W)] \Delta W - W \Delta (w_c/W) \tag{A2}$$

where Δ denotes "a change in".

Among the lavas observed in Etna, u_c varied by about 100, W by about 3/2, and (w_c/W) by about 4/3. Given the absolute values of W and (w_c/W) (see main text), Eq. (A2) gives a maximum Δw_s of about 3. Hence, for a difference in u_c of 100, the smallest change in $\dot{\varepsilon}_{av}$ must be greater than 100/3, i.e. greater than an order of magnitude.

References

Armienti P, Clocchiatti R, Innocenti F, Pompilio M, Villari L (1987) 1984–1985 Mount Etna effusive activity. Rend Soc It Min Petr 42:225–236

Blake S (1989) Viscoplastic models of lava domes. IAVCEI Proc Volcanol 2:88–126 (this vol)

Brandeis G, Jaupart C (1987) The kinetics of nucleation and crystal growth and scaling laws for magmatic crystallisation. Contrib Mineral Petrol 96:24–34

Bullard FM (1947) Studies on Paricutin volcano, Michoacan, Mexico. Geol Soc Am Bull 58:433–449

Chalmers B (1964) Principles of solidification. John Wiley, New York

Chester DK, Duncan AM, Guest JE, Kilburn CRJ (1985) Mount Etna. The anatomy of a volcano. Chapman and Hall, London, pp 1–404

Corrigan GM (1982) Supercooling and the crystallization of plagioclase, olivine, and clinopyroxene from basaltic magmas. Min Mag 46:31–42

Daly RA (1911) The nature of volcanic action. Proc Am Acad Arts Sci 47:47–122

Daly RA (1914) Igneous rocks and their origin. McGraw-Hill, New York

Dana JD (1890) Characteristics of volcanoes. Dodd, Mead and Co, New York

Dawson JB (1962) The geology of Oldoinyo Lengai. Bull Volcanol 24:349–389

Day AL, Shepherd ES (1913) Water and volcanic activity. Geol Soc Am Bull 24:573–606

Dowty E (1980) Crystal growth and nucleation theory and the numerical simulation of igneous crystallisation. In: Hargraves RB (ed) Physics of magmatic processes. Princeton University Press, New Jersey, pp 419–485

Dragoni M, Bonafede M, Boschi E (1986) Downslope flow models of a Bingham liquid: implications for lava flows. J Volcanol Geotherm Res 30:305–325

Dutton CE (1884) Hawaiian volcanoes. US Geol Surv 4th Annu Rep: pp 75–219

Einarsson T (1949) The eruption of Hekla 1947–1948: IV, 3. The flowing lava. Studies of its main physical and chemical properties. Soc Scientarium Islandica, Reykjavik, pp 1–70

Emerson OH (1926) The formation of aa and pahoehoe. Am J Sci 12 (5th ser):109–114

Fink JH (1984) Structural geologic constraints on the rheology of rhyolitic obsidian. J Non-Crystalline Solids 67:135–146

Fitton JG, Kilburn CRJ, Thirlwall MF, Hughes DJ (1983) 1982 Eruption of Mount Cameroon, West Africa. Nature 306:327–332

Foster HL, Mason AC (1955) 1950 and 1951 Eruptions of Mihara yama, Oshima volcano, Japan. Geol Soc Am Bull 66:731–762

Gamond JF (1983) Displacement features associated with fault zones: a comparison between observed examples and experimental models. J Struct Geol 5:33–45

Geikie A (1903) Textbook of geology, Vol 1 (4th edn). Macmillan, London

Gibb FGF (1974) Supercooling and the crystallisation of plagioclase from a basaltic magma. Min Mag 39:641–653

Guest JE, Kilburn CRJ, Pinkerton H, Duncan AM (1987) The evolution of lava flow-fields: observations of the 1981 and 1983 eruptions of Mount Etna, Sicily. Bull Volcanol 49:527–540

Hulme G (1974) The interpretation of lava flow morphology. Geophys J R Astr Soc 39:361–383

Jaggar TA (1917) On the terms aphrolith and dermolith. J Wash Acad Sci 7:277–281

Jaggar TA (1930) Distinction between pahoehoe and aa or block lava. The Volcano Letter 281:1–3

Jones AE (1943) Classification of lava surfaces. Am Geophys Union Trans part 1:265–268

Kieffer G (1979) Pahoehoe et aa: problèmes de morphologie superficielle des coulées volcaniques fraîches. 4éme Colloque de Géomorphologie Volcanique, Publ Inst Geog Faculté de Clermont-Ferrand 57:33–46

Kilburn CRJ (1981) Pahoehoe and aa lavas: a discussion and continuation of the model of Peterson and Tilling. J Volcanol Geotherm Res 11:373–389

Kilburn CRJ (1983) Studies of lava flow development. In: Tazieff H, Sabroux J-C (eds) Forecasting volcanic events. Elsevier, Amsterdam, pp 83–98

Kilburn CRJ, Lopes RMC (1988) The growth of aa flow-fields on Mount Etna, Sicily. J Geophys Res 93:14,759–14,772

Kirkpatrick RJ (1976) Towards a kinetic model for the crystallisation of magma bodies. J Geophys Res 81:2565–2571

Kouchi A, Tsuchiyama A, Sunagawa I (1986) Effect of stirring on crystallization kinetics of basalt: texture and element partitioning. Contrib Mineral Petrol 93:429–438

Krauskopf KB (1948) Lava movement at Paricutin volcano, Mexico. Geol Soc Am Bull 59:1267–1283

Lipman PW, Banks NG (1987) Aa flow dynamics, Mauna Loa 1984. US Geol Surv Prof Pap 1350:1527–1567

Lipman PW, Banks NG, Rhodes JM (1985) Gas-release induced crystallization of 1984 Mauna Loa magma, Hawaii, and effects on lava rheology. Nature 317:604–607

Macdonald GA (1953) Pahoehoe, aa and block lava. Am J Sci 251:169–191

Macdonald GA (1967) Forms and structures of extrusive basaltic rocks. In: Poldervaart AA, Hess HH (eds) Basalts: the Poldervaart treatise on rocks of basaltic composition, Vol 1. Wiley-Interscience, New York, pp 1–61

Macdonald GA (1972) Volcanoes. Prentice-Hall, New Jersey, pp 1–150

Mandelbrot BB (1977) Fractals. Form, chance, and dimension. WH Freeman, San Francisco, pp 1–365

Mercalli G (1907) I vulcani attivi della terra. Ulrico Hoepli, Milan, pp 1–421

Mullin JW, Raven KD (1962) Influence of mechanical agitation on the nucleation of aqueous salt solutions. Nature 195:35–38

Nichols RL (1938) Grooved lava. J Geol 46:601–614

Nye JF (1952) The mechanics of glacier flow. J Glaciol 2:82–93

Perret FA (1950) Volcanological observations. Carnegie Inst Wash Publ 549:1–151

Peterson DW, Tilling RI (1980) Transition of basaltic lava from pahoehoe to aa, Kilauea volcano, Hawaii: field observations and key factors. J Volcanol Geotherm Res 7:271–293

Pieri DC, Baloga SM (1986) Eruption rate, area and length relationships for some Hawaiian lava flows. J Volcanol Geotherm Res 30:29–45

Pinkerton H, Sparks RSJ (1976) The 1975 sub-terminal lavas, Mount Etna: a case history of the formation of a compound lava field. J Volcanol Geotherm Res 1:167–182

Pinkerton H, Sparks RSJ (1978) Field measurements on the rheology of lava. Nature
    276:383–385
Pinkerton H, Wilson L, Factors controlling the lengths of channel-fed lava flows. Bull
    Volcanol (in press)
Rittmann A (1962) Volcanoes and their activity. Interscience, New York, pp 1–305
Robson GR (1967) Thickness of Etnean lavas. Nature 216:251–252
Rowland SK, Walker GPL (1987) Toothpaste lava: characteristics and origin of a lava struc-
    tural type transitional between pahoehoe and aa. Bull Volcanol 49:631–641
Rowland SK, Walker GPL (in press) Pahoehoe and a'a in Hawaii: volumetric flow rate con-
    trols the lava structure. Bull Volcanol
Scrope GP (1856) On the formation of craters, and the nature of the liquidity of lavas. Q
    J Geol Soc Lond 12:326–350
Shaw HR (1969) Rheology of basalt in the melting range. J Petrol 10:510–535
Shaw HR, Wright TL, Peck DL, Okamura R (1968) The viscosity of basaltic magma: an
    analysis of field measurements in Makaopuhi lava lake, Hawaii. Am J Sci 261:255–264
Sparks RSJ, Pinkerton H (1978) Effect of degassing on the rheology of basaltic lava. Nature
    276:385–386
Sparks RSJ, Pinkerton H, Hulme G (1976) Classification and formation of lava levees on
    Mount Etna, Sicily. Geology 4:269–271
Steele JH Jr, Lentz DF (1978) Application of fractographic-microstructural correlations in
    evaluating failure mechanisms in two types of steels. In: Strauss BM, Cullen WH Jr (eds)
    Fractography in failure analysis ASTM STP 645, Am Soc for Testing and Materials, pp
    5–31
Walker GPL (1967) Thickness and viscosity of Etnean lavas. Nature 213:484–485
Washington HS (1923) Petrology of the Hawaiian Islands; IV. The formation of aa and
    pahoehoe. Am J Sci 6 (5th series):409–423
Watanabe T (1940) Eruptions of molten sulphur from the Siretoko-Iosan volcano, Hok-
    kaido, Japan. Jap J Geol Geog Trans Abs 17:289–310
Wentworth CK, Macdonald GA (1953) Structures and forms of basaltic rocks in Hawaii.
    US Geol Surv Bull 994, pp 1–98
Wood HO (1917) Notes on the 1916 eruption of Mauna Loa. J Geol 25:334–335
Zingg T (1935) Beitrag zur Schotteranalyse. Schweiz Mineral Petrog Mitt 15:39–140

Longitudinal Variations in Rheological Properties of Lavas: Puu Oo Basalt Flows, Kilauea Volcano, Hawaii

J.H. Fink and J. Zimbelman

Abstract

Four well-documented lava flows from the 1983 Puu Oo eruption in Hawaii reveal exponential increases of both yield strength and viscosity with distance from vent and with time from onset of eruption. Rheologic descriptors were determined from observations of flow-margin thickness, flow-front position as a function of time, and underlying topographic slope. Longitudinal gradients of calculated strength and viscosity progressively increase through the sequence of four flows and correlate with systematic increases in eruptive temperature and mafic character. If this suggested trend of increasing rheologic gradient with increasing mafic content is found to hold for a wider chemical range, it may prove to be a diagnostic tool for remote evaluation of flow composition.

1 Introduction

Most lava flow studies have been motivated by the need to estimate how far and fast individual flows might travel or by attempts to deduce the compositions of flows on other planets. Models for predicting flow length and for remotely determining composition require knowledge or assumptions about a flow's rheologic and cooling behavior. Hazard studies can be based on either active or fully emplaced flows, whereas planetary modeling must rely on measurements of solidified flows.

Laboratory studies have demonstrated the profound effects of cooling on both the viscosity and yield strength of magmas (e.g., Murase and McBirney 1973; McBirney and Murase 1984). Qualitative field observations and simple calculations have also shown that these properties may increase markedly downstream (e.g., Minikami 1951; Booth and Self 1973; Moore et al. 1980). Few investigations of flow morphology have systematically incorporated cooling behavior, primarily due to a lack of sufficiently accurate field measurements of lava temperature.

In this chapter we present observations of flow morphology and simple calculations of rheologic parameters for four exceptionally well-monitored lava flows emplaced in Hawaii between February and July 1983. We show that for all of the flows, both yield strength and viscosity increase exponentially downstream. Using observations of flow-front position (Wolfe et al. 1989), we find that changes in strength and viscosity correlate well with time as well as distance. This sequence of flows showed a progressive decrease in silica content and corresponding increase in eruptive temperature (Neal et al. 1989). We use these data to demonstrate that the *rates* of increase for both strength and viscosity depend on composition and eruption temperature.

Our results illustrate that: (1) statistically significant rheologic trends can be obtained using a large number of relatively crude morphological measurements; (2) comparisons of yield strengths or viscosities calculated for two different flows require that the morphological measurements be taken from equivalent distances along the flow paths; and (3) longitudinal variations in rheologic properties inferred from simple morphological data may reflect the rate of cooling, which in turn depends on the eruption temperature and composition.

2 Background

Yield strengths and viscosities of lava have been measured both directly and indirectly in the field and the laboratory. Penetrometers (Sparks and Pinkerton 1978; Pinkerton and Sparks 1978) and shear vanes (e.g., Shaw et al. 1968) that relate applied stresses to resultant strain rates have been used to determine directly rheologic properties of Hawaiian and Etnaen lavas, whereas falling sphere and concentric cylinder viscometers have been employed in laboratory measurements (e.g., Murase and McBirney 1973; Spera et al. 1982; McBirney and Murase 1984). These studies have shown that the actual rheologic properties of magma include time- and shear rate-dependent effects which are not taken into account by the two-parameter Bingham model. Nonetheless, the assumption of Bingham behavior has proven useful for interpretations of flow morphology (e.g., Hulme 1974).

Indirect determinations of viscosity of active flows have been made from observations of velocity, density, and slope using the Jeffrey's equation which assumes Newtonian rheology (e.g., Nichols 1939, Macdonald 1963) or modified Navier-Stokes equations based on the Bingham model (e.g., Moore 1987). Most yield strength estimates are based on the idea that Bingham materials come to rest when basal shear stresses drop below the yield value. Since basal stress is the product of thickness, unit weight, and underlying slope, yield strength may be computed from geometric measurements made on emplaced flows, even if they were never observed while active. The ability of this model to estimate yield strengths of prehistoric flows has led to widespread application in extraterrestrial settings (e.g., Hulme 1974; Moore et al. 1978; Zimbelman 1985; Head and Wilson 1986; Cattermole 1987).

Yield strength and viscosity values calculated indirectly from flow morphology or velocities are commonly up to several orders of magnitude higher than those determined for pure melt phase in the laboratory (e.g., Moore et al. 1978; Huppert et al. 1982; Fink and Zimbelman 1986). This mismatch has been attributed to several factors including the influence of a solidified crust, and the resistance to flow provided by underlying topographic roughness and breccia within the flow-front talus pile. Crystallization also causes a marked increase in both rheologic parameters, as has been demonstrated in several laboratory studies (e.g., Shaw 1969).

Although estimating lava rheology from flow geometry is conceptually straightforward, in practice there are many questions about which measurements to make. In an effort to reduce these ambiguities, Fink and Zimbelman (1986) compared calculations of yield strength and viscosity based on several different geometric parameters for the distal 2 km of the well-documented, Episode 5 lava flow from the 1963 Puu Oo eruption of Kilauea Volcano. (In this previous paper, individual flows were referred to as Phases; these are now described as Episodes to conform with USGS usage). Using carefully surveyed cross-sections and direct observations of flow velocity and temperature (Wolfe et al. 1989; Neal et al. 1989), they found that strength and viscosity both increased downstream. They also determined that strength estimates based on the heights of flow margins showed increases more consistent with measured temperature values than estimates based on levee dimensions, central channel depths, or the overall height to width ratio of the flow. These results indicated that for flows not observed during emplacement (including those on other planets), yield strength calculations should be based on flow margin heights rather than other morphological parameters.

Fink and Zimbelman (1986) also found that reasonable viscosity estimates required observations of velocity obtained from many longitudinal positions along the active flow, rather than average values computed from vent effusion rates, overall flow dimensions, or empirical relationships between viscosity and strength. This stipulation led them to conclude that accurate viscosity estimates are generally not possible for flows not observed during emplacement.

3 Observations

3.1 The Puu Oo Lavas

The Puu Oo eruption along the East Rift of Kilauea Volcano began with 47 distinct channel-fed aa episodes from January 1983–July 1986. In mid-1986, the style of eruption changed to a more steady effusion of pahoehoe from a lava lake through an extensive tube system. Throughout the eruption, the U.S. Geological Survey's Hawaiian Volcano Observatory maintained a monitoring program that included hand-leveled measurement of thicknesses at many locations along the margins of the emplaced flows. These flows were mapped and thicknesses recorded on a topographic base with 20-ft contours, from which preflow slopes could later be determined. For some of the lobes, successive flow front positions were recorded, which enabled average velocities to be calculated as a function of position and time. For nearly all eruptive episodes, the USGS measured temperatures near the vent and, whenever possible, at distal locations as well. Samples of lava from each episode were also collected for chemical analysis.

The first 20 episodes of the eruption have been summarized by Wolfe et al. (1987) and documented in detail by Wolfe et al. (1989). For the first 11 episodes the composition became progressively more mafic (e.g., CaO increased from 10.1 to 10.9 wt%; $Na_2O + K_2O$ decreased from 3.3 to 2.7 wt%), and the eruption temperatures showed a corresponding steady increase (1117° to 1138°C) (Neal et al. 1989). From Episodes 12 to 20, the compositions and temperatures showed much less variation.

During the first 11 episodes, several flows passed over the cliffs (palis) to the south and advanced through rain forests or along roads of the Royal Gardens subdivision. The earliest of these southern flows traversed older preexisting topography and did not cover lavas from earlier episodes of the eruption. In contrast, the lobes that moved to the north and east, and later flows that traveled south, formed an anastamosing network for which the boundaries and thicknesses of individual units could not be as easily distinguished on later topographic maps.

3.2 Procedure

In our earlier study of the distal 2 km of the Episode 5 flow (Fink and Zimbelman 1986), we found systematic downstream increases in both yield strength and viscosity. Similar trends for calculated strengths of other lava flows had been reported by Moore et al. (1980) and Moore (1987), and for viscosities by Booth and Self (1973), Baloga and Pieri (1986), and Moore (1987). In order to test the statistical and physical significance of these trends for the Puu Oo flows, we needed many more measurements per flow, as well as data from several different flows. The painstaking surveying procedures which we had used earlier to construct cross-sections did not seem a practical way to obtain flow margin thicknesses along the lengths of a suite of flows. In contrast, the USGS measurements provide readily available thickness data along the entire lengths of most of the Puu Oo flows.

For the present study we focused on flows that traveled south and that did not fill preexisting channels for most of their length; these had the best-defined boundaries, and underlying topographic information was available. We concentrated on relatively long flows, because we felt these would best show longitudinal trends, and we restricted our attention to Episodes 1 – 20, which had the best topographic, chronologic, and temperature data available. For episodes that produced more than one lava flow, we selected the first flow formed, so that we would always be comparing the earliest erupted lavas. These constraints left us four flows to compare (Fig. 1): the southeast flow from the 1123 vent of Episode 2, the flow from Episode 3 which issued from the O-Vent (3-O), the Episode 4 flow, and the eastern branch of the Episode 5 flow (5-E), hereafter referred to as Episodes 2, 3, 4, and 5.

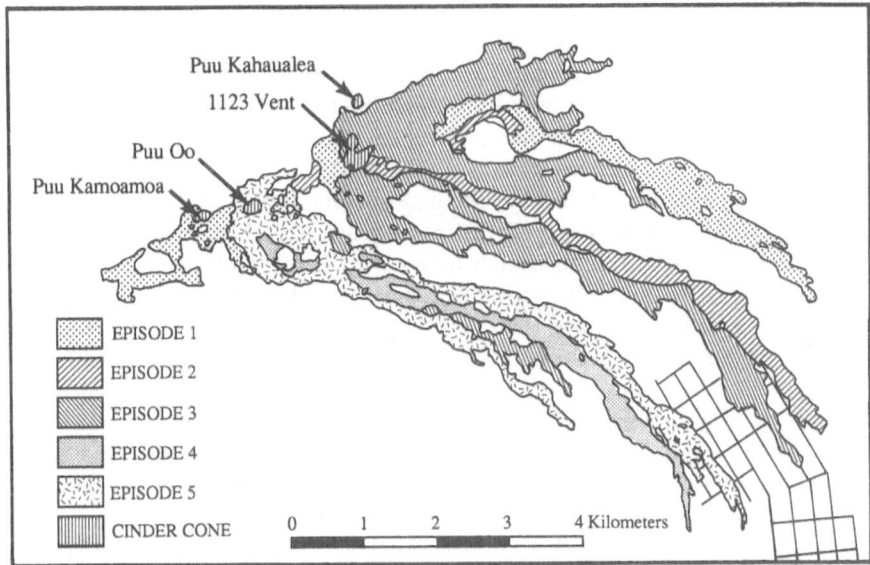

Fig. 1. Map showing the outlines of the Puu Oo lava flows after emplacement of Episode 5. *Straight lines* represent roads in upper portion of Royal Gardens subdivison (After Wolfe et al. 1989)

Observed thicknesses were combined with slopes derived from topographic maps and an assumed density of 2.5 g cm^{-3} to arrive at yield strengths, τ, for the Episode 2−5 flows, using the equation (Johnson 1970):

$$\tau = \varrho \, \mathrm{g h} \sin \theta \, , \tag{1}$$

where ϱ = density, g = acceleration due to gravity, h = flow margin thickness, and θ = underlying slope. Flow margin thicknesses usually reflect the yield strength of the lava as it initially passes a particular location, as opposed to midchannel dimensions which change throughout the course of a flow's development. A constant density of 2.5 g cm^{-3} was assumed, because information about longitudinal variations in specific gravity of the Puu Oo flows was not available at the time of this study. The actual value of 2.5 g cm^{-3} was selected to be consistent with out earlier calculations (Fink and Zimbelman 1986). Lipman and Banks (1987) measured densities in the proximal section of the 1984 Mauna Loa lava-flow channel as low as 0.8 g cm^{-3}, although in the levees, flow margins, and distal portions of the channel, values in the range 2.0−2.6 g cm^{-3} were more typical.

Flow front velocity (v) data combined with thickness estimates and slopes also allowed us to calculate Newtonian viscosities, η, using the Jeffrey's equation:

$$\eta = (\varrho \, \mathrm{g h}^2)/(3 \, \mathrm{v}) \, . \tag{2}$$

The Jeffrey's equation does not take into account yield strength, which should cause our Newtonian estimates to be somewhat higher than Bingham viscosity values, and it assumes an infinitely wide channel (Johnson 1970; Moore 1987). It also ignores any possible pseudoplastic behavior (e.g., Pinkerton and Sparks 1978) which means that our viscosity values may be too high at higher strain rates.

Although several studies have pointed out the limited applicability of this equation to lava flows (e.g., McBirney and Murase 1984; Baloga and Pieri 1986), and despite the inconsistency inherent in assuming Newtonian rheology in one calculation and Bingham rheology in another, the Jeffrey's equation does offer a separate means of estimating relative changes in flow rheology. Comparing apparent viscosities calculated from flow front velocities with yield strengths based on thicknesses of nearby flow margins provides two measures of lava rheology taken at roughly the same time and place.

Flow front velocity data also allowed us to estimate when the front passed those sites where margin thickness was later measured. Variations in calculated yield strength and viscosity could thus be compared not only with distance from the vent but also with time since the flow began to advance. In these calculations, time and distance were both measured from the closest observation point to the vent, rather than from the vent itself. These most proximal points were generally within 500 m of the vent.

Rheologic values were independently estimated by combining observed temperatures with laboratory data. Neal et al. (1989) report that average vent (eruption) temperatures for Episodes 1 – 11 increased steadily from around 1117° to 1140°, but they remained nearly constant at around 1138 °C from Episodes 12 through 20. The few distal measurements made in the interiors of southward advancing flows showed values of about 1095 °C within 1 km of their termini. These are considered minimum temperatures (Neal et al. 1989).

Laboratory measurements of yield strengths for basalts with compositions similar to those from Puu Oo (McBirney and Murase 1984) show rapid increases in strength (more than two orders of magnitude) as the temperature drops from 1200° to 1150 °C, but that it remains nearly constant upon further cooling (Fig. 2). Similar strength versus cooling relationships were found by Minikami (1951) for the somewhat more silicic basalts of Oshima Volcano in Japan and by McBirney and Murase (1984) for andesite and dacite compositions. Thus we expect a steadily cooling basalt to show a rapid initial rise in strength followed by little or no later increase. If the cooling rate decreased with time, this shallowing of the strength versus time profile would be accentuated. Conversely, if the cooling rate increased, then a more linear strength profile might develop. For the temperature range of 1127° to 1115 °C measured near the vents of Episodes 2 – 5, the laboratory data of McBirney and Murase (1984) would predict a limited strength range of only 5900 to 6100 Pa ($\log \tau = 3.77$ to 3.79). Even for the maximum range of reported temperatures from the vent and flow fronts of Episodes 1 – 11 (1144° to 1094 °C), the predicted strengths would only vary from about 5200 to 6300 Pa ($\log \tau = 3.72$ to 3.80).

Fig. 2. Logarithm of laboratory-determined yield strength versus temperature plots for three different compositions (data replotted form McBirney and Murase 1984). *MLB* Mauna Loa Basalt; *CRB* Columbia River Basalt; *MHA* Mount Hood Andesite. Also shown are near-vent strength values for Episodes *2–5* obtained by projecting the strength versus distance curves to the origin. Temperatures are from Neal et al. (1989)

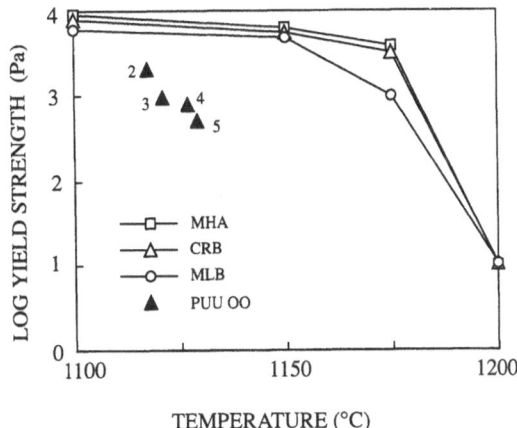

TEMPERATURE (°C)

Laboratory data for the viscosity of basalt (e.g., Shaw 1969) show a more continuous exponential variation with temperature than the yield strength data. Thus we expect steadily cooling basalts to show exponential increases in viscosity. Deviations from linear cooling rates should lead to corresponding variations in viscosity. As with yield strength, we can use laboratory data and observed temperatures to predict viscosity values for the Puu Oo flows. Using Shaw's (1969) empirical relationship for temperature-dependent viscosities of crystal-rich magmas, we can extrapolate from an observed value of 630 Pa-s (log η = 2.80) measured in the laboratory at 1100 °C by Murase and McBirney (1973) both upward and downward to estimates of 0.3 and 1600 Pa-s (log η = −0.52 to 3.20) for the maximum and minimum observed Puu Oo temperatures of 1144° and 1094 °C, respectively.

4 Results and Discussion

In this section we present a series of graphs showing longitudinal variations of calculated yield strength and viscosity for the Episode 2–5 flows. Distances were measured along the central axes of the flows using a planimeter and a 1:24000 scale topographic base map. Emplacement times were computed from flow front advance data by interpolation, assuming a constant velocity between observation points. Slopes were calculated by measuring the distances between topographic contours adjacent to sites where flow margin heights were determined.

For all four flows, calculated yield strength increased exponentially with distance (Fig. 3) and with time (Fig. 4). Several different correlations were investigated; exponential fits were better than various power-law fits (including linear) for nearly all cases. Correlations were also stronger for those cases with more observations and when strength was plotted against distance rather than time, probably reflecting the fact that times had to be interpolated from flow

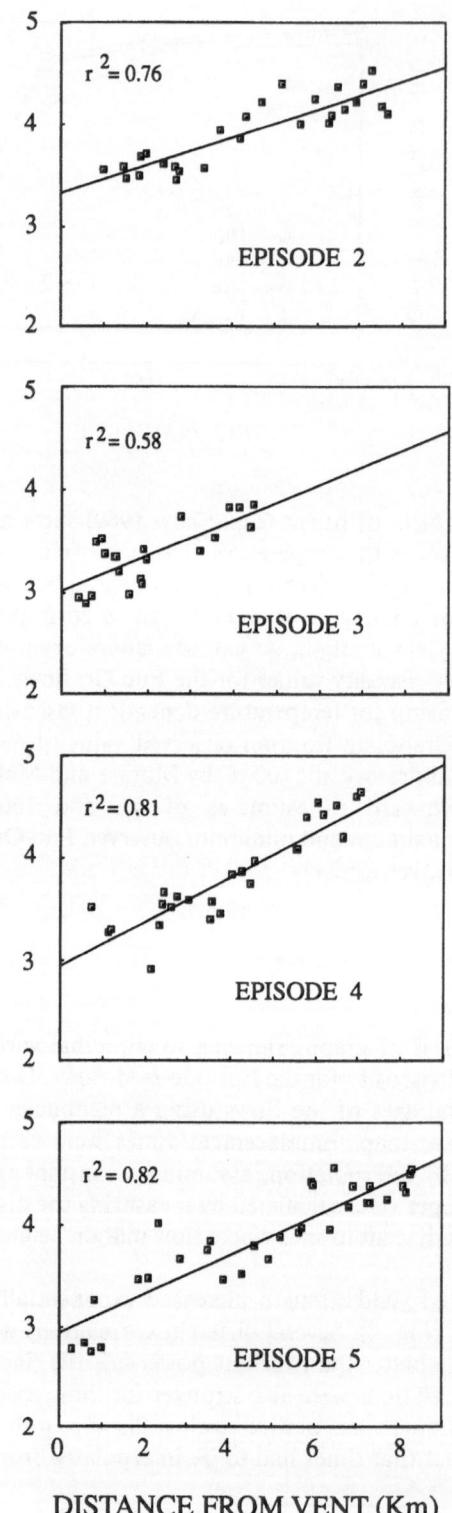

Fig. 3. Plots of yield strength versus distance for flows of Episodes *2–5*. Note increase in slopes of successive regression lines. Correlation coefficients shown in each plot

Fig. 4. Plots of yield strength
versus time calculated from
flow front position data for
Episodes *2−5*

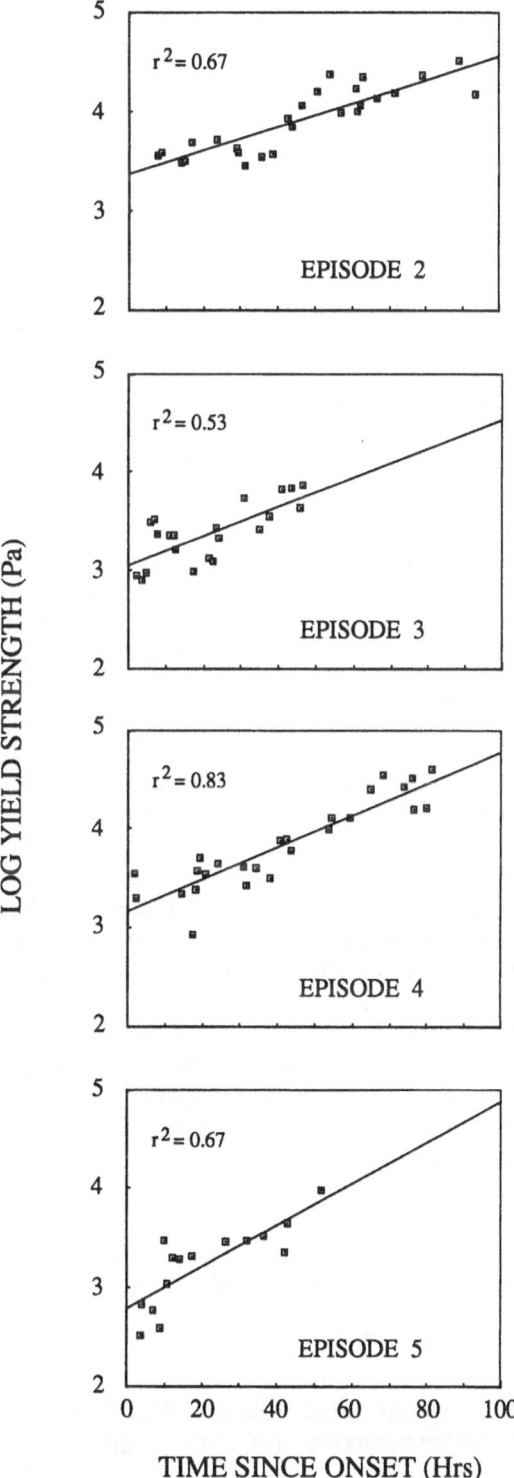

LOG YIELD STRENGTH (Pa)

TIME SINCE ONSET (Hrs)

front position data with an assumption of constant velocity between points, whereas distances were measured directly. The minimum and maximum calculated values of log τ for Episodes 2−5 were 2.51 and 4.60, respectively. These widely bracket the limited range that would be expected (3.75 to 3.80) based on the reported temperatures (1144° to 1094 °C), and McBirney and Murase's (1984) laboratory data.

Calculated viscosities also increased exponentially with distance (Fig. 5) and with time (Fig. 6) for each of the flows. Once again, viscosity correlated slightly better with distance than with time. Estimated log η varied from about 1.78 to 5.51, which again exceed the range expected from laboratory determinations of basalt for the observed temperature range.

The calculated strengths and viscosities shown in Figs. 3 through 6 are roughly consistent with field determinations for basalts by earlier workers (e.g., Pinkerton and Sparks 1978; Moore et al. 1978; Moore 1987). However, along each individual flow we found yield strengths that varied by nearly two orders of magnitude, and viscosities that covered ranges of almost three orders of magnitude. Clearly, a single calculated strength or viscosity for a given flow will not be diagnostic of composition. Taking into account the longitudinal position of a measurement site leads to a more restricted range of values; there still may be a large variation in calculated properties. For example, near-vent log τ values in Episodes 2 through 5 ranged from 2.48 to 3.41 whereas distal values ranged from 3.56 to 4.56. Log η similarly varied from 1.78 to 3.30 near the vent, and 4.08 to 5.51 near the flow fronts.

One of the motivations for calculating these rheologic parameters was to evaluate means of remotely estimating lava composition. Given the scatter of rheologic values calculated for these four flows and the imprecision of the original data on which the calculations were based, the exponential forms of the longitudinal trends are surprisingly consistent. Since similar profiles can in principle be obtained for remotely observed flows, we investigated whether the longitudinal gradient in the rheologic properties might prove to be more diagnostic of flow composition than individual measures of strength or viscosity.

In Fig. 7a we have plotted the slopes of the four strength versus distance regression lines from Fig. 3 against the average of near-vent temperatures T_e, for each episode (here considered to be close to the actual eruption temperatures). The slopes of the regression lines represent the exponents α for equations of the form

$$\tau = A \exp(\alpha L) , \tag{3}$$

where A and α are constants. The four points in Fig. 7a define a linear trend represented by the equation

$$\alpha = m T_e + b , \tag{4}$$

where m and b are constants. Similar results arise from plotting the slopes of the viscosity versus distance (Fig. 7b), strength versus time (Fig. 7c), and viscosity versus time (Fig. 7d) graphs. All four of these plots indicate that the

Fig. 5. Plots of viscosity versus distance for flows of Episodes 2−5

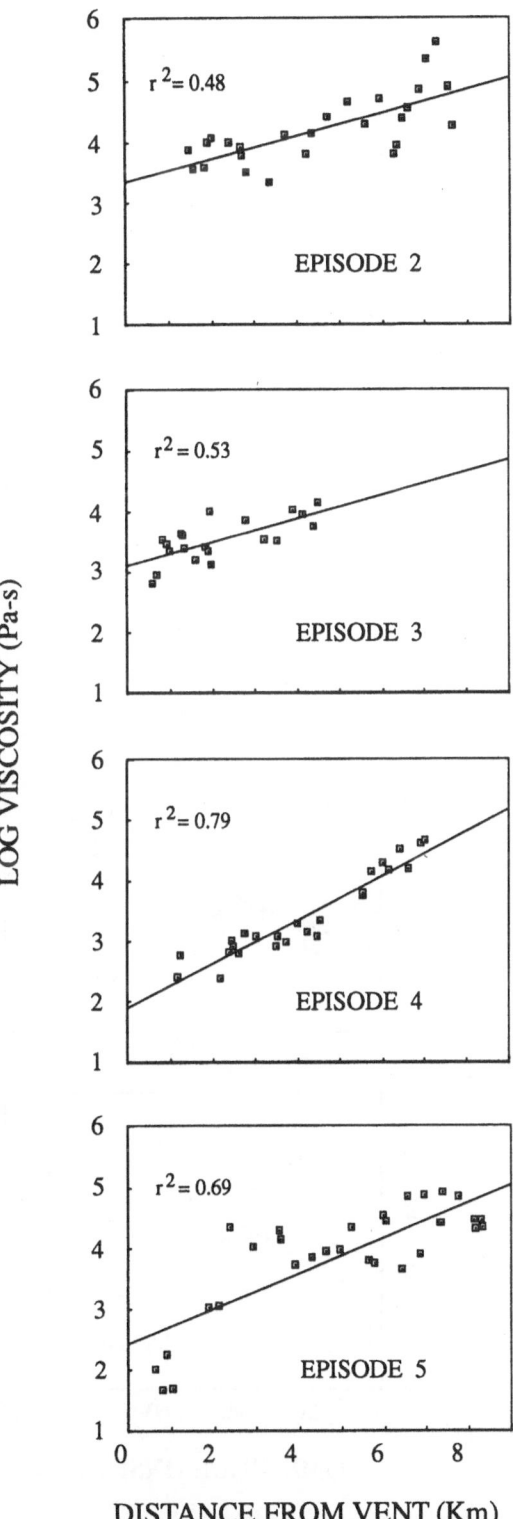

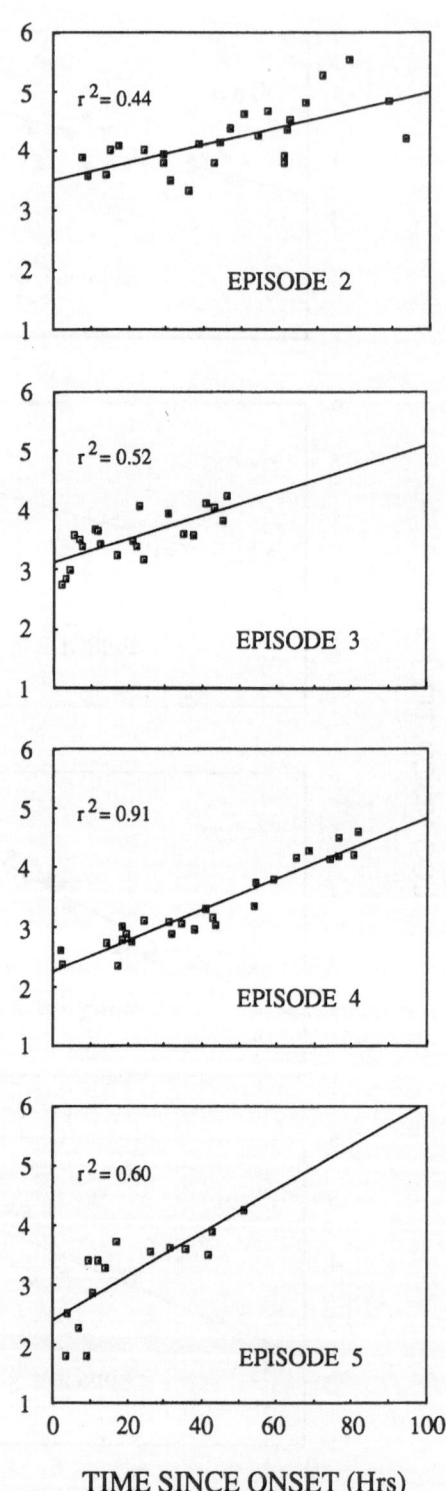

Fig. 6. Plots of viscosity versus time for flows of Episodes 2–5

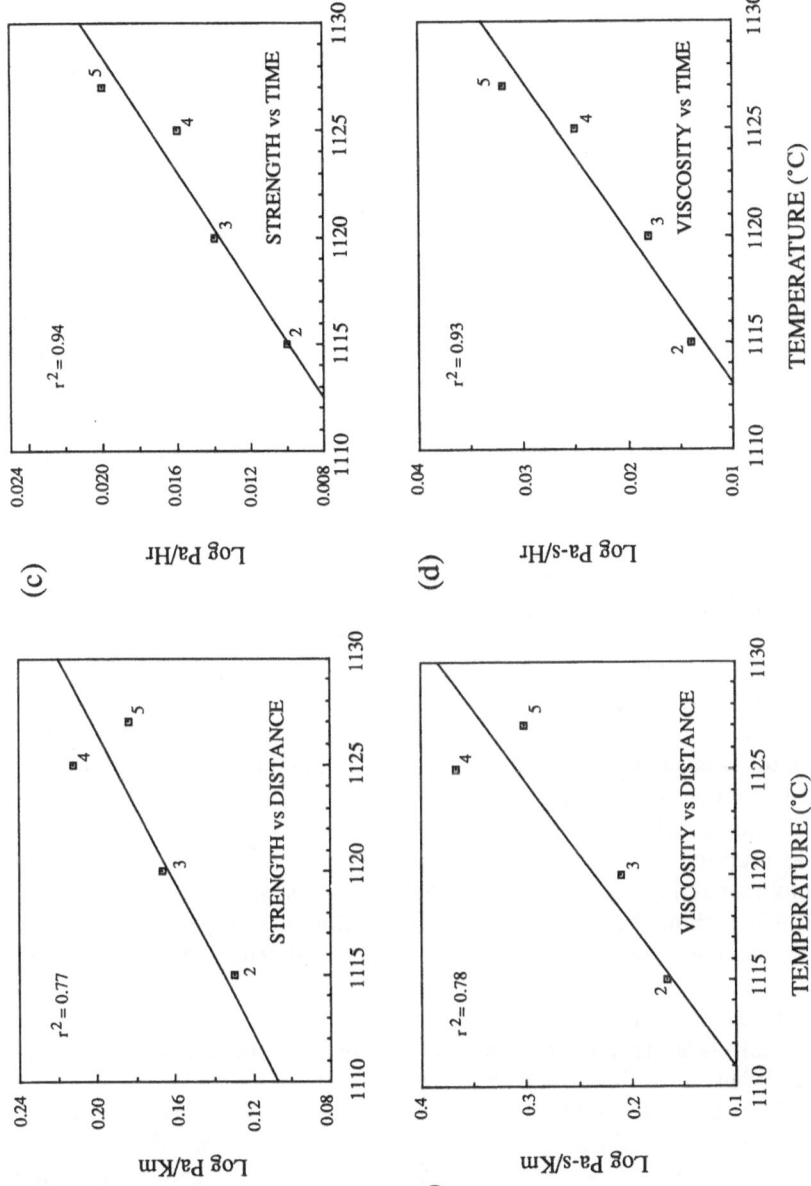

Fig. 7a–d. Slopes of regression lines from Figs. 3–6 plotted against eruption temperature. Note that in all cases there is an increase of rheologic gradient with increasing eruptive temperature. *Numbers* next to data points refer to eruptive Episode

longitudinal gradients of the rheologic parameters τ and η over distance or time are proportional to eruption temperature, T_e. Similar correlations are obtained if the longitudinal gradients of rheologic properties (as represented by the exponential parameters, α) are plotted against the concentration of either SiO_2 or MgO, rather than temperature (C. A. Neal, pers. commun. 1988).

Before attempting to interpret Fig. 7, it is necessary to restate the assumptions that it incorporates. Each data point represents the slope of the exponential regression of a rheologic parameter (yield strength or viscosity) versus distance or time. The rheologic estimates are based on flow margin heights, flow front positions, and preflow topographic slopes. Temperatures for a given flow are averages of several measurements taken within a few kilometers of the vent. Finally, we have not directly taken into account variations in effusion rate at the vent with time, or the effects of surges on the observed velocities.

There are two possible explanations for the increased longitudinal gradients with higher eruption temperature: either the flow length (or duration) drops with increasing temperature (and increasing mafic character of the flow), or the difference between the maximum and minimum values of the rheologic parameters increases with temperature. As there is no apparent reason why higher extrusion temperatures should result in shorter flows (and since the flow lengths do not show such a correlation), we will assume that it is the contrasts between the near-vent and distal rheologic values that increase. Larger rheologic contrasts in turn require either the near-vent values to decrease or the distal values to go up. Experimental data (e.g., McBirney and Murase 1984) indicate that neither the yield strength nor the viscosity should *increase* anywhere along a flow when eruption temperature increases; rather the values should everywhere decrease or remain the same. Thus, the observed larger rheologic contrasts require that strengths and viscosities decrease more rapidly near the vents than near the flow fronts as eruption temperatures increase.

A more rapid decrease of near-vent values is consistent with laboratory data (Fig. 2) which show that for a range of compositions, yield strength does not change as rapidly at low temperatures as it does at high temperatures. Laboratory data for viscosity do not show the same relationship; viscosity increases exponentially as temperature decreases over nearly the entire melting range of basalt. Crystallization upon cooling should actually cause the rate of viscosity increase to be higher at lower temperatures.

An alternative explanation for the observed trend is that brecciation has a stronger influence on rheologic properties near the flow front than near the vent. All four of these flows underwent transitions from pahoehoe to aa within a few kilometers of the vent, and became progressively more brecciated as they advanced downslope. In the near-vent region where breccia was poorly developed, the bulk rheologic properties of the flow would more closely reflect those of the erupting magma, which in turn would be most sensitive to the eruption temperature. Near the toe of an aa flow, the bulk properties would be dominated by those of the thickening breccia blanket. This suggestion is consistent with earlier field studies of basalt rheology which generally found higher values of yield strength and viscosity than those found in laboratory measurements.

Two additional factors which may affect calculated strength and viscosity values are variations in density and in crystal content. Systematic downstream increases like those measured at Mauna Loa for specific gravity by Lipman and Banks (1987), and for crystallinity by Lipman et al. (1985) could both result in the types of viscosity and yield strength gradients we calculate. If the rates of increase in either density or crystal content can be shown to correlate with eruptive temperature, then these could provide additional explanation for the observed rheologic gradients.

If the longitudinal rheologic trends observed in these four basalt flows can be demonstrated to occur over a wider range of compositions, then they suggest a possible means for remotely estimating composition. Flows of increasing silica content (which in some cases correlates with decreasing eruptive temperature) should show progressively lower longitudinal gradients in yield strength or viscosity. This conclusion is consistent with Baloga and Pieri's (1986) theoretical deduction that the rate of downstream flow thickening should decrease with increasing silica content. Additional field studies of flows with a range of compositions are needed to determine how diagnostic of composition the gradient of strength or viscosity can be. If, as is the case for these Puu Oo flows, the slope of a rheologic profile (i.e., the exponential parameter, α) is nearly constant, then one only need obtain rheologic measurements along part of the length of a flow to estimate its composition. This would be helpful, because it is commonly difficult to identify the vent areas or fronts of remotely observed flows.

5 Conclusions

We have presented morphologically based calculations of lava yield strength and viscosity for four very well-documented flows in Hawaii. Computed yield strengths for all of the flows ranged from 3×10^2 to 4×10^4 Pa, and viscosities ranged from 60 to 3×10^5 Pa-s, which are slightly lower than those reported by earlier workers for other Kilauean basalts. Individual flows show up to nearly two orders of magnitude variation of yield strength and nearly three orders of magnitude in viscosity. Strengths and viscosities increase exponentially with distance and time for all of the flows.

The slopes of regression curves, representing the exponential coefficient, were plotted against eruption temperature for the four flows. These show a steady increase with eruptive temperature and composition. We suggest that the gradient of yield strength or viscosity, as determined from the sorts of measurements discussed above, may prove to be an effective diagnostic aid in the remote identification of flow compositions. Additional studies of flows with a wider range of compositions are needed.

Acknowledgments. The authors thank members of the U.S. Geological Survey's Hawaiian Volcano Observatory for logistic assistance in the field and for access to unpublished data. We are particularly grateful to Ed Wolfe, Tina Neal, George Ulrich, and Christina Heliker.

Thanks also to Mike Malin, Steve Baloga, and Andrea Borgia for stimulating discussions. Robert Anderson, Tina Neal, and Ed Wolfe provided helpful reviews. This research was supported by NASA grant NAGW 529 from the Planetary Geology Program.

References

Baloga S, Pieri D (1986) Time-dependent profiles of lava flows. J Geophys Res 91:9543–9552

Booth B, Self S (1973) Rheological features of the 1971 Mount Etna lavas. Philos Trans R Soc Lond A 274:99–106

Cattermole P (1987) Sequence, rheological properties, and effusion rates of volcanic flows at Alba Patera, Mars. Proc Lun Planet Sci Conf 17, J Geophys Res 92:E553–E560

Fink JH, Zimbelman JR (1986) Rheology of the 1983 Royal Gardens basalt flows, Kilauea Volcano, Hawaii. Bull Volcanol 48:87–96

Head JW, Wilson L (1986) Volcanic processes and landforms on Venus: theory, predictions, and observations. J Geophys Res 91:9407–9446

Huppert HE, Shepherd JB, Sigurdsson H, Sparks RSJ (1982) On lava dome growth, with application to the 1979 lava extrusion of the Soufriere of St. Vincent. J Volcanol Geotherm Res 14:199–222

Hulme G (1974) The interpretation of lava flow morphology. Geophys J R Astr Soc 39:361–383

Johnson AM (1970) Physical processes in geology. Freeman, Cooper, San Francisco, p 577

Lipman PW, Banks NG (1987) Aa flow dynamics, Mauna Loa 1984: In: Decker RW, Wright TL, Stauffer PH (eds) US Geol Surv Prof Pap 1350:1527–1567

Lipman PW, Banks NJ, Rhodes JM (1985) Degassing-induced crystallization of basaltic magma and effects on lava rheology. Nature 317:604–607

Macdonald GA (1963) Physical properties of erupting Hawaiian magmas. Bull Geol Soc Am 74:1071–1078

McBirney AR, Murase T (1984) Rheological properties of magmas. Annu Rev Earth Planet Sci 12:337–357

Minikami T (1951) On the temperature and viscosity of the fresh lava extruded in the 1951 Oosima eruption. Bull Earthq Res Inst 29:487–498

Moore HJ (1987) Preliminary estimates of rheologic properties of 1984 Mauna Loa lava. In: Decker RW, Wright TL, Stauffer PH (eds) US Geol Surv Prof Pap 1350:1569–1588

Moore HJ, Arthur DWG, Schaber GG (1978) Yield strengths of flows on the Earth, Mars and Moon. Proc Lunar Planet Sci Conf 9:3351–3378

Moore HJ, Kachadoorian R, Moore RB (1980) Estimates of lava flow velocities in channels of the Pu'u Kia'i flow. Hawai'i. Rept Planet Geol Prog NASA TM82582:269–271

Murase T, McBirney AR (1973) Properties of some common igneous rocks and their melts at high temperatures. Bull Geol Soc Am 84:3563–3592

Neal CA, Duggan TJ, Wolf EW, Brandt EL (1989) Lava samples, temperatures, and compositions. Chapter 2. In: Wolfe EW (ed) The Puu Oo eruption of Kilauea Volcano, Hawaii, Episodes 1–20, January 3, 1983 through June 8, 1984. USGS Prof Pap 1463:99–126

Nichols RL (1939) Viscosity of lava. J Geol 47:290–302

Pinkerton H, Sparks RSJ (1978) Field measurements of the rheology of lava. Nature 276:383–385

Shaw HR (1969) Rheology of basalt in the melting range. J Petrol 10:510–535

Shaw HR, Wright TL, Peck DL, Okamura R (1968) The viscosity of basaltic magma: an analysis of field measurements in Makaopuhi lava lake, Hawaii. Am J Sci 266:225–264

Sparks RSJ, Pinkerton H (1978) Effects of degassing on rheology of basaltic lava. Nature 276:385–386

Spera FJ, Yuen DA, Kirschvink SJ (1982) Thermal boundary layer convection in silicic magma chambers: effects of temperature-dependent rheology and implications for thermogravitational chemical fractionation. J Geophys Res 87:8755–8767

Wolfe EW, Garcia MO, Jackson DB, Koyanagi RY, Neal CA, Okamura AT (1987) The Puu Oo eruption of Kilauea Volcano, Episodes 1–20, January 3, 1983, to June 8, 1984. USGS Prof Pap 1350:471–508

Wolfe EW, Neal CA, Banks NG, Duggan TJ (1989) Geologic observations and chronology of eruptive events. In: Wolfe EW (ed) The Puu Oo eruption of Kilauea Volcano, Hawaii, Episodes 1–20, January 3, 1983 through June 8, 1984. USGS Prof Pap 1463:1–98

Zimbelman JR (1985) Estimates of rheologic properties of flows on the martian volcano Ascraeus Mons. Proc Lunar Planet Sci Conf 16, J Geophys Res 90:D157–D162

Numerical Simulation of Lava Flows on Some Volcanoes in Japan

K. Ishihara, M. Iguchi, and K. Kamo

Abstract

The present chapter proposes a numerical calculation for simulating the macroscopic movements of lava flows on actual topography of volcanoes and tests it for three recent Japanese eruptions. The topography of the volcanoes as shown on pre-eruption maps was digitized using a grid in which the sampling interval was smaller than the width of the main stream of lava. The flux of lava among meshes is calculated from the steady-state solution of the Navier-Stokes equation for a Bingham fluid flowing due to gravity on an inclined plane. The calculation evaluates temperature change due to cooling by radiation. Viscosity is then estimated using an empirical relationship between viscosity and magma temperature and composition. The effusion rate and the duration of extrusion were estimated from the records of the eruptions. Three lava flows with different chemical compositions were simulated to check the validity of the method: the 1983 Miyakejima lava flows ($53-54\%$ SiO_2), the 1986 Izu-Oshima lava flows ($54-58\%$), and the 1914 Sakurajima lava flows ($59-62\%$). The simulations match fairly well the progress of flow fronts and actual thickness of the lavas. Some discrepancies in inundation area were found at the branches of lava streams and at the margins of main streams. These are largely due to insufficient precision of maps of the pre-eruptive topography and to the size of the sampling interval in the horizontal dimension being large relative to the widths of lava streams.

1 Introduction

Many geophysicists and geologists have studied the physical properties of lava flows by field observations. Hagiwara (1941) estimated viscosity of the 1940 lava flow at Miyakejima from the velocity of the front. Hagiwara et al. (1946) and Nagata et al. (1946) observed the movements of the lava flows at Sakurajima and estimated the viscosities. Minakami (1951) observed the movements of the lava flows in detail along the streams at Izu-Oshima Volcano in 1951. He derived the empirical relationship between temperature and viscosity. Shaw et al. (1968) measured the viscosity of the lava at Makaopuhi lake with a rotational viscometer and discussed the rheology of lava from the relation between the shear rate and the stress. Pinkerton and Sparks (1978) measured temperatures and yield strengths of lava flows at Mount Etna. Lipman and Banks (1987) measured temperature change of lavas with time and distance from the eruptive vent in the 1984 Mauna Loa eruption. Walker (1967) determined the relationship between the thickness of lava and the angle of slope at Mount Etna. Fink and Zimbelman (1986) surveyed the topography of a lava flow from the 1983 Kilauea eruption and estimated viscosity and yield strength of the lava from the analysis of the topographic profiles.

The effect of rheological parameters and extrusion rates on the movements and the morphologies of lava flows have also been discussed theoretically. Shaw and Swanson (1970) estimated the effects of some parameters on the profile of flood basalts. Baloga and Pieri (1986) investigated the combined effects of a time-dependent effusion rate and a spatially varying viscosity on the thickness profile of a flowing lava.

The three-dimensional numerical simulation of a lava flow was proposed by Crisci et al. (1986). They experimented with the method on imaginary topography for several cases of different locations of vents. However, they did not assure that the method was applicable to lava flows on actual topographies.

The purpose of this chapter is to present a method to simulate the macroscopic movements of lava flows on actual topographies, based on the previous studies mentioned above. The macroscopic movements imply flow velocity, thickness, and inundation area. Three recent lava flows in Japan (the 1914 Sakurajima, the 1983 Miyakejima, and the 1986 Izu-Oshima lava flow) are reproduced by the numerical calculation. The simulated lava flows are compared with the actual ones concerning the flow velocity, the thickness, and the inundation area in order to prove the applicability of the method.

2 Method

Lava flows progress when the thickness attains a critical value and the basal stress exceeds the yield strength. Robson (1967) first proposed a Bingham fluid model from the relation between the thickness of lava and the inclination angle of the slope at Mount Etna. The laboratory experiments of Hulme (1974) verified that the morphology of a lava flow was approximate to that of Bingham fluid. In our method of calculation, lava flows are regarded as Bingham fluids and modeled as illustrated schematically in Fig. 1. First, a group of two-dimensional square meshes are provided. The mesh size is chosen so that divergent or convergent flow within a mesh can be neglected. Each mesh has an altitude corresponding to the actual topography at its center. At the initial state, the thickness of lava at each grid of the mesh is set at zero. The zero thickness means that no lava flows cover the mesh. The lava starts to be discharged at a certain rate from a mesh corresponding to a vent. The thickness of lava at the vent mesh increases by a rate calculated from the volume of lava extruded during each time interval. When the thickness at the vent mesh reaches a critical level, outflow to the adjacent meshes occurs. In this way, the lava flow advances to the adjacent meshes, whenever the thickness at a mesh exceeds the critical level.

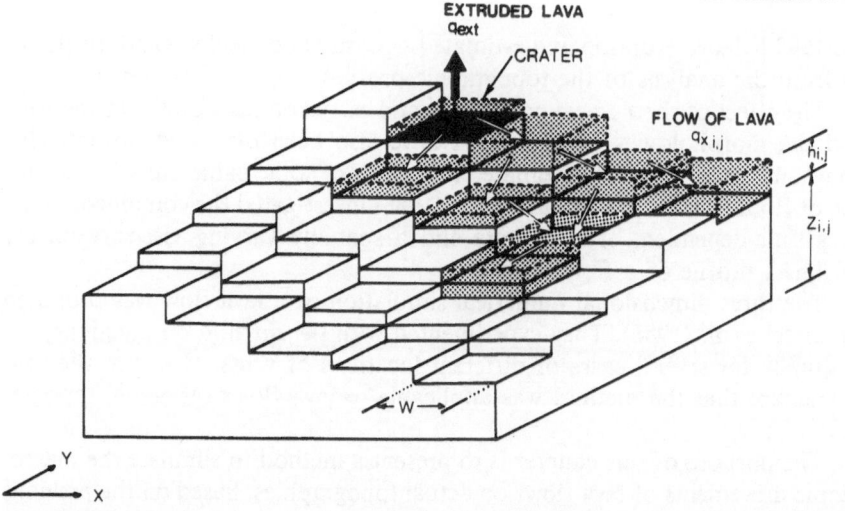

Fig. 1. Schematic illustration of numerical simulation of lava flows on a digital topography

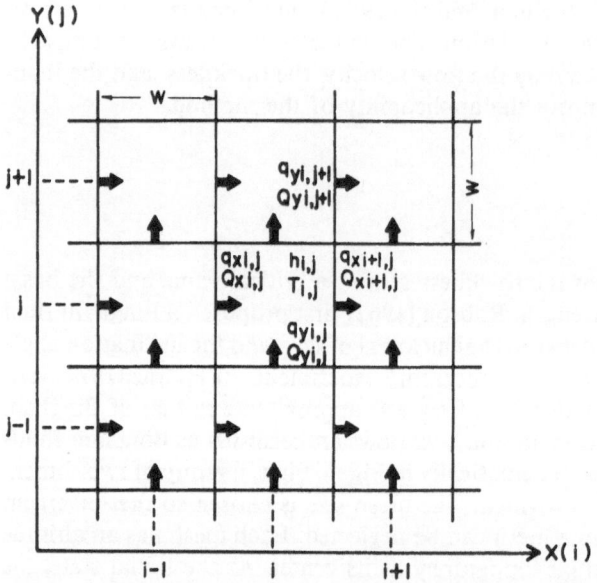

Fig. 2. Definition of inflow and outflow of lava at the mesh (i,j)

2.1 Increase in Volume

The inflow and the outflow among meshes are illustrated in Fig. 2. The
$X(i)$-axis, $Y(j)$-axis, and Z-axis are taken in the direction of the east, the
north, and vertically upward, respectively. A volcanic body is divided into
square meshes, the size of which is $w \times w$. Each mesh has the altitude $Z_{i,j}$.
The meshes are numbered from $(1,1)$ to (M, N). The symbol with subscript (i,j)

Table 1. Explanation of symbols

a	Angle of inclination of slope
g	Acceleration due to gravity
h	Thickness of lava flow
h_{cr}	Critical thickness
h_e	Effective thickness
k	Parameters defined in Eq. (32)
q_{ext}	Extrusion rate
q_u	Flux through the area of width
$q_{xi,j}$	Flux from the mesh $(i-1, j)$ to the mesh (i, j)
$q_{yi,j}$	Flux from the mesh $(i, j-1)$ to the mesh (i, j)
$Q_{i,j}$	Heat in the mesh (i, j)
$Q_{xi,j}$	Heat flux from the mesh $(i-1, j)$, to the mesh (i, j)
$Q_{yi,j}$	Heat flux from the mesh $(i, j-1)$ to the mesh (i, j)
$\Delta Q_{i,j}$	Increase in heat in the mesh (i, j)
$\Delta Q_{mi,j}$	Increase in heat in the mesh (i, j) due to flowing-in and flowing-out
$\Delta Q_{ri,j}$	Decrease in heat in the mesh (i, j) due to radiative cooling
S_b	Shear stress at the bottom of lava flows
S_y	Yield strength
t	Time
Δt	Time interval
T_{ext}	Absolute temperature at the time of extrusion
$T_{i,j}$	Absolute temperature in the mesh (i, j)
$T_{si,j}$	Absolute temperature on the surface in the mesh (i, j)
$V_{i,j}$	Volume of lava in the mesh (i, j)
$\Delta V_{i,j}$	Increase in volume in the mesh (i, j)
v_p	Velocity of plug
w	Width of a mesh
$Z_{i,j}$	Altitude in the mesh (i,j)
ΔZ	Difference in height between two meshes
ε	Emissivity of lava
η	Viscosity of lava
θ	Temperature in centigrade
ϱ	Density of lava
σ	Stefan-Boltzmann's constant

represent the values at the *i*-th and *j*-th mesh in the x- and y-direction, respectively. The definitions of symbols are summarized in Table 1.

Now the mesh (i,j) is considered. Let $q_{xi,j}$ and $q_{yi,j}$ be flux into the mesh (i,j) from the mesh $(i-1,j)$ and from the mesh $(i,j-1)$, respectively. Let $q_{xi+1,j}$ and $q_{yi,j+1}$ be flux from the mesh (i,j) into the mesh $(i+1,j)$ and into the mesh $(i,j+1)$, respectively. The volume in a mesh is changed by the inflow and the outflow. It is assumed that the flux between the mesh (i,j) and the four surrounding meshes are calculated independently and the flux of lava is constant during a time interval. The increase in the volume at the mesh (i,j) during a time interval Δt is

$$\Delta V_{i,j} = (q_{xi,j} + q_{yi,j} - q_{xi+1,j} - q_{yi,j+1}) \Delta t \ . \tag{1}$$

Adding the volume of lava extruded in a time interval, the increase in volume at a mesh which corresponds to a vent is obtained as

$$\Delta V_{i,j} = (q_{xi,j} + q_{yi,j} - q_{xi+1,j} - q_{yi,j+1} + q_{ext}) \Delta t . \tag{2}$$

Here, the extrusion rate is also assumed to be constant during the time interval. Dividing increase in volume $\Delta V_{i,j}$ by the area of a mesh w^2, the increase in thickness at the mesh (i,j) during the time interval is obtained as

$$\Delta h_{i,j} = \Delta V_{i,j}/w^2 . \tag{3}$$

Adding the increase in thickness, the thickness at the time $t + \Delta t$ is obtained as

$$h_{i,j}(t + \Delta t) = h_{i,j}(t) + \Delta h_{i,j} . \tag{4}$$

The thickness with time at each mesh is calculated successively by iterating the procedure represented by Eqs. (1)–(4).

2.2 Flux of Lava

The basic formula to calculate the flux between two adjacent meshes is the steady state solution of the Navier-Stokes' equation for the Bingham fluid with a constant thickness which flows downward due to gravity on an inclined plane. According to Dragoni et al. (1986), the flux per unit width of flow (q_u) is

$$q_u = \frac{S_y h_{cr}^2}{3 \eta} [a^3 - (3/2) a^2 + 1/2] \quad (h > h_{cr}) , \tag{5}$$

$$q_u = 0 \qquad\qquad (h \leq h_{cr}) , \tag{6}$$

where

$$a = h/h_{cr} . \tag{7}$$

The critical thickness (h_{cr}) depends on the yield strength (S_y) and the angle of the slope (A) as described

$$h_{cr} = S_y/(\varrho g \sin A) , \tag{8}$$

where the symbols ϱ and g represent density of lava and acceleration due to gravity, respectively.

As the altitudes are discrete in the digital topography, the function $\sin A$ is represented by the difference in altitude between two adjacent meshes (ΔZ) and the width of a mesh (w) as follows:

$$\sin A = \Delta Z/(\Delta Z^2 + w^2)^{1/2} . \tag{9}$$

Eqs. (5)–(8) are applied to the numerical calculation of flux of lava between two adjacent meshes. The flux from the mesh ($i-1,j$) to the mesh (i,j) through the width (w) is written as follows.

$$q_{xi,j} = \frac{S_y h_{cr}^2 w}{3 \eta} [a^3 - (3/2) a^2 + 1/2] \quad (h_e > h_{cr}) , \tag{10}$$

$$q_{xi,j} = 0 \quad (h_e \leq h_{cr}) , \tag{11}$$

where

$$a = h_e/h_{cr} . \tag{12}$$

Similarly, the flux from the mesh $(i, j-1)$ to the mesh (i, j) through the width (w) is written as follows,

$$q_{yi,j} = \frac{S_y h_{cr}^2 w}{3 \eta} [a^3 - (3/2) a^2 + 1/2] \quad (h_e > h_{cr}) , \tag{13}$$

$$q_{yi,j} = 0 \quad (h_e \leq h_{cr}) , \tag{14}$$

and

$$a = h_e/h_{cr} , \tag{15}$$

where the viscosity coefficient (η) and the yield strength (S_y) are the values at the mesh from which the lava flows out. The symbol h_e is the effective thickness defined according to the direction of flow and the relation of heights between two adjacent meshes, as shown in Fig. 3. Case 1 corresponds to a normal

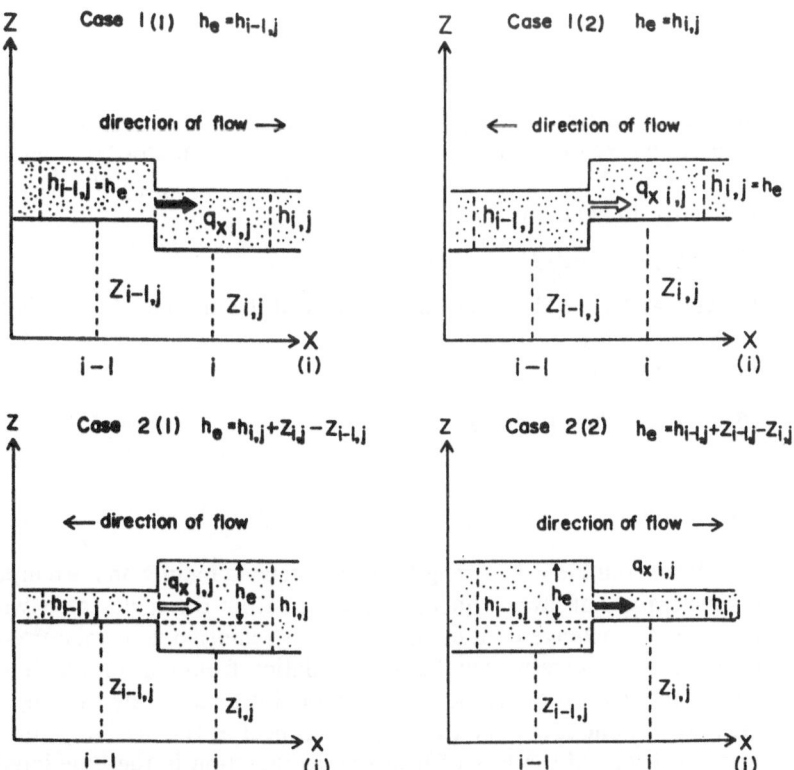

Fig. 3. Definition of the effective thickness of lava flow

flow from a higher position to a lower one, and Case 2 applies to an overflow from a lower position to a higher one.

Case 1: The effective thickness h_e is defined by the thickness at the higher mesh.

1. In cases for which the altitude at the mesh $(i-1,j)$ is higher than that in the mesh (i,j), that is, $Z_{i,j} < Z_{i-1,j}$ and $Z_{i,j} + h_{i,j} < Z_{i-1,j} + h_{i-1,j}$, the effective thickness is

$$h_e = h_{i-1,j} \ . \tag{16}$$

2. In cases for which the altitude at the mesh (i,j) is higher than that at the mesh $(i-1,j)$, that is, $Z_{i,j} \geqq Z_{i-1,j}$ and $Z_{i,j} + h_{i,j} \geqq Z_{i-1,j} + h_{i-1,j}$, the effective thickness is

$$h_e = h_{i,j} \ . \tag{17}$$

The difference in altitude (ΔZ) is defined by the difference of ground surface between the higher and lower meshes. That is

$$\Delta Z = Z_{i-1,j} - Z_{i,j} \ . \tag{18}$$

Case 2: Outflow from the lower mesh to the higher one may occur when the level of lava at a basin or a narrow channel is rising gradually. The parameter ΔZ is defined by the difference in altitude of lava surface. That is

$$\Delta Z = (Z_{i-1,j} + h_{i-1,j}) - (Z_{i,j} + h_{i,j}) \ . \tag{19}$$

The effective thickness is defined by the difference in altitude between the lava surface in the mesh exited and the ground surface in the mesh entered.

1. In cases where the altitude of lava surface at the mesh (i,j) is higher than that at the mesh $(i-1,j)$, that is, $Z_{i,j} + h_{i,j} \geqq Z_{i-1,j} + h_{i-1,j}$ and $Z_{i,j} < Z_{i-1,j}$, the effective thickness is

$$h_e = Z_{i,j} + h_{i,j} - Z_{i-1,j} \ . \tag{20}$$

2. In cases where the altitude of lava surface at the mesh $(i-1,j)$ is higher than that at the mesh (i,j), that is, $Z_{i,j} + h_{i,j} < Z_{i-1,j} + h_{i-1,j}$ and $Z_{i,j} \geqq Z_{i-1,j}$, the effective thickness is

$$h_e = Z_{i-1,j} + h_{i-1,j} - Z_{i,j} \ . \tag{21}$$

2.3 Temperature

The viscosity and the yield strength of lavas depend mainly on the temperature. The temperature of lavas which are extruded from the vents varies with time and space. The change of the temperature is mainly due to mixture of lavas with different temperatures and due to radiation from the surface. In our calculation, the change of temperature is modeled as illustrated in Fig. 4.

Let $\Delta Q_{mi,j}$ and $\Delta Q_{ri,j}$ be the increase in heat at the mesh (i,j) due to the mixture of lava and the loss of heat due to radiation in the time interval Δt, respectively. The heat at the time $t + \Delta t$ is represented as follows,

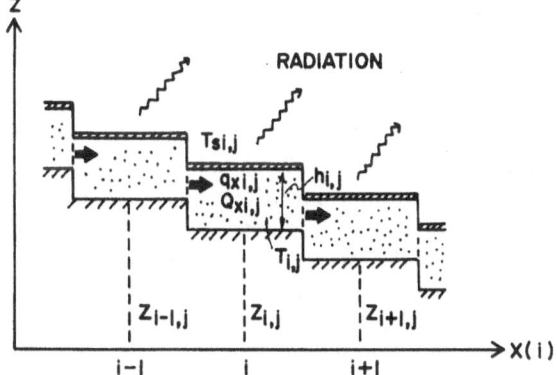

Fig. 4. Change of temperature due to mixture of lava and cooling by radiation

$$Q_{i,j}(t+\Delta t) = Q_{i,j}(t) + \Delta Q_{mi,j} - \Delta Q_{ri,j} \ . \tag{22}$$

Then the absolute temperature $(T_{i,j})$ at the mesh (i,j) at the time $t + \Delta t$ can be written as

$$T_{i,j}(t+\Delta t) = \frac{Q_{i,j}(t+\Delta t)}{c\varrho\,h_{i,j}(t+\Delta t)\,w^2} \ . \tag{23}$$

The change of heat due to the inflow and the outflow is obtained in the same way as illustrated in Fig. 2. Replacing $q_{xi,j}$, $q_{yi,j}$, and q_{ext} in Eqs. (1) and (2) by $Q_{xi,j}$, $Q_{yi,j}$ and $c\varrho\,q_{ext}\,T_{ext}$, the equation for conservation of heat is obtained as

$$\Delta Q_{mi,j} = (Q_{xi,j} + Q_{yi,j} - Q_{xi+1,j} - Q_{yi,j+1})\,\Delta t \ . \tag{24}$$

Adding the heat of extruded lava, the increase in heat at a mesh corresponding to a vent is

$$\Delta Q_{mi,j} = (Q_{xi,j} + Q_{yi,j} - Q_{xi+1,j} - Q_{yi,j+1} + c\varrho\,q_{ext}\,T_{ext})\Delta t \ , \tag{25}$$

where

$$Q_{xi,j} = c\varrho\,q_{xi,j}\,T \ , \tag{26}$$

$$Q_{yi,j} = c\varrho\,q_{yi,j}\,T \ . \tag{27}$$

The symbol T denotes absolute temperature at the mesh where the lava flows out. c is the specific heat.

The heat of lava is lost mainly due to radiation from the surface. Conduction to the air and the ground surface is negligible when the temperature is high. The heat loss due to radiation is estimated from the Stefan-Boltzmann's law (Shaw and Swanson 1970; Danes 1972). The loss of heat in the time interval Δt is written as

$$\Delta Q_{ri,j} = \varepsilon\sigma\,T_{si,j}^4\,\Delta t \ , \tag{28}$$

where the symbol ε, σ, and $T_{si,j}$ represent emissivity of lava, Stefan-Boltzmann's constant, and the absolute temperature on the surface of lava at the mesh (i, j), respectively.

When the surface of lava is covered by a layer of clinkers, the difference in temperature between the surface and the molten lava under the layer of clinkers becomes large. Archambault and Tanguy (1976) measured the temperature distribution of lava flows on Mount Etna. The measurement showed that the temperature on the surface was 300 °C lower than that of the internal molten lava. Minakami (1951) observed that the lava began to be covered by clinkers at a temperature of 1000°– 1030 °C. Considering these observations, two cases of relationships of temperature between the surface and the internal lava are assumed as follows.

1. When clinkers are not present, the temperature on the surface is equal to that of internal molten lava $[T_{i,j}(t) > 1303 \text{ K}]$

$$T_{si,j}(t) = T_{i,j}(t) \ . \tag{29}$$

2. When a layer of clinkers covers a molten lava flow, the temperature on the surface decreases by 300 K $[T_{i,j}(t) \leq 1303 \text{ K}]$

$$T_{si,j}(t) = T_{i,j}(t) - 300 \text{ K} \ . \tag{30}$$

While the lava is flowing, the size of clinkers is negligibly thin compared with the thickness of molten lava. Therefore, it is assumed that the layer of clinkers does not affect the mechanical movements of lava flows.

The heat and the temperature at each mesh are calculated by putting Eqs. (24) – (30) into Eqs. (22) and (23).

2.4 Viscosity

The viscosity of lava depends strongly on the temperature. The viscosities of lavas are calculated by the empirical relationship between the temperature and viscosity derived from the observation at Izu-Oshima in 1951 by Minakami (1951). The silica content was 52 – 53% (Tsuya et al. 1952). The relation between the temperature in centigrade (θ) and the viscosity (η) in poise is shown in Fig. 5 and described as follows,

$$\log \eta = 6.115 - 0.0181 \, (\theta - 1000) \quad 1000 \leq \theta \leq 1150 °C \ . \tag{31}$$

The viscosity of lava depends not only on the temperature but also on the chemical composition (Murase and McBirney 1973) and the gas content (Sparks and Pinkerton 1978). Equation (31) is not directly applicable to the other lava flows. In our method, Eq. (31) is generalized to apply it to the other lava flows whose chemical properties are different. Introducing a parameter k, Eq. (31) is generalized to

$$\log \eta = k - 0.0181 \, \theta \ , \tag{32}$$

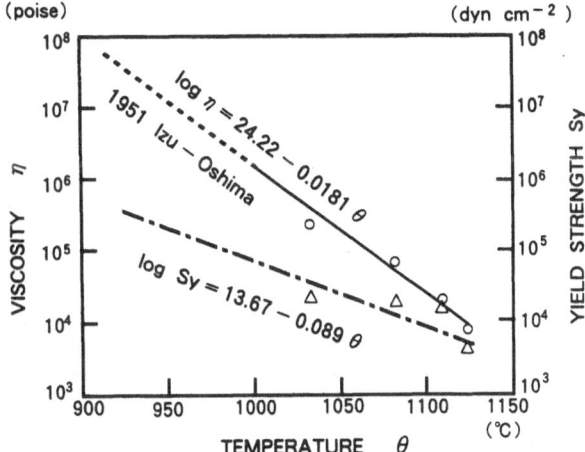

Fig. 5. Relationships between temperature and viscosity, and temperature and yield stress for the 1951 Izu-Oshima lava flow. *Circles* and *triangles* denote the viscosity and the yield stress, respectively. Estimated from the data obtained by Minakami (1951)

under the assumption that the ratio of the increase in the viscosity to the decrease in temperature is constant. The parameter k which depends on chemcial composition of lava and gas content, should be determined for each lava flow. As it is difficult to estimate the value k considering gas content, it is estimated empirically from the temperature and viscosity by field measurements or from the silica content. In the calculation, the viscosity of the lava flow at each mesh is calculated from

$$\log \eta_{i,j} = k - 0.0181 \, (T_{i,j} - 273) \ . \tag{33}$$

2.5 Yield Strength

The yield strength of lava also depends on temperature (Pinkerton and Sparks 1978). The flowing velocity at the surface of a Bingham fluid is described by (Dragoni et al. 1986)

$$v_p = \frac{S_b h}{2 \eta} \, [1 - S_y/S_b]^2 \ , \tag{34}$$

where the symbol S_b denotes the shear stress at the bottom of lava flows, and is represented as:

$$S_b = \varrho g h \sin A \ . \tag{35}$$

From the above equation, yield strength is derived as follows,

$$S_y = S_b - (2 \eta v_p S_b/h)^{1/2} \ . \tag{36}$$

Here, yield strength was estimated also from the measurements by Minakami (1951). The relation between the temperature in centigrade and the yield strength in dyn cm^{-2} is shown in Fig. 5. The relationship is formulated as follows,

$$\log S_y = 13.67 - 0.089\,\theta \ . \tag{37}$$

Hulme (1974) discussed the relationship between yield strength and silica contents but there are few data on how yield strength depends on both temperature and silica content. In our calculation, it is assumed that the yield strength at each mesh will be calculated from the following formula:

$$\log S_{yi,j} = 13.67 - 0.089\,(T_{i,j} - 273) \ . \tag{38}$$

In our simulation, the flux of lavas is calculated in an X- and Y-component, respectively. The critical thickness of lava flows (h_{cr}) is evaluated for the each component according to Eqs. (8) and (38).

2.6 Computation

The flow chart of the numerical calculation is illustrated in Fig. 6. At first, the following data, which are necessary for the numerical simulation, are put in:

1. Altitude at each mesh;
2. Conditions for extrusion of lava (the positions of assumed craters, the extrusion rate, and the duration of extrusion);
3. Physical constants and properties of lava (acceleration due to gravity, Stefan-Boltzmann's constant, emissivity, specific heat capacity, density of lava, initial temperature, and parameter k);
4. Time interval of the calculation.

Then, the calculation for the following parameters are iterated according to the flow chart until the thickness is not changed at any meshes.

1. Viscosity and yield strength [Eqs. (33) and (38)];
2. Critical thickness [Eq. (8)];
3. Flux between two adjacent meshes [Eqs. (10)–(15)]. If a thickness is less than the critical one, the flux vanishes.
4. Increase in volume [Eqs. (1) and (2)];
5. Increase in thickness [Eq. (3)];
6. Heat flux between two adjacent meshes [Eqs. (24)–(27)];
7. Decrease in heat due to cooling by radiation [Eq. (28)];
8. Change of heat [Eq. (22)];
9. Conversion to temperature from the heat [Eq. (23)].

The calculated results are written by a line printer and recorded on a magnetic tape with the proper time interval. The operative program written in FORTRAN is shown in the Appendix 1. It is shown in the Appendix 2 how much the effusion rate and the initial temperature of lava affect the progress of simulated lava flows in the case of a simple topography.

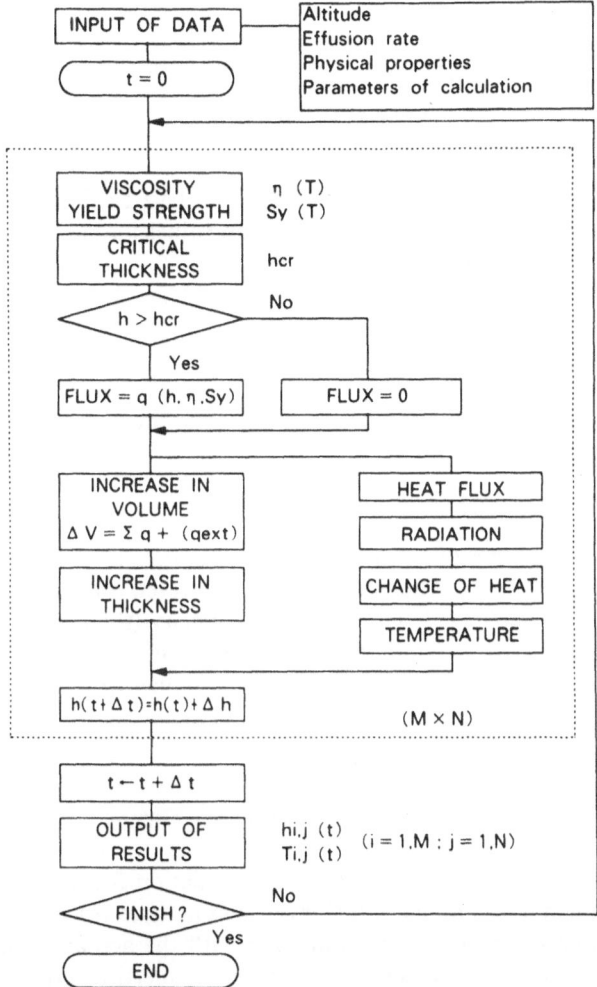

Fig. 6. Flow chart for the numerical calculation. The calculations enclosed by broken lines are done for each mesh (i,j) $[i = 1, M; j = 1, N]$

When the calculation is put into practice, the proper choice of the time interval Δt and the width of mesh w is important to achieve the calculation effectively. Small values of Δt and w are favorable to obtain a stable solution and to simulate lava flow exactly. However it takes a long time to achieve the calculation. In contrast, larger values of Δt and w result in an unstable solution and an inaccurate simulation, although the calculation is completed in a short time. It is assumed in our method proposed here that the thickness at each mesh is approximately constant during a time interval. Therefore, the change of the volume of lava during every time interval must be small compared with the volume at the mesh. The condition can be represented as follows,

$$|\Delta V_{i,j}| \ll V_{i,j} . \tag{39}$$

Putting Eqs. (1) and (10) into Eq. (39), the conditions can be written as

$$\Delta t \ll \frac{3\eta wh}{S_y h_{cr}^2} [(h/h_{cr})^3 - (3/2)(h/h_{cr})^2 + 1/2] . \tag{40}$$

In case of the 1983 Miyakejima lava flow, inequality (40) is estimated to be $\Delta t \ll 100$ s by putting the mean values into the right term of inequality (40), $\eta = 2 \times 10^5$ poise, $S_y = 1 \times 10^4$ dyn cm^{-2}, $h = 2.5$ m and $w = 25$ m. The experimental calculations were carried out for several values of time interval. For time intervals greater than 2 s, the calculated thickness at a steep slope oscillates or becomes negative. For time intervals less than 1 s, the calculated thickness at each mesh converges to the same value. Therefore, approximately, the inequality (40) can be rewritten as follows,

$$\Delta t < \frac{3\eta wh}{100 S_y h_{cr}^2} [(h/h_{cr})^3 - (3/2)(h/h_{cr})^2 + 1/2] . \tag{41}$$

The size of the mesh should be sufficiently small compared with the width of lava flows. The size of meshes, however, is restricted by the precision of original topographic maps and the capacity (the memory size and the processing speed) of the computer operating in the simulation. In practical calculations, the size of the mesh should be determined from the narrowest width of noticeable streams of lava to be simulated.

3 Application to Actual Lava Flows

The method of the numerical simulation for lava flows proposed here is applied to reproduce actual lava flows at three volcanoes in Japan, and the reproduced lava flows are compared with the actual ones. These lava flows differ from each other in their chemical properties. The locations of volcanoes are shown in Fig. 7. Miyakejima Volcano and Izu-Oshima Volcano belong to Izu-Mariana Islands. Sakurajima Volcano is located in south Kyushu.

The 1983 lava at Miyakejima was basalt, and the silica content was 53−54% (Fujii et al. 1984). The 1986 lava extruded by the fissure eruption at Izu-Oshima Volcano was richer in silica than the other historical lava flows at the island. The lava was basaltic andesite, and the silica content was 54−58% (Aramaki et al. 1987). The 1914 Sakurajima lava flows were more viscous and more voluminous than these two lava flows. The lava was andesite, and the silica content was 59−62% (Aramaki et al. 1981).

3.1 1983 Miyakejima Lava Flow

The fissure eruption occurred on the southwestern flank of the island at 15:20 (JST) on October 3, 1983. Lavas began to pour out of the vents along the fis-

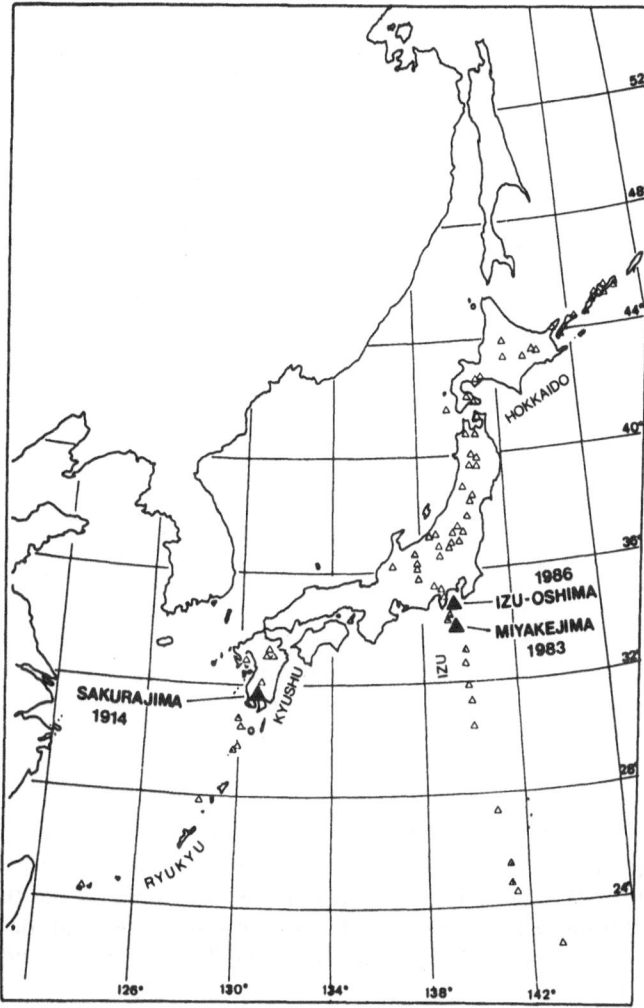

Fig. 7. Locations of Miyakejima, Izu-Oshima, and Sakurajima Volcanoes

sure almost simultaneously at the onset of the eruption. The eruption ceased the next morning. The distribution of the lava flows and their thicknesses are illustrated in Fig. 8. The northern lava flow streamed down to Ako village and the southern one covered Awabe village. The area covered by these lava flows was 1.6 km². The mean thickness and the total volume of lava were estimated to be 2.5 m and 4.0×10^6 m³, respectively (Ishihara et al. 1984b).

Concerning the lava flowing toward Ako village, Soya et al. (1984) summarized the arrival times of the flow front at three positions. These positions are also shown with the times in Fig. 8. The front of the lava flow reached Ako village approximately 2 h after the start of the extrusion.

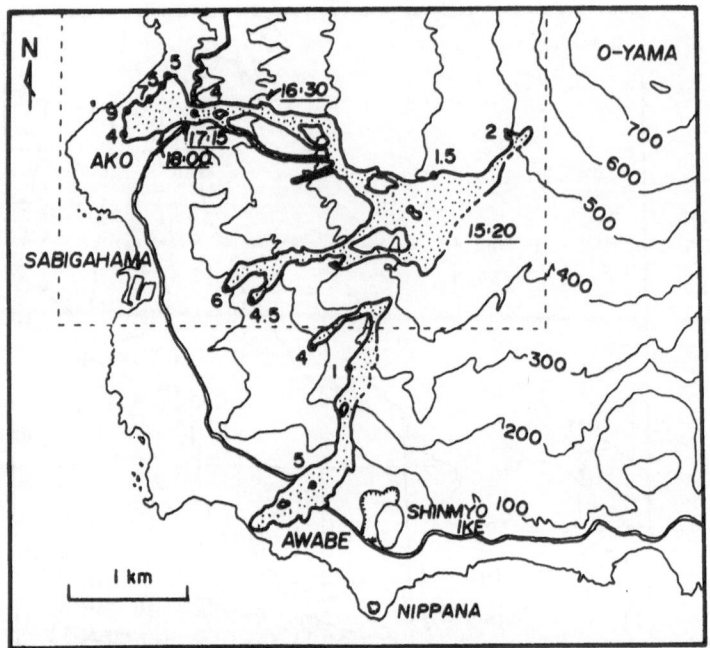

Fig. 8. Distribution of the 1983 Miyakejima lava flow. The *numbers* beside *dots* represent the thickness in meters. The arrival time of the flow front at each position is shown by the *underlined number.* The lava flows enclosed with *broken lines* were reproduced

The thickness of the lava flow was measured at the fronts and the margins (Ishihara et al. 1984a). The lava flow gradually became thicker with distance from the vents. The thickness of the lava flow near the fissure, 400–500 m in altitude, was 1–2 m. The thickness on the prefectural road was 4 m. The average thickness at Ako village was 5 m. The lava flow beside the school at Ako village is the deepest and the thickness was 9 m.

The northern part of the lava flow was simulated because the flow front was observed better. Aramaki and Hayakawa (1984) summarized the process of the eruption. The extrusion from the fissure continued at a high discharge rate until 16:38 (JST). Thus, it was assumed that the extrusion of lava continued for 90 min at the constant rate of 540 m^3 s^{-1}. Ten vents were assumed along the fissure. The extrusion rate at each vent was determined from the preliminary results of the simulation (Ishihara et al. 1984a, b).

The relationship between temperature and viscosity assumed in the calculation is shown in Fig. 9. The silica content of the 1983 Miyakejima lava (53–54%) was regarded as almost the same as that of the 1951 Izu-Oshima lava flow (52–53%). In our calculation, it was assumed that the relationship between viscosity and temperature of the 1951 Izu-Oshima lava was applicable to that of the 1983 lava flow at Miyakejima. Thus the parameter k was assumed to be 24.22. The initial viscosity of the lava was estimated to be 5×10^4

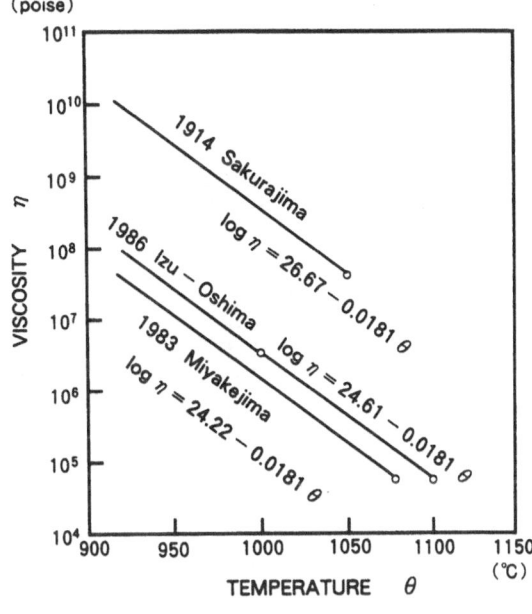

Fig. 9. Relation between temperature and viscosity assumed in the calculation. *Circles* denote the initial viscosities and temperatures of lavas extruded from the craters

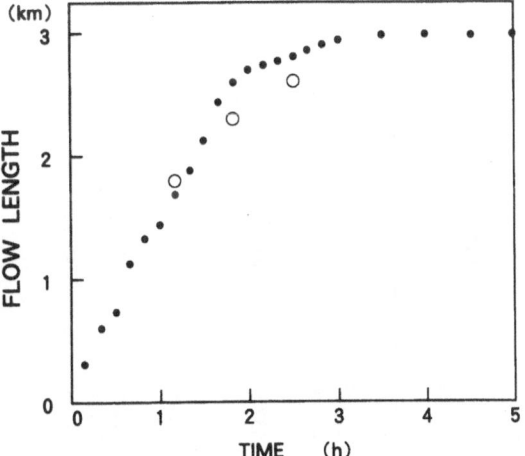

Fig. 10. Flow length of the 1983 Miyakejima lava versus time from the beginning of the extrusion. *Solid circles* denote flow lengths by the simulation. *Open circles* represent the times when the flow front reached the three points illustrated in Fig. 8

poise (Ishihara et al. 1984b). Therefore, the initial temperature was assumed to be 1078 °C.

The conditions provided for the simulation are summarized in Table 2.

The simulated lava flow was compared with the actual flow. The change of the flow lengths from the fissure with time are shown in Fig. 10. The arrival time of the simulated lava flow at the position 500 m east of the prefectural road coincided with that of the actual one. The simulated flow front arrived at the prefectural road and at Ako village about 20 min earlier than the actual one.

Table 2. Conditions for simulation

Lava flow	1983 Miyakejima	1986 Izu-Oshima			1914 Sakurajima
		LCI	LBI	LBIII	
Topography					
Scale of original map	1:5000	1:5000	1:5000		1:50000
Size of mesh	25×25 m	10×10 m	25×25 m		50×50 m
Extrusion of lava					
Rate	537 m³ s⁻¹	44 m³ s⁻¹	212 m³ s⁻¹:0–40 min 478 m³ s⁻¹:40–70 min 212 m³ s⁻¹:70–300 min	231 m³ s⁻¹:0–40 min 521 m³ s⁻¹:40–70 min 231 m³ s⁻¹:70–300 min	2380 m³ s⁻¹:0–14 h 171 m³ s⁻¹:14–40 h 84 m³ s⁻¹:40–66 h 459 m³ s⁻¹:66–100 h
Duration	1.5 h	1 h	5 h	5 h	100 h
Physical constants of lava					
Temperature	1078 °C	1100 °C	1000 °C	1000 °C	970 °C 1050 °C
Viscosity	5×10⁴ P (k = 24.22)	5×10⁴ P (k = 24.61)	3×10⁶ P (k = 24.61)	3×10⁶ P (k = 24.61)	1×10⁹ P 5×10⁷ P (k = 26.67)
Density			2.5 g cm⁻³		
Emissivity			0.9		
Specific heat capacity			8.4×10⁶ erg g⁻¹ K⁻¹		
Acceleration due to gravity			980 cm s⁻²		
Stefan-Boltzmann's constant			5.67×10⁻⁵ erg s⁻¹ cm⁻¹ K⁻⁴		
Time interval of calculation	0.5 s	0.1 s	0.4 s	0.2 s	20 s

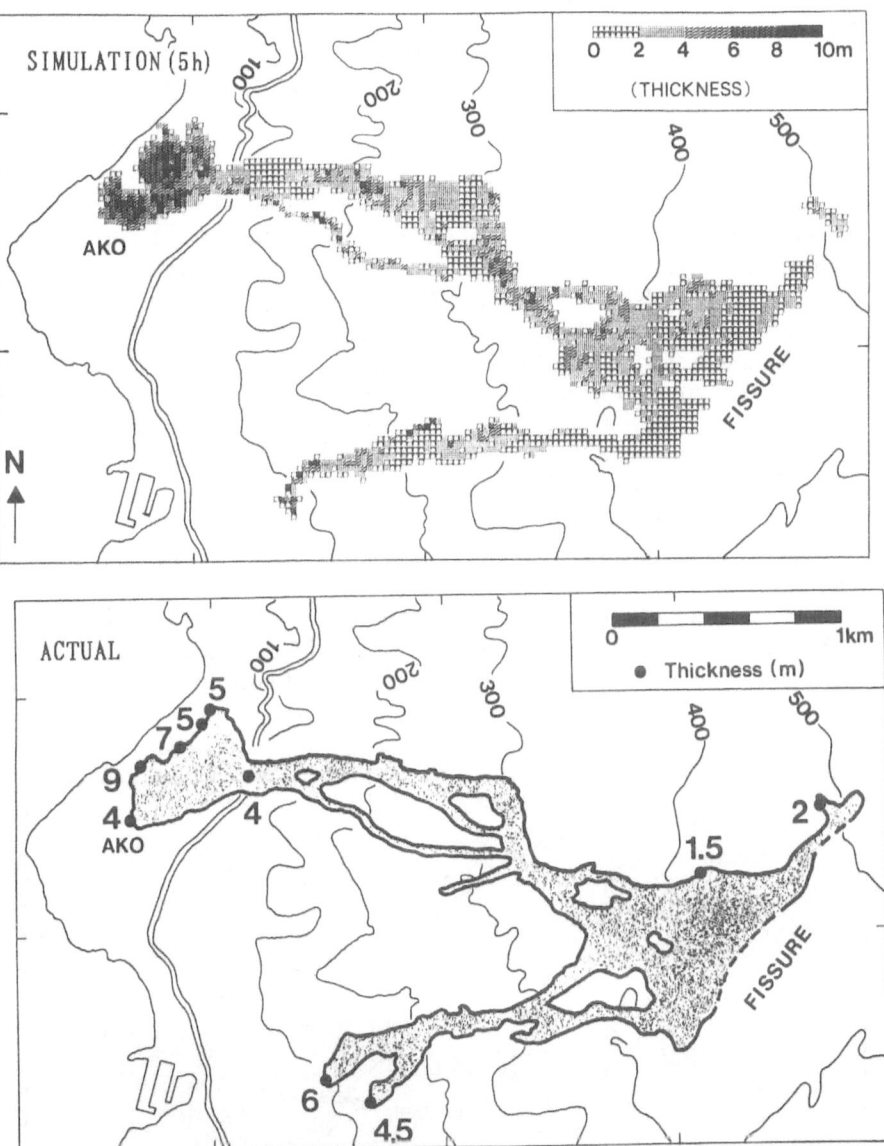

Fig. 11. Comparison of the simulated lava flow with the actual one at Miyakejima in October 1983. The *numbers* in the *lower figure* represent thickness of the actual lava flow

After 5 h, the simulated lava did not extend further. The thickness and the inundation area of the simulated lava flow after 5 h were compared with those of the actual one, as shown in Fig. 11. The thickness of the simulated lava flow almost coincided with that of the actual one. Although the narrow branches of the lava flow could not be reproduced, the outline of the simulated lava flow almost agreed with that of the actual one. The excess area of the simulated lava flow was 4% of the actual one and the area deficient in the simulated lava flow was 8%.

3.2 1986 Izu-Oshima Lava Flow

A summit eruption began on November 15, 1986 at Izu-Oshima Volcano and the lava overflowed from the summit crater of Miharayama (LA). After a week of intermittent eruption from the summit, a fissure eruption commenced in the caldera at 16:15 (JST) on November 21. The fissure extended northwestward opening several vents in the caldera and on the northwestern flank of the volcano. Although most of the lavas poured out into the caldera (lava flow LBI and LBIII), a stream of lava flowed down the western flank and approached Motomachi town (lava flow LCI). The distribution of these lava flows is shown in Fig. 12. The areas of the lava flow LBI, LBIII and LCI were 6.9×10^5, 4.2×10^5, and 6.1×10^4 m^2, respectively. The volumes of these lavas were estimated to be 4.3×10^6, 4.7×10^6, and 1.6×10^5 m^3 respectively, from the elevation change of the area covered by the lava flows (Nagaoka and Ogawa 1987; Ishihara et al. 1988).

The activities of these vents with time were summarized by Hayakawa (1987). It was assumed that the extrusion of lava LCI continued for 1 h at the constant rate of 44 m^3 s^{-1}, and that the extrusion of lava flow LBI and LBIII

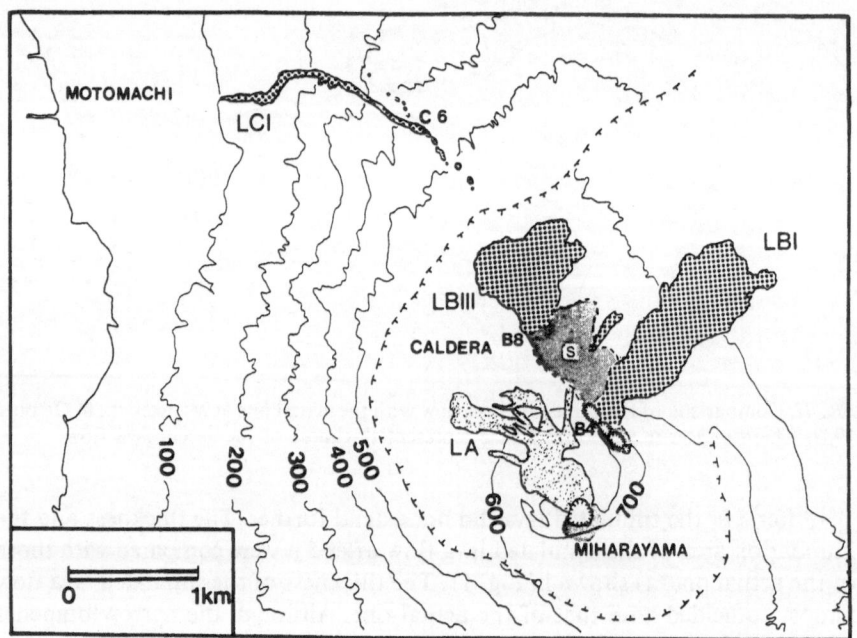

Fig. 12. Distribution of the 1986 Izu-Oshima lava flows. *S* denotes scoria cones

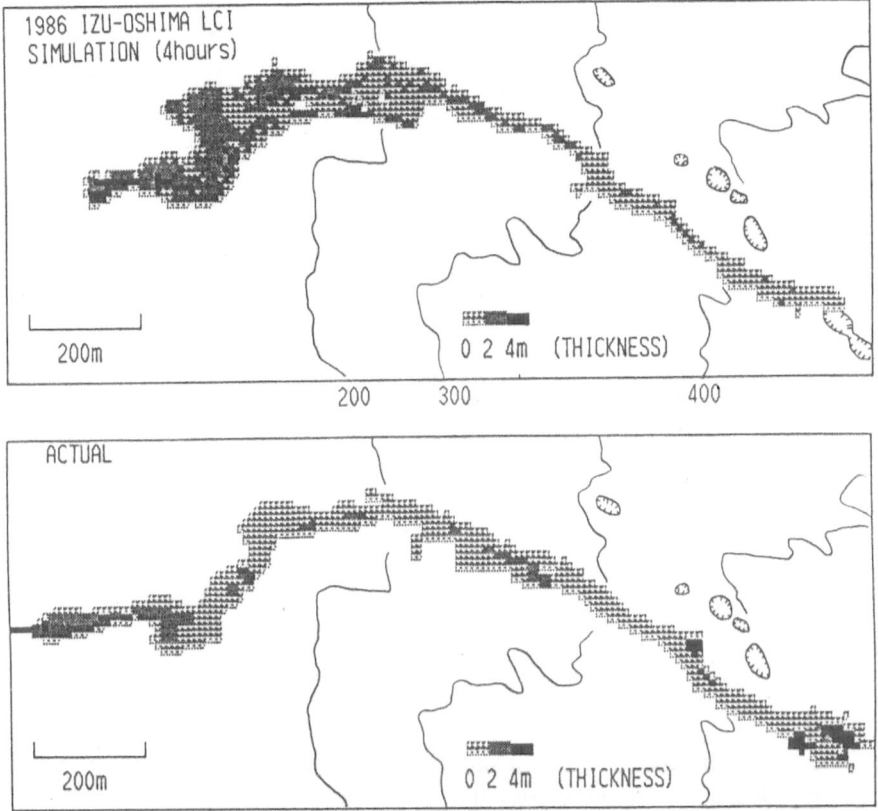

Fig. 13. Comparison of the simulated lava flow with the actual one, LCI, at Izu-Oshima in November 1986

continued for 5 h at variable rates of $212-478$ and $231-521$ m^3 s^{-1}, respectively.

The temperature and viscosity of the lavas were not observed directly but geochemical analysis (Fujii et al. 1988) indicated that the silica contents of these lavas were higher than those of either historical lava flows at the volcano or the 1983 Miyakejima lava flow. Therefore, the new lava flows were considered to be more viscous than the 1951 lava flow. Thus, the parameter k was assumed to be 24.61 as illustrated in Fig. 9.

In the preliminary calculation, lava flows were calculated assuming the initial temperatures of lava to be 1000 °C and 1100 °C, and the simulated lava flows were compared with the actual ones. The length of the simulated lava flow LCI on the western flank and LBIII in the caldera became 1.3 and 1.9 km, respectively, when the initial temperatures were assumed to be 1100 °C. When initial temperatures were assumed to be 1000 °C, the lengths of simulated lava flow LCI and LBIII decreased to 1.0 and 1.0 km, respectively. The lengths of the actual lava flow LCI and LBIII were 1.6 and 1.0 km, respectively. Inferred

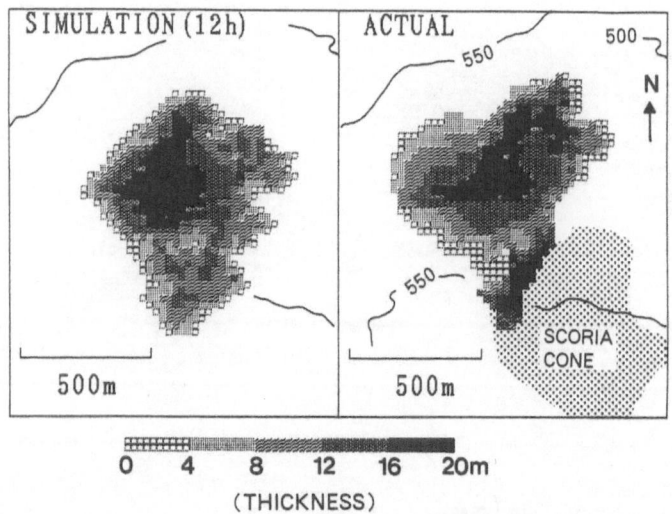

Fig. 14. Comparison of the simulated lava flow with the actual one, LBIII, at Izu-Oshima in November 1986

from the lengths of actual and simulated lava flows, the initial temperatures of the lava flow LBIII were considered to be lower than that of the LCI (Ishihara et al. 1988).

Here, it was assumed that the initial temperature of the LBI and the LBIII in the caldera were 1000 °C and that of the LCI on the western flank was 1100 °C. The conditions provided for the simulations are summarized in Table 2.

The distribution of the actual lava flow and that of the simulated lava flow LCI after 4 h are shown in Fig. 13. The narrow flow front with a width of less than 20 m could not be reproduced. The simulated lava flow was 300 m shorter than the actual one. However, the outline of the simulated lava flow agreed well with the actual one and the simulated ($6.3 \times 10^4 \, m^2$) is almost equal to the actual value ($6.1 \times 10^4 \, m^2$). The thickness of the simulated lava also coincided with that of the actual flow.

The distribution of the actual lava flow and that of the simulated lava flow LBIII after 12 h are shown in Fig. 14. The inundation area of the actual lava flow LBIII and the simulated one were 4.2×10^5 and $5.4 \times 10^5 \, m^2$, respectively. The excess area of the simulated lava flow was 37% of the actual one and the area deficient in the simulated lava flow was 9%. The actual thickness of the flow front was 16−20 m and the simulated one was 12−16 m. The thickness and the inundation area of the simulated lava flow thus coincided closely with those of the actual flow.

The distribution of the actual lava flow and that of the simulated lava flow LBI after 12 h are shown in Fig. 15. In this case the pattern of the simulated lava flow clearly differed from that of the actual one. The simulated lava flow

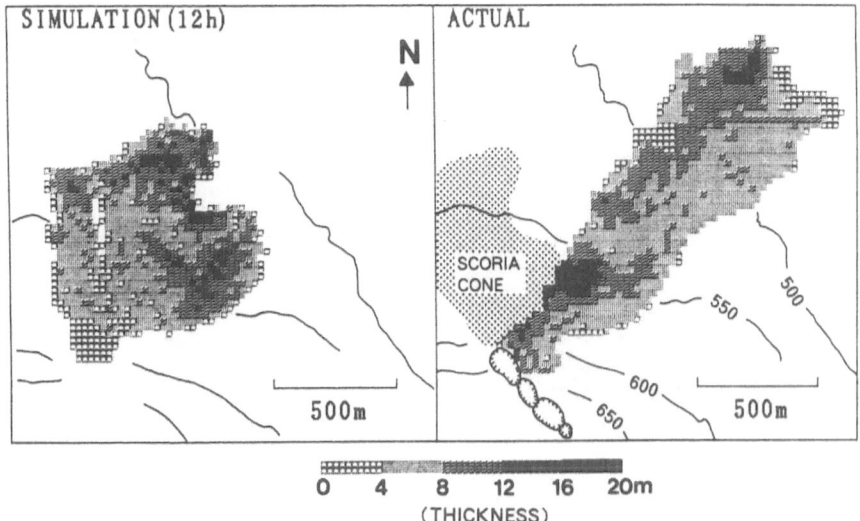

Fig. 15. Comparison of the simulated lava flow with the actual one, LBI, at Izu-Oshima in November 1986

deviated northwestward from the actual flow. In particular, on the northern side of the vent, a false branch of the lava flow was produced by the numerical simulation. Although there were steep slopes on the northern side before the eruption, it was later determined that scoria cones were formed at the position shown in Fig. 12 (Endo et al. 1987). This fact suggests that the accumulation of scoria blocked the formation of a branch stream to the north (Ishihara et al. 1988).

3.3 1914 Sakurajima Lava Flow

The 1914 eruption at Sakurajima Volcano was the largest in Japan in this century. The eruptive activity was summarized by Omori (1916) and Koto (1916). The lava extruded from parasitic craters formed on the eastern and the western flank. The total volume of the lava was estimated to be 1.34×10^9 m^3 (Ishihara et al. 1981).

The eruption on the western flank began at 10:05 (JST) on January 12, 1914. The extrusion of lava started at about 8 P.M. on January 13, 34 h after the onset of the eruption. The advance of the flow front was observed from Kagoshima city and recorded on photographs. Ishihara et al. (1985) summarized the extent of the lava flow according to Koto (1916), Omori (1916) and the photographs, as illustrated in Fig. 16. The flow front extended to a distance of about 1 km from the coast line on the morning of January 14. The height of the front was 70 m. More than half of the volume of the lava was extruded in the first 14 h. The total volume extruded on the western flank was estimated to be 2.5×10^8 m^3.

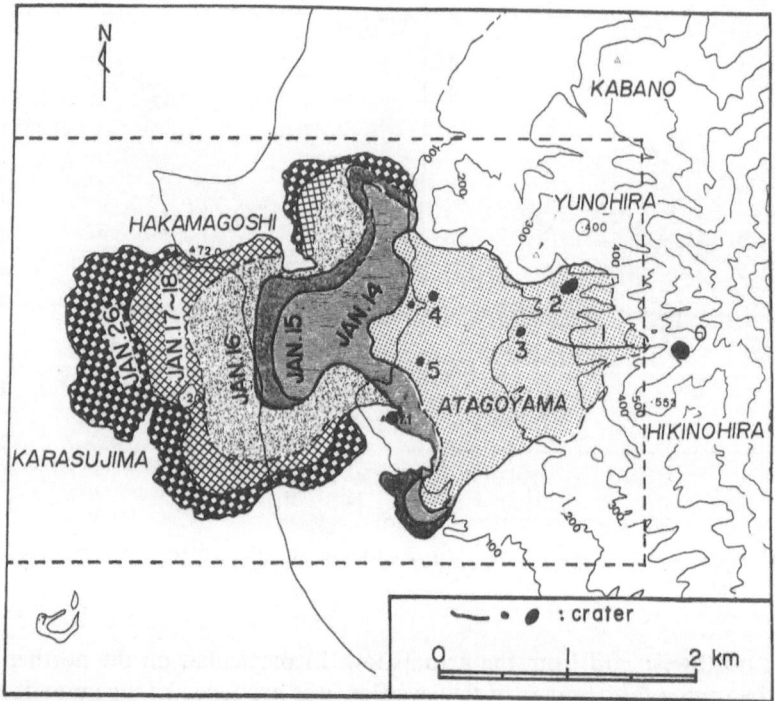

Fig. 16. Extent of the 1914 Sakurajima lava field with time on the western flank (Ishihara et al. 1985)

In the calculation, it was assumed that the upper craters (No. 1 and No. 2) produced lava at a rate of 2380 $m^3 s^{-1}$ during the first 14 h of the extrusion, and the lower craters (Nos. 3−5) extruded lava at variable rates of 87−461 $m^3 s^{-1}$ from 14 to 100 h after the onset.

Although the extrusion continued at a low discharge rate after 100 h from the beginning, the calculation was terminated after 100 h, because the flow front had already streamed into the sea after 66 h and the method presented here cannot reproduce a subaqueous lava flow. Moreover, submarine topography is poorly defined.

The relationship between temperature and viscosity assumed in the calculation is shown in Fig. 9. As the temperature and viscosity of the 1914 lava were not measured directly, the temperature and the parameter k were determined by analogy to measurements carried out in the 1946 eruptions because the silica content of the 1914 lava is the same as that of the 1946 lava flow (Aramaki et al. 1981). The temperature and viscosity of the 1946 lava flow were 950 °C and 3×10^9 poise near the coast line, about 3 km from the vent (Hagiwara et al. 1946). From Eq. (32), the parameter k was estimated to be 26.67. It was assumed that the value of k was applicable to the 1914 lava.

The preliminary calculation (Ishihara et al. 1985) revealed that the rapid progress of the lava flow during the period from January 16 to 18 could not

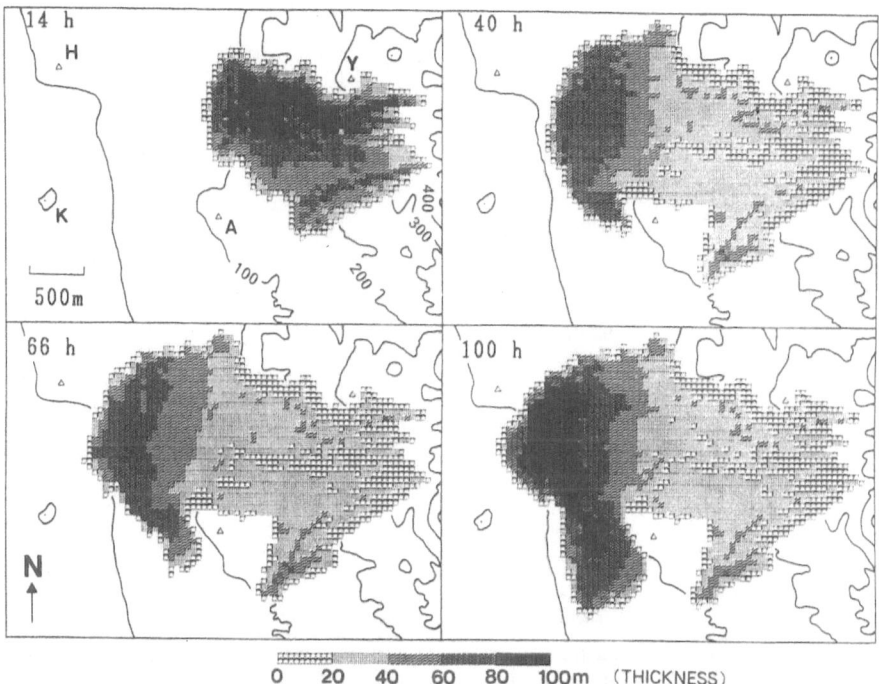

Fig. 17. Distribution of the simulated lava flow at Sakurajima

be explained by the assumption that lavas with the same viscosity were extruded from the upper and lower craters. Therefore, it was assumed that the initial temperatures of the upper and lower craters were 970° and 1050°C, respectively.

The conditions provided for the simulations are summarized in Table 2.

The simulated lava flow was compared with the actual one. The simulated distribution of thickness 14, 40, 66, and 100 h after the beginning are shown in Fig. 17. The thickness of the lava flow estimated from the photographs was 60–70, 40–50, and 40–50 m at 14, 40, and 66 h after the beginning, respectively. The thickness of the simulated lava at these times was 80–90, 60–70, and 50–60 m respectively. The differences between observed and calculated thicknesses were less than 20 m.

The increase of the area and length of the simulated and actual lava flows with time is shown in Fig. 18. The area covered by the simulated lava flow almost coincided with that of the actual one until 100 h after the beginning of the extrusion. The length of the simulated lava flow coincided with that of the actual one after 14 h. The lengths of the simulated lava flow after 40, 66, and 100 h were shorter by 0.3, 0.4, and 0.5 km than the actual ones. Although the positions of the simulated flow front at each time coincided with those of the actual flow, the simulated flow extended too widely at the northern margin below 50 m elevation. The excess area of the simulated lava flow at 100 h after

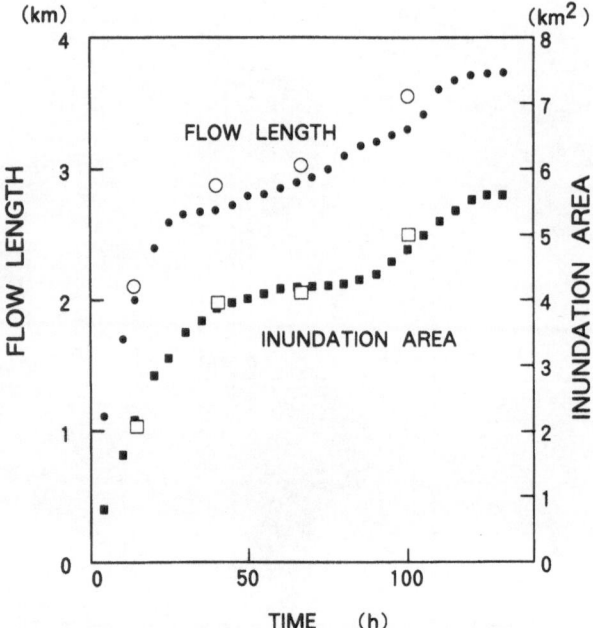

Fig. 18. Inundation area and flow length of the simulated lava flow. *Open symbols* denote actual values. *Solid symbols* denote the calculated values

the beginning was 12% of the actual one and the area deficient in the simulated lava flow was 16%.

4 Discussion

The method proposed here was applied to actual lava flows to examine its applicability and limits. The pattern and the thickness of the main streams of the simulated lava flows agreed with those of the actual lava flows except LBI at Izu-Oshima. For some paths and branches, however, the simulated lava flows disagreed with the actual ones.

The conditions provided for the simulation were examined before the execution of the calculations. The initial temperature and the parameter k, the extrusion rate, and the extrusion duration were determined from field observations, chemical analysis, and preliminary calculations, and were examined before the execution of calculations. The change of extrusion rate of the 1914 Sakurajima lava flow, whose duration was longer than the other two flows, was estimated in more detail using photographs taken at that time. Therefore, we conclude that the discrepancies were caused mainly by the imprecision of the topographic data.

The simulated lava flow at Sakurajima Volcano spread too widely at the northern margin, below 50 m elevation. The topographic map of Sakurajima

Volcano before the 1914 eruption was published in 1903. The reduced scale of the original map is 1 : 50 000 and the interval of the contours of the altitudes is 20 m. Therefore, the digital topography made from this old map may not describe the actual surface precisely, especially, at the gentle slopes on the lower flanks. The poor correspondence may be caused by the imprecision of the original maps.

The distribution of the simulated lava flow LBI at Izu-Oshima Volcano deviated northwestward from the actual one. The actual lava flow covered the 1951 lava flow where contours of elevation are not plotted on the topographic map. Thus, the topographic map is not as precise on the 1951 lava flows as on other parts where the contours are plotted.

The small branch of the 1983 Miyakejima lava and the narrow front of the lava flow LCI at Izu-Oshima could not be reproduced well. These small lava flows streamed into narrow valleys, whose channel positions and the fine-scale topographies were not represented on the digital topographic maps used here. It is possible to reproduce these small lava flows more exactly if maps are digitized with a shorter sampling interval. However, the digital topographic map made from a group of finer meshes could not represent the actual topography accurately if the sampling interval exceeded the resolution of the original maps.

The results of the application reveal that our method of numerical simulation of lava flows is able to reproduce the macroscopic movements of lava flows. However, our modeling does not include either the estimation of heat loss from the sides of lava flows or the topographic change of the solidified lavas which lost their fluidity.

At their margins, lava flows cool rapidly because the surface exposed to the atmosphere is larger than in the channel zone. Especially for thick, highly viscous flows, heat loss from the steep margins cannot be neglected. The thickness at the margin of the 1914 Sakurajima lava flow is 40 – 50 m and is almost as large as the length of a mesh. The simulated lava flow spread more widely than the actual one at their margins. This result suggests that the effect of cooling at their margins is underestimated in our simulation. In the case of the 1914 Sakurajima lava flow, better results might be obtained by including the loss of heat from the sides. In the case of less viscous lava flows which are thin compared with the length of a mesh, the cooling effect at the sides may be neglected.

The method cannot be applied to lava flows which are extruded intermittently, as obseved at Arenal Volcano (Wadge 1983) and in the Pu'u O'o eruption of Kilauea Volcano (Wolfe et al. 1987). These lavas flowed on top of previous lava flows which had already lost their fluidity. However, in our simulation, it is assumed that a mixture of lavas with different temperatures is possible even in the region where the temperature of lavas decreases below the solidus. Therefore, our method may be applicable only for simulating lava flows which are extruded continuously from vents.

To simulate lava flows more exactly, it is necessary to modify our method to include both the evaluation of heat loss from the sides of lava flows and the topographic changes due to the accumulation of solidified lavas.

5 Conclusion

A method of numerical simulation for lava flows on actual topography is presented. The 1983 Miyakejima, the 1986 Izu-Oshima and the 1914 Sakurajima lava flows were reproduced by the method using boundary conditions defined by field observations. Comparing the reproduced lava flows with actual ones in terms of inundation area, flow length, and thickness, it is concluded that these three lava flows are reproduced fairly well. The method is effective for reproducing basaltic, basaltic andesitic, and andesitic lava flows whose viscosities range from 10^4 to 10^9 poise. The method may be used to predict the paths of lava flows and to develop plans for diversion of flows or evacuation of populations.

Acknowledgment. We thank Raymond Dibble of Victoria University of Wellington who read our manuscript and gave us helpful suggestions. Jonathan Fink encouraged us to contribute to this proceeding. Michael Sheridan, Michael Malin and Gail A Mahood reviewed our manuscript and gave us helpful comments. We presented this paper at IUGG meeting in Vancouver, Canada, in 1987. Thomas Pierson of U.S. Geological Survey helped us to prepare for the presentation.

References

Aramaki S, Hayakawa Y (1984) Sequence and mode of eruption of the October 3–4, 1983 eruption of Miyakejima. Bull Volcanol Soc Jpn 29 Special Issue The 1983 Eruption of Miyakejima: S24–S35

Aramaki S, Fukuyama H, Kamo K, Kamada M (1981) Symposium on arc volcanism Tokyo and Hakone, Japan, Field excursion guide to Sakurajima, Kirishima and Aso Volcanoes. Volcanol Soc Jpn pp 1–17

Aramaki S, Fujii T, Kaneko T, Ishii T, Ozawa K, Fukuoka T (1987) Petrography of the ejecta and lavas of the 1986 eruption of Oshima volcano, Izu. Bull Volcanol Soc Jpn 32:182–183 (in Japanese)

Archambault C, Tanguy JC (1976) Comparative temperature measurements on Mount Etna lavas: problems and techniques. J Volcanol Geotherm Res 1:113–125

Baloga S, Pieri D (1986) Time-dependent profiles of lava flows. J Geophys Res 91:9543–9552

Crisci GM, Gregorio SD, Pindaro O, Ranieri GA (1986) Lava flow simulation by a discrete cellular model: first implementation. Int J Model Simul 6:137–140

Danes ZF (1972) Dynamics of lava flows. J Geophys Res 77:1430–1432

Dragoni M, Bonafede M, Boschi E (1986) Downslope flow models of a Bingham liquid: implications for lava flows. J Volcanol Geotherm Res 30:305–325

Endo K, Chiba T, Miyaji N, Sumita M, Uno R, Miyahara T, Tachikawa S (1987) Eruptive products and the eruption type of the 1986 Izu-Oshima eruption. Bull Volcanol Soc Jpn 32:169 (in Japanese)

Fink JH, Zimbelman JR (1986) Rheology of the 1983 Royal Gardens basalt flows, Kilauea Volcano, Hawaii. Bull Volcanol 48:87–96

Fujii T, Aramaki S, Fukuoka T, Chiba T (1984) Petrology of the ejecta and lavas of the 1983 eruption of Miyakejima. Bull Volcanol Soc Jpn 29 Special Issue The 1983 Eruption of Miyakejima: S266–S282 (in Japanese)

Fujii T, Aramaki S, Kaneko T, Ozawa K, Kawanabe Y, Fukuoka T (1988) Petrology of the lavas and ejecta of the November 1986 eruption of Izu-Oshima Volcano. Bull Volcanol

Soc Jpn 33 Special Number The 1986 Eruption of Izu-Oshima: S234–S254 (in Japanese)

Hagiwara T (1941) Viscosity of Akabakkyo lava. Bull Earthq Res Inst 19:299–303

Hagiwara T, Omote S, Murauchi S, Akashi K, Yamada Z (1946) The eruption of Mt. Sakura-Zima in 1946. Bull Earthq Res Inst 24:143–159 (in Japanese)

Hayakawa Y (1987) Mass and magma discharge rate of Izu Oshima 1986 eruption. Bull Volcanol Soc Jpn 32:183 (in Japanese)

Hulme G (1974) The interpretation of lava flow morphology. Geophys J R Astron Soc 39:361–383

Ishihara K, Takayama T, Tanaka Y, Hirabayashi J (1981) Lava flows at Sakurajima Volcano (I) – Volume of the Historical Lava Flows –. Ann Disast Prev Res Inst Kyoto Univ 24B-1:1–10 (in Japanese)

Ishihara K, Iguchi M, Kamo K (1984a) A numerical simulation of the 1983 lava flows at Miyakejima. Ann Disast Prev Res Inst Kyoto Univ 27B-1:1–14 (in Japanese)

Ishihara K, Iguchi M, Kamo K (1984b) A numerical simulation of basaltic lava flows and its application to the 1983 lava flows at Miyakejima. Bull Volcanol Soc Jpn 29 Special Issue The 1983 Eruption of Miyakejima: S242–S252 (in Japanese)

Ishihara K, Iguchi M, Kamo K (1985) Lava flows at Sakurajima Volcano (II) – Numerical Simulation of the 1914 Lava Flows on the Western Side of Sakurajima –. Ann Disast Prev Res Inst Kyoto Univ 28B-1:1–11 (in Japanese)

Ishihara K, Iguchi M, Kamo K (1988) Reproduction of the 1986 Izu-Oshima lava flows. Bull Volcanol Soc Jpn 33 Special Number The 1986 Eruption of Izu-Oshima: S64–S76 (in Japanese)

Koto B (1916) The great eruption of Sakura-jima in 1914. J Coll Science Imp Univ Tokyo 38:1–237

Lipman PW, Banks NG (1987) Aa flow dynamics, Mauna Loa 1984. Volcanism in Hawaii, U.S. Geol Surv Prof Pap 1350:1527–1567

Minakami T (1951) On the temperature and viscosity of the fresh lava extruded in the 1951 Oo-sima eruption. Bull Earthq Res Inst 29:487–498

Murase T, McBirney AR (1973) Properties of some common igneous rocks and their melts at high temperatures. Bull Geol Soc Am 84:3563–3592

Nagaoka M, Ogawa K (1987) Forms and structures of volcanic products of the 1986 eruption of Izu-Oshima volcano and its volume measured by photogrammetry. Bull Volcanol Soc Jpn 32:171–172 (in Japanese)

Nagata T, Sakuma S, Fukushima N (1946) On the lava flow newly ejected from Sakura-jima Volcano. Bull Earthq Res Inst 24:161–169 (in Japanese)

Omori F (1916) The Sakura-jima eruptions and earthquakes, III. Imp Earthq Invest Com 8:234–251

Pinkerton H, Sparks RSJ (1978) Field measurements of the rheology of lava. Nature 276:383–385

Robson GR (1967) Thickness of Etnean lavas. Nature 216:251–252

Shaw HR, Swanson DA (1970) Eruption and flow rate of flood basalts, Proceeding, Second Columbia River Basalt Syposium, 1969. Eastern Washington State College Press, Cheney, pp 271–299

Shaw HR, Wright TL, Peck DL, Okamura R (1968) The viscosity of basaltic magma: an analysis of field measurements in Makaopuhi Lava Lake, Hawaii. Am J Sci 266:225–264

Soya T, Uto K, Suto S (1984) The products of the 1983 eruption of the Miyakejima Volcano – with special reference to the lava. Bull Volcanol Soc Jpn 29 Special Issue The 1983 Eruption of Miyakejima:S230–S241 (in Japanese)

Sparks RSJ, Pinkerton H (1978) Effect of degassing on rheology of basaltic lava. Nature 276:385–386

Tsuya H, Morimoto R, Ossaka J (1952) Chemical composition of the 1951 – lavas of Oshima Volcano, Seven Izu-Islands, Japan. Bull Earthq Res Inst 30:231–236

Wadge G (1983) The magma budget of Volcano Arenal, Costa Rica, from 1968 to 1980. J Volcanol Geotherm Res 19:281–302

Walker GPL (1967) Thickness and viscosity of Etnean lavas. Nature 213:484–485
Wolfe EW, Garcia MO, Jackson DB, Koyanagi RY, Neal CA, Okamura AT (1987) The Puu
    Oo eruption of Kilauea Volcano, episodes 1–20, January 3, 1983, to June 8, 1984. U.S.
    Geol Surv Prof Pap 1350:471–508

Appendix 1: Program for Simulation

The operable program of the numerical simulation is written in FORTRAN.
The program is shown in Fig. 19. This program was operated by the mini-computer OKITAC 50 model 60, manufactured by Oki Electric Industry Co. Ltd.
Topographic data are read from a magnetic tape driver in the statements from
200 to 220. The subroutine programs for transferring calculated results to magnetic tape drivers are omitted.

Appendix 2: Effect of Initial Temperature and Effusion Rate of Lava on the Progress of Simulated Lava Flows

To evaluate the effects of the initial temperature and the effusion rate of lava
in our simulation, the progress of a simulated lava flowing on an inclined plane
are examined. The basic conditions in the simulation are as follows:

1. The angle of the inclined plane is 0.04 rad,
2. The size of mesh is 25×25 m,
3. The total volume of extruded lavas is $1.125 \times 10^6 \, m^3$,
4. The value of the parameter k is 24.61.

The conditions for a standard case (case A) are as follows:

1. Extrusion rate and duration are $62.5 \, m^3 \, s^{-1}$ and 5 h, respectively,
2. Initial temperature of extruded lava is 1100 °C ($\eta = 5.0 \times 10^4$ poise).

The other four cases were calculated, and compared with case A; cases B
and C: initial temperature of lava is the same as in case A, but the effusion
rate is twice (duration: a half) in case B, and a half (duration: twice) in case
C. The differences in the final length and area between cases B and C, and case
A are approximately $\pm 10\%$ and $\pm 20\%$, respectively, as illustrated in the left
side of Fig. 20. The rate of extension of a simulated lava flow is more sensitive
to the changes in effusion rate.

Cases D and E: the effusion rate is the same as in case A, but the initial
temperature of lava is changed to be 1050° ($\eta = 4.0 \times 10^5$ poise) in case D, and
1000 °C ($\eta = 3.2 \times 10^6$ poise) in case E. The flow lengths of simulated lava
flows in cases D and E decrease to 70 and 38% of the simulated flow in case
A, respectively, as illustrated in the right side of Fig. 20.

Fig. 19. Computational program for the simulation

```
C     LAVA FLOW SIMULATION  FOR BINGHAM FLUID
C                                                      87/06/12
C     ****************FROM CARD READER****************************
C     READ AREA OF MAP
C     INTERVAL OF LP OUT (MINUTES)
C     COMMENT FOR SIMULATION
C     PHYSICAL CONSTANT OF LAVA
C     POSITION, EXTRUSION RATE AND DURATION, TEMPERATURE OF CRATER
C     ***********************************************************
      COMMON /BM/MTB(1600)
      COMMON /PAR/W, W2, DLVF, G, DT
      DIMENSION MDT(40, 40), ISBM(100), NSYM(40)
      DIMENSION IFX(20), IFY(20), PRDC(20), EXST(20), EXED(20), ITP(20)
      DIMENSION HA(90, 70), HB(90, 70)
      DIMENSION IZ(90, 70)
      DIMENSION TH(90, 70), TA(90, 70), CC(90, 70), SS(90, 70)
      EMIS(H, T, DT)=T/(1.0+7.0E-14*T*T*T*DT/H)**(1.0/3.0)
      VIS(T)=3.551E+28*EXP(-4.1677E-2*T)
      YLD(T)=1.2449E+15*EXP(-2.047E-2*T)
      DATA IEND, JEND/90, 70/
      DATA ICN/3200/
      DATA NSYM/1H0, 1H1, 1H2, 1H3, 1H4, 1H5, 1H6, 1H7, 1H8, 1H9, 1HA, 1HB, 1HC, 1HD,
     11HE, 1HF, 1HG, 1HH, 1HI, 1HJ, 1HK, 1HL, 1HM, 1HN, 1HO, 1HP, 1HQ, 1HR, 1HS, 1HT,
     11HU, 1HV, 1HW, 1HX, 1HY, 1HZ, 1H , 1H*, 1H., 1H-/
      KEND=(IEND-1)/100+1
C *** READ ORIGIN OF COORDINATE
      READ(9, 119, END=900) NOCS, MSTX, MSTY, MEDX, MEDY, NSTX, NSTY
  119 FORMAT(7I5)
      READ(9, 129) IWRTDT
  129 FORMAT(I5)
      WRTDT=IWRTDT*60.0
C *** READ HEIGHT DATA FROM MT
  200 CALL OPENMT(3, ICN, IER)
      IF(IER.NE.0) GO TO 900
  210 CALL READMT(MSTX, MSTY, MEDX, MEDY, NOCS, MDT, IOX, IOY, ICN, IER)
      IF(IER.NE.0) GO TO 220
      MDLX=(IOX-MSTX)*40-NSTX+1
      MDLY=(IOY-MSTY)*40-NSTY+1
      DO 230 I=1, 40
      DO 231 J=1, 40
          MX=MDLX+I
          MY=MDLY+J
            IF(MX.LT.1) GO TO 231
            IF(MY.LT.1) GO TO 231
            IF(MX.GT.IEND) GO TO 231
            IF(MY.GT.JEND) GO TO 231
            IZ(MX, MY)=MDT(I, J)
  231 CONTINUE
  230 CONTINUE
      GO TO 210
  220 CALL CLOSMT(ICN)
      IF(IER.EQ.2) GO TO 900
      READ(9, 418) IT, MVT
      T=FLOAT(IT)*60.0
  418 FORMAT(2I5)
      CALL EOS5G(0, 1, MTB, ICN, I, J)
      CALL LVI(C, TLVF, PRDC, EXST, EXED, IFX, IFY, NOCR, MSTX, MSTY,
     1          NSTX, NSTY, TC, MDT, ITP, TCR, TDF)
C     *** INITIAL PARAMETER AND AREA SET
      DO 441 I=1, IEND
      DO 442 J=1, JEND
          TA(I, J)=FLOAT(ITP(1))
          CC(I, J)=VIS(TA(I, J))
          SS(I, J)=YLD(TA(I, J))
  442 CONTINUE
  441 CONTINUE
C *** DETERMINATION OF INITIAL CALCULATION AREA
      IMIN=IFX(1)
      IMAX=IFX(1)
      JMIN=IFY(1)
      JMAX=IFY(1)
      DO 443 ICR=1, NOCR
          IF(IFX(ICR).LT.IMIN) IMIN=IFX(ICR)
          IF(IFX(ICR).GT.IMAX) IMAX=IFX(ICR)
          IF(IFY(ICR).LT.JMIN) JMIN=IFY(ICR)
          IF(IFY(ICR).GT.JMAX) JMAX=IFY(ICR)
  443 CONTINUE
C     *   CALCULATION OF VELOCITY OF LAVA FLOW START
  510 CONTINUE
      IDT=IFIX(WRTDT/DT)
      DO 500 ICT=1, IDT
      T=T+DT
C     ***   DETERMIN THE CALCULATION AREA
```

```
      IST=IMIN-1
      IED=IMAX+1
      JST=JMIN-1
      JED=JMAX+1
      IF(IST.LT.1) IST=1
      IF(JST.LT.1) JST=1
      IF(IED.GT.IEND) IED=IEND
      IF(JED.GT.JEND) JED=JEND
      IEDD=IED-1
C *** CALCULATE HB ( EAST OR WEST DIRECTION )
      DO 540 J=JST,JED
          DO 541 I=IST,IEDD
              M=I+1
              IF(HA(I,J).EQ.0.0.AND.HA(M,J).EQ.0.0) GO TO 541
              CALL HGTM(IZ(I,J),IZ(M,J),HA(I,J),HA(M,J),
     1        CC(I,J),CC(M,J),TA(I,J),TA(M,J),SS(I,J),SS(M,J),
     2        HIO,THIO)
              HB(I,J)=HB(I,J)+HIO
              HB(M,J)=HB(M,J)-HIO
              TH(I,J)=TH(I,J)+THIO
              TH(M,J)=TH(M,J)-THIO
  541     CONTINUE
  540 CONTINUE
C *** CALCULATE HB ( NORTH OR SOUTH DIRECTION )
      JEDD=JED-1
      DO 542 I=IST,IED
          DO 543 J=JST,JEDD
              M=J+1
              IF(HA(I,J).EQ.0.0.AND.HA(I,M).EQ.0.0) GO TO 543
              CALL HGTM(IZ(I,J),IZ(I,M),HA(I,J),HA(I,M),
     1        CC(I,J),CC(I,M),TA(I,J),TA(I,M),SS(I,J),SS(I,M),
     2        HIO,THIO)
              HB(I,J)=HB(I,J)+HIO
              HB(I,M)=HB(I,M)-HIO
              TH(I,J)=TH(I,J)+THIO
              TH(I,M)=TH(I,M)-THIO
  543         CONTINUE
  542 CONTINUE
C     ***   ADD LAVA TO CRATER MESH   ***
      DO 520 ICR=1,NOCR
          IF(T.GT.EXED(ICR)) GO TO 520
          IF(T.LT.EXST(ICR)) GO TO 520
          I=IFX(ICR)
          J=IFY(ICR)
          HC=PRDC(ICR)*DT
          IF(T.GE.2400.0.AND.T.LT.4200.0) HC=HC*2.0
          HB(I,J)=HB(I,J)+HC
          TH(I,J)=TH(I,J)+HC*ITP(ICR)
  520 CONTINUE
C     ***   H'=H+DH   ***
      DO 550 I=IST,IED
      DO 551 J=JST,JED
          HD=HA(I,J)+HB(I,J)
          IF(HD.EQ.0.0) GO TO 551
          THD=TA(I,J)*HA(I,J)+TH(I,J)
          TD=THD/HD
C         *** SURFACE TEMPERATURE
          IF(TD.GE.TCR) TDIFF=0.0
          IF(TD.LT.TCR) TDIFF=TDF
          TD=TD-TDIFF
          TD=EMIS(HD,TD,DT)
          TD=TD+TDIFF
          CC(I,J)=VIS(TD)
          SS(I,J)=YLD(TD)
          HA(I,J)=HD
          TA(I,J)=TD
          HB(I,J)=0.0
          TH(I,J)=0.0
  551 CONTINUE
  550 CONTINUE
C     ***   ENLARGE OF CALCULATION AREA   ***
      DO 544 I=IST,IED
          IF(HA(I,JST).NE.0.0) JMIN=JST
          IF(HA(I,JED).NE.0.0) JMAX=JED
  544 CONTINUE
      DO 545 J=JST,JED
          IF(HA(IST,J).NE.0.0) IMIN=IST
          IF(HA(IED,J).NE.0.0) IMAX=IED
  545 CONTINUE
  500 CONTINUE
C     *** CALCULATION OF TOTAL HEIGHT OF LAVA FLOW.
      TOTALV=0.0
      NMESH=0
```

```
              DO 571 I=IST,IED
              DO 572 J=JST,JED
                 TOTALV=TOTALV+HA(I,J)
                 IF(HA(I,J).NE.0.0) NMESH=NMESH+1
       572 CONTINUE
       571 CONTINUE
C *** WRITE THICKNESS OF LAVA FLOW ON LP
           IT=IFIX(T/60.0+0.5)
           ITH=IT/60
           ITM=IT-ITH*60
           DO 569 LPOUT=1,2
           IF(LPOUT.EQ.1) UNIT=1.0
           IF(LPOUT.EQ.2) UNIT=10.0
           DO 562 K=1,KEND
              LST=1
              LED=IEND
              WRITE(6,567) ITH,ITM
              DO 560 JJ=1,JEND
              J=JEND+1-JJ
                 DO 565 I=LST,LED
                    IF(LPOUT.EQ.1) AA=HA(I,J)
                    IF(LPOUT.EQ.2) AA=ITP(1)-TA(I,J)
                    IS=I-LST+1
                    IF(I.GT.IEND) IBM=37
                     IF(I.GT.IEND) GO TO 564
                    IBM=IFIX(AA/UNIT)+1
                    IF(IBM.GT.36) IBM=38
                    IF(AA.EQ.0.0) IBM=39
                    IF(AA.LT.0.0) IBM=40
                    IF(IZ(I,J).LE.0 .AND.AA.EQ.0.0) IBM=37
       564          ISBM(IS)=NSYM(IBM)
       565          CONTINUE
                 WRITE(6,566) J,(ISBM(I),I=1,100),J
       560      CONTINUE
       562 CONTINUE
           WRITE(6,568) UNIT
       568 FORMAT(1H ,100X,'UNIT=',F5.1)
       569 CONTINUE
       573 FORMAT(1H ,'TOTAL VOLUME=',F15.5,5X,'AREA=',I5,5X,
          1         10X,4I6)
       574 FORMAT(1H ,4F15.7)
           WRITE(6,573) TOTALV,NMESH,IST,IED,JST,JED
           IF(T.GE.TLVF) GO TO 900
           GO TO 510
       900 CONTINUE
       566 FORMAT(1H ,5X,I3,2X,100A1,2X,I3)
       567 FORMAT(1H1,10X,I5,'H',I4,'M',/,
          17X,'NO',                         11X,'1',9X,'2',9X,'3',9X,'4',9X,'5',
          29X,'6',9X,'7',9X,'8',9X,'9',9X,'0',/)
           STOP      999
           END
C
C
           SUBROUTINE HGTM(MO,MA,HO,HA,CO,CA,TO,TA,SO,SA,DH,DTH)
C      FLUX
           COMMON /PAR/W,W2,DLVF,G,DT
           ZA=FLOAT(MA)
           ZO=FLOAT(MO)
           HRA=ZA+HA
           HRO=ZO+HO
           IF(HRA.GT.HRO) GO TO 100
           C=CO
           T=TO
           S=SO
           IF(ZO.GT.ZA) GO TO 20
        10 DZ=HRA-HRO
           H=HRO-ZA
           GO TO 200
        20 DZ=ZA-ZO
           H=HO
           GO TO 200
       100 C=CA
           T=TA
           S=SA
           IF(ZA.GT.ZO) GO TO 120
       110 DZ=HRA-HRO
           H=HRA-ZO
           GO TO 200
       120 DZ=ZA-ZO
           H=HA
           GO TO 200
       200 CONTINUE
           DR=SQRT(DZ*DZ+W2)
```

```
      SINA=DZ/DR
      IF(SINA.EQ.0.0) GO TO 300
      IF(SINA.LT.0.0) SGN=-1.0
      IF(SINA.GT.0.0) SGN=1.0
      SINA=ABS(SINA)
      HC=S/(DLVF*G*SINA)
      IF(H.LE.HC) GO TO 300
      H1=H/HC
      Q=S*HC*HC/(3.0*C)*(H1*H1*H1-1.5*H1*H1+0.5)*SGN
      GO TO 400
  300 Q=0.0
  400 DH=Q/W*DT
      DTH=DH*T
      RETURN
      END
C
C
      SUBROUTINE LVI(C, TLVF, PRDC, EXST, EXED, IFX, IFY, NOCR, MSTX, MSTY,
     1              NSTX, NSTY, TC, MDT, ITP, TCR, TDF)
C *** READ SIMULATION PARAMETER FROM SYS009*******************************
C     DLVF    : DENSITY
C     ETA     : VISCOSITY
C     DELT    : DT(SEC)
C     TLVF    : SIMULATION TIME (HOUR)
C     IFX     : LOCATION OF CRATER   (ICRX * 100X)
C     IFY     : LOCATION OF CRATER   (ICRY * 100Y)
C     PRDCT   : EXTRUSION RATE (M/SEC)
C     EXSTM   : START TIME OF EXTRUSION
C     EXEDM   : END TIME OF EXTRUSION
C     TCR     : SOLIDUS TEMPERATURE
C     TDF     : DIFFERENCE BETWEEN SURFACE AND INSIDE
C     ITP     : INITIAL TEMPARATURE
C     ITOP    : WRITE POSITION OF MT
C ***********************************************************************
      COMMON /PAR/W, W2, DLVF, G, DELT
      DIMENSION PRDC(20), EXST(20), EXED(20), IFX(20), IFY(20), ITP(20)
      DIMENSION MCN(40), MDT(40, 40)
      VIS(T)=3.551E+28*EXP(-4.1677E-2*T)
      G=9.8
      W=25.0
      W2=W*W
      READ(9, 103) (MCN(I), I=1, 40)
  103 FORMAT(40A2)
      READ(9, 100) DLVF, TCR, TDF, DELT, TLVF, NOCR, ITOP
  100 FORMAT(F10.5, 2F5.0, 2F10.5, 2I5)
      DO 10 I=1, NOCR
          READ(9, 101) NOCRT, ICRX, ICRY, I00X, I00Y, PRDCT, EXSTM, EXEDM, INITT
C
          PRDC(NOCRT)=PRDCT
          IFX(NOCRT)=ICRX+(I00X-MSTX)*40-NSTX+1
          IFY(NOCRT)=ICRY+(I00Y-MSTY)*40-NSTY+1
          EXED(NOCRT)=EXEDM*60.0
          EXST(NOCRT)=EXSTM*60.0
          ITP(NOCRT)=INITT+273
   10 CONTINUE
  101 FORMAT(5I5, 3F10.5, I5)
      TCR=TCR+273.0
      WRITE(6, 200)
      WRITE(6, 201) DLVF, DELT, TLVF, TCR, TDF
  200 FORMAT(1H1, ///, 11X, '*   LAVA FLOW SIMULATION (IZU-OSHIMA)   *', /
     1//)
  201 FORMAT(1H , 10X, 'DENSITY(G/CC)        ', F10.5, /,
     2             11X, 'DT(SEC)             ', F10.5, /,
     3             11X, 'CALCULATION(HOUR)   ', F10.5, /,
     4             11X, 'SOLIDUS TEMP(K)     ', F10.5, /,
     5             11X, 'DIFFERRENCE TEMP(K) ', F10.5, ///)
      DO 20 I=1, NOCR
          WRITE(6, 202) I, PRDC(I), EXST(I), EXED(I), IFX(I), IFY(I), ITP(I)
   20 CONTINUE
  202 FORMAT(1H , 10X, I2, 5X, 'RATE(M/S)=', F7.5, 5X, 'TIME(S)=', F8.1, '->',
     1F8.1, 3X, 2I5, 5X, 'TEMP(K)=', I4, /)
      WRITE(6, 203) (MCN(I), I=1, 40)
  203 FORMAT(1H , ///, '###', 5X, 40A2)
      MDT(1, 1)=IFIX(TLVF*60.0)
      DLVF=DLVF*1000.0
      TLVF=TLVF*3600.0
      C=VIS(FLOAT(ITP(1)))
      WRITE(6, 204) C
  204 FORMAT(1H , //, 11X, 'VISCOSITY(PA.S)      ', E10.2)
  900 RETURN
      END
```

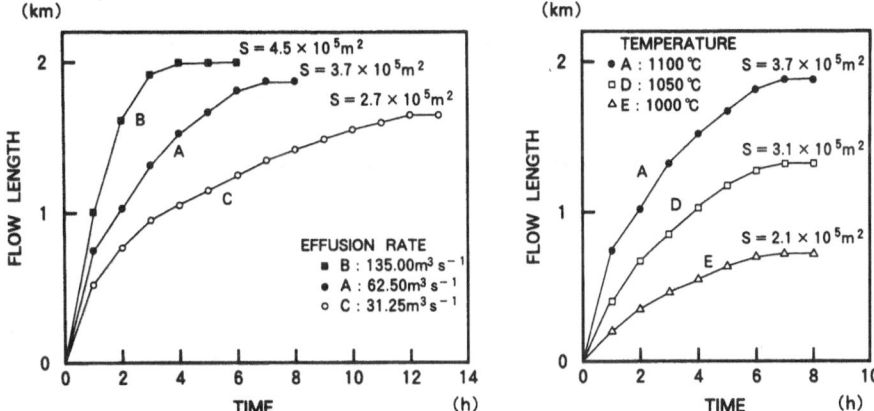

Fig. 20. Flow lengths of simulated lava flowing on a plane inclined with an angle of 0.04 rad. *A* is the standard case [effusion rate = 62.5 m^3 s^{-1}, duration = 5 h, and initial temperature = 1100 °C ($\eta = 5.0 \times 10^4$ poise)]. *B* and *C* are the cases when the effusion rate is changed, and *D* and *E* are the cases when the initial temperature is changed. *S* denotes the final area of each simulated lava flow

On the Mechanisms of Lava Flow Emplacement and Volcano Growth: Arenal, Costa Rica

A. Borgia and S. R. Linneman

Abstract

Arenal Volcano is composed of a hierarchical series of geologic units: unit flow, composite flow, lava field, and lava armor. Volume-limited unit flows are emplaced at short time intervals to make up composite flows. Composite flows form lava fields, and lava fields in turn, constitute the lava armor (the volcano). Tephra and lava breccias are selectively eroded from the steep slopes of the volcano by heavy rains and contribute little to the actual shape of the cone. This constructive process has important consequences for the distribution of the age of lava on a composite cone. We show that lava of significantly different ages may be juxtaposed at all scales from the unit flow, to the composite flow, to the lava field, and to the lava armor. These relations are applicable to the time sequential sampling of a composite volcano.

Detailed observations of the dynamic behavior of unit flows indicate that two dimensionless parameters determine the distribution of lava between an active, flowing component and a passive, stationary component. The first parameter, f, is a measure of how much lava the front uses to advance relative to how much it uses to build up levees. The second parameter, q, is the fraction of lava that is able to drain out of the channel when no more lava from the vent feeds the flow. Both parameters have a primary role in determining the final dimensions of a lava flow. These parameters may be calculated from observations of lava flowing onto different topography and at different times after effusion. This data set may allow the prediction of f and q for future flows, and as a consequence, the final flow length along possible flow paths is also predictable.

The development of a thermal structure within the flow plays a critical role in the dynamic evolution of a unit flow. The weight of a cold, highly viscous crust at the surface of the flow actively modifies the stress distribution in the flow and controls the rate of processes such as front velocity, levee formation, and growth of surges. We propose that for a given flux of lava there is a critical channel length beyond which the flow accelerates triggering the separation of the flow from its source near the vent. Thus, the unit flows are volume-limited. Based on this hypothesis we derive a relation for the velocity and position of the flow front at any time after effusion has started, assuming the time functions of f, q, and flow rate are known. We find that the length of a unit flow is directly proportional to f, q, and the flow rate and it is inversely proportional to the cross-sectional area of the channel and to the sine of the slope. These relations also hold for composite flows.

Finally, by making the approximation that a composite flow grows to a constant slope we derive equations for the evolution of lava fields and the growth of the volcanic structure. These relations explain the asymmetric distribution, areal extent, and slope of the various lava fields at Arenal and allow us to infer the position of buried craters and contacts. Remarkably, our model is based on mass conservation and makes no assumption about rheology. With comparable observations this method may be applicable to other volcanoes similar to Arenal.

1 Introduction

The study of the dynamic processes that determine the geometrical parameters of lava flows, especially length, is relevant in determining the potential hazard that a particular flow may have on human and natural environments. Several studies have made substantial contributions to the understanding of these processes. Walker (1973), and later Wadge (1978), show that the length of Etnean lava flows is directly proportional to the rate of effusion, while Malin (1980) suggests a direct relation between total volume and length of Hawaiian lava flows. Pieri and Baloga (1986) present a relation between the planimetric area of lava fields and the eruption rate. Baloga and Pieri (1986) suggest that variation in flow depth at the source and the form of the viscosity variation with distance from the vent have significant influence on the morphology and dimensions of lava flows. Dragoni (1987), assuming a temperature-dependent Bingham rheology, radiative cooling, and a negligible thermal gradient orthogonal to the flow, derives analytical relations for flow front velocity and flow depth as a function of time. Finally, Pinkerton and Wilson (1987) and Guest et al. (1987) illustrate the extreme complexity of lava flow emplacement and recognize that the length of flows may be controlled by either the volume of lava erupted (volume-limited flows) or by cooling (temperature-limited flows).

For more purely scientific and aesthetic reasons, the shape of volcanoes has also long attracted scientific attention. Milne (1878, 1879) and Becker (1885) derived a logarithmic and an exponential topographic profile, respectively, for an ideal volcano composed of a uniform mass of constant strength. Shteynberg and Solov'yev (1976) used a similar assumption to argue that the upper part of a volcano, where stresses do not overcome the strength of the rocks, retains the shape of a cone, while its lower part deforms under the load of the upper cone acquiring a logarithmic profile. Recently, Lacey et al. (1981) and Angevine et al. (1984) modeled a volcano as an uniform porous medium in which magma flows to the surface following the path of least resistance. This model predicts that a volcano grows by maintaining its topography tangent to an equipotential surface determined by the hydraulic head of the magma. Despite strong criticisms (Wood 1982; Wadge and Francis 1982), this model emphasizes two very important, but formerly neglected, concepts: the constructive process in controlling the volcanic form, and the potential surface of magma pressure in controlling vent location and, indirectly, the constructive process.

We believe that the principal weakness of these models is the assumption that volcanoes have similar shapes. When considered in greater detail, the shape of volcanoes varies considerably, but the differences may be appreciated only on a geologic and structural basis. Milne (1878) observed during a visit to Icelandic volcanoes, "The wilderness of form presented to us by such mountains as these is so evidently the combined effect of many and varied causes, that it would be vain to seek a simple explanation for the formation of the whole." Williams and McBirney (1979) point out that the principal factors controlling the shape of a volcano are magma composition, mechanism of emplacement, relative abundance of lava and pyroclastic ejecta, distribution of vents, stage of growth, and degree of erosion of the edifice. All these controlling factors usually result in a complex morphology which cannot be explained by a simple model. In contrast, at Arenal Volcano, the constructive process is remarkably simple: (1) erupted lavas maintained approximately the same composition through most of its life history; (2) the geometry of lava fields is controlled by the dynamics of emplacement of volume-limited unit flows erupted during a steady, low rate of effusion; (3) heavy rains rapidly erode pyroclastic ejecta and lava breccias from the steep flanks of the cone and as a consequence the geometry of the volcanic edifice is mainly controlled by the topography of lava fields.

In this chapter we present a model for the growth of Arenal Volcano developed on the basis of the geology and the mechanisms of lava flow emplacement and lava field growth during the current eruption. Our model differ from previous attempts to model volcanic form in that it emphasizes the eminent role of geology and structure in the construction of the volcanic edifice, a relationship ignored by earlier models. The mathematical model utilizes the field observation that Arenal is made up of a hierarchical series of geologic units: the volcano is built by the superposition of successive lava fields; each lava field contains a number of composite flows; and each of these, in turn, includes a number of unit flows. Our report bridges the gap between lava flow dynamics and volcano growth and is a logical extension of the paper by Borgia et al. (1983). In the discussion below we assume that the reader is familiar with that paper.

2 Description of Natural Phenomena

Arenal Volcano, in northern Costa Rica (Fig. 1), is a small ($\approx 1.5 \times 10^{10}\,\mathrm{m}^3$), young ($\approx 3000\,\mathrm{yr}\ BP$) cone made predominantly of high viscosity ($\approx 10^6\,\mathrm{Pa\,s}$) basaltic andesite ($\approx 54\,\mathrm{wt\%}\ SiO_2$) lava flows (Wadge 1983;

Fig. 1. Topographic and geologic map of Arenal Volcano with the areal distribution of the various lava fields. Note the break in slope at the contacts and the parabolic intersections between the lava fields. Topography based on the 1 : 10000 map by Juan Bravo Chacon, map drawn by A. H. Arce and M. Ch. Barboza, contours every 10 m. Geology after Borgia et al. (1988)

Lava field	erupted from crater
Al_l	A beginning in 1968
Al_h	C beginning in 1973
$A2_l$	E (position inferred)
$A2_h$	D beginning less then 300 years ago
$A3$	F (position inferred)
$A4$	G (position inferred)

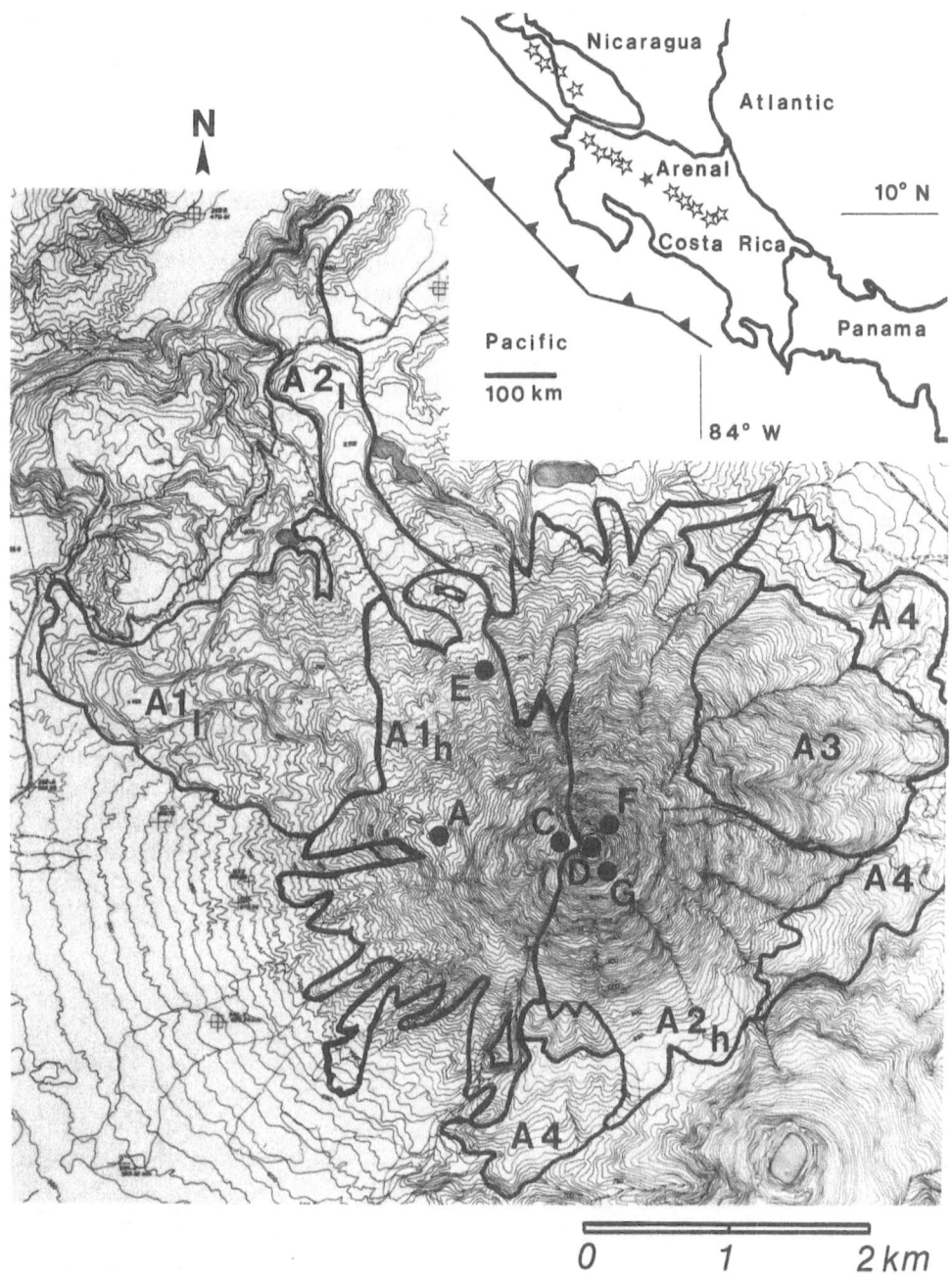

Cigolini et al. 1984; Reagan et al. 1987; Borgia et al. 1988). The cone was built by emplacement of flows during a few (five to eight), large volume ($\approx 15\%$ of present volume of the cone) eruptive episodes (Borgia et al. 1988). The most recent episode has produced blocky basaltic andesite lava flows almost continuously since September, 1968. This effusive activity has emanated from two of the three craters created during a brief explosive period in July, 1968 (Melson and Saenz 1968, 1973). Lava flows were initially erupted from a lower crater (A in Fig. 1). After a short interruption in 1974, the effusion shifted to the near summit crater (C in Fig. 1; Bennett and Raccichini 1977). Wadge (1983) reports a total volume of erupted magma of 0.3 km³ for the period 1968–1980. After an initial decrease (3 to 0.3 m³ s⁻¹), the effusion rate has remained nearly constant since 1973 (Wadge 1983).

The following sections describe the important constructive features of Arenal Volcano during its recent and continuing eruptive phase. These descriptions apply specifically to the period of continuous effusion of lava flows observed between January 1980 and June 1983. Explosive activity beginning in 1984 prohibited any further study of the development of lava flows.

2.1 The Active Crater

The active crater (crater C in Figs. 1 and 2) is located near the summit of Arenal and is bounded on the east by a horseshoe-shaped scarp carved into the summit during the 1968 explosive phase. Since the beginning of effusive activity at crater C in 1974 (Bennett and Raccichini 1977), an approximately 200-m-high lava cone has grown by lava emplacement. Accordingly, the active vent area has moved up and by 1986 was only a few tens of meters below the summit of the volcano. The vent area consists of an active vent (where effusion is taking place) and, occasionally, older vents, which may still contain partially molten lava.

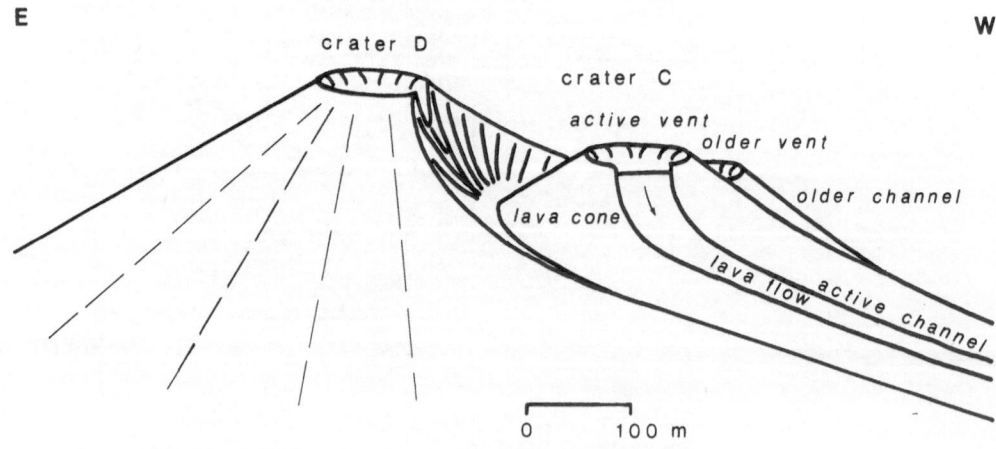

Fig. 2. Sketch of Arenal's summit area in January–March 1983

During effusive activity, lava rises from the conduit into the active vent and flows through one side of the lava cone into the channel proper (see below). Changes in lava level during continuous effusion produce a series of small concentric lava ridges within the vent. Heat loss at the vent area is mainly due to radiation since a glowing surface is ubiquitous. Temperature on the surface of cracks is $>1000\,°C$ (Cigolini et al. 1984).

As the height of the lava cone increases, its steep flanks ($>40°$ near the summit) become unstable and, due to erosion (principally landsliding) and magma pressure, may collapse outward. The new breach drains the flow in a different direction. This direction is most likely toward the side of the cone which has the steepest (most unstable) slope. The change in direction of outflow is controlled by continuous constructive factors such as lava effusion rate and cone growth rate, and discontinuous destructive factors such as explosions and earthquakes. Hence, these changes tend to occur at roughly regular intervals as observed in 1980–1983, when flow direction changed every 6–8 months. Moreover, these changes affect the evolution of the lava flow field because they cause the end of one composite lava flow and the beginning of a new one.

2.2 The Unit Flows

A *unit flow* (Fig. 3a) is a finite quantity of lava, which is emplaced during continuous effusion, moves downhill by gravity away from the vent, and maintains fluid continuity throughout its length. It consists of a continuous fluid lava body enveloped by debris. A vertical cross-section shows top debris, a vesiculated crust, an unvesiculated core, and bottom debris (Cigolini and Borgia 1980). A unit flow may be divided into a channel zone and a frontal zone. The channel extends from the vent down to the frontal zone and is characterized by parallel streamlines and increasing length with time. The channel is further subdivided into a channel proper, where the lava actually flows downhill, and the levees which are the stationary lateral boundaries of the flow (equivalent to the static levees of Guest et al. 1987) and are constructed at the flow front.

The frontal zone is the most distal part of the flow. It constitutes the hydraulic connection between the channel proper and the levees: the lava flows from the channel proper through the frontal zone into the levees. In the frontal zone, streamlines diverge showing a three-dimensional fountain flow velocity distribution (Rose 1961; Mavridis et al. 1986). Lava elements decelerate and stretch laterally as they approach the slower moving lava-air interface at the flow front. They move outward and downward toward the side and bottom regions where they are quenched in the levees and flow bottom, respectively.

The path of the lava, then, is from the vent, through the channel proper, to the front, and from the front to the levees. It follows that the lava increases in age from the vent to the front in the channel proper, but from the front to the vent in the levees (Fig. 3a). The same is true in a vertical cross-section of

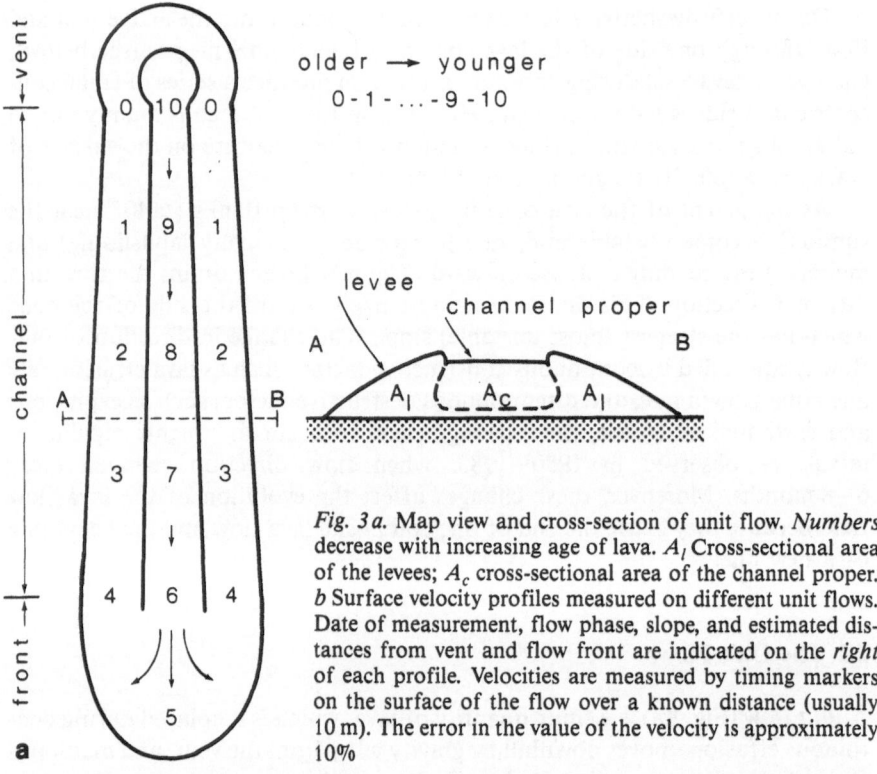

Fig. 3 a. Map view and cross-section of unit flow. *Numbers* decrease with increasing age of lava. A_l Cross-sectional area of the levees; A_c cross-sectional area of the channel proper. *b* Surface velocity profiles measured on different unit flows. Date of measurement, flow phase, slope, and estimated distances from vent and flow front are indicated on the *right* of each profile. Velocities are measured by timing markers on the surface of the flow over a known distance (usually 10 m). The error in the value of the velocity is approximately 10%

the flow: since the lava and debris at the bottom of the flow are deposited by the flow front, the age of lava increases from the vent to the front in the upper part of the flow and from the front to the vent at the flow bottom. The validity of this relation has also been observed in the age of deposits produced by the emplacement of wax flows under water (Hallworth et al. 1987).

A unit flow shows a developing phase and a collapsing phase. During the developing phase lava is continuously supplied by the vent to the flow. Once the effusion of lava is cut at the vent, the unit flow ends its developing phase and enters the collapsing phase. The fluid lava still remaining in the channel is drained out by gravity into the flow front; a collapsed and extended crust is left behind filling the channel proper floor (Borgia et al. 1983; Cigolini et al. 1984).

The cooling of the lava in the channel is by radiation near the vent only. As soon as a crust has formed, conduction and convection of heat into the air is prevalent. Cooling is thereby greatly reduced within the channel proper. Cooling becomes again very effective at the flow front where the glowing nucleus of the flow is exposed to the air. In a qualitative sense, the cooling of the unit flow may be approximated by a two-step process. A first sharp drop in temperature at the crater to form a cooler, semirigid crust, and a second sharp drop at the front where the lava is finally quenched to form the levees.

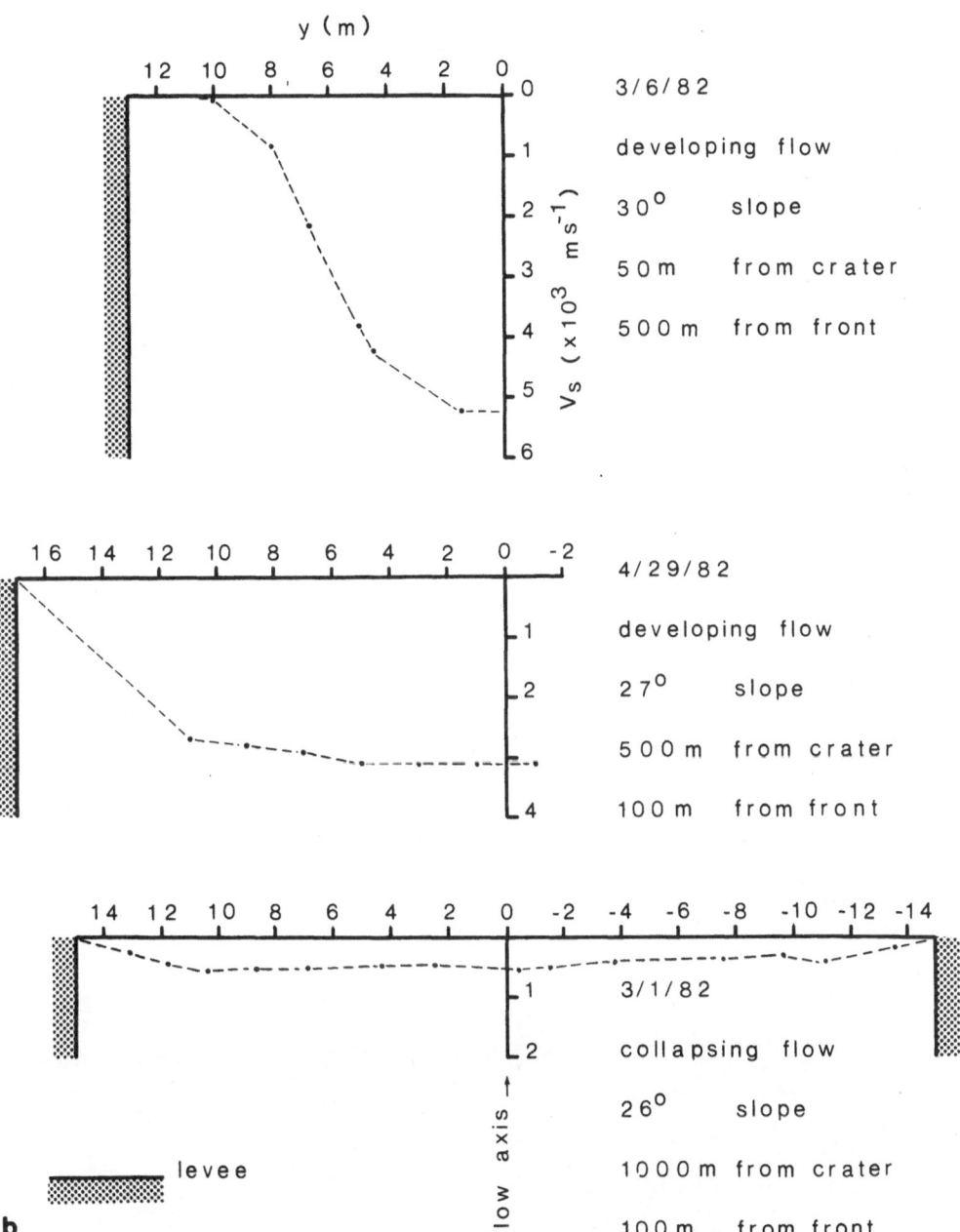

The growth of this thermal structure within the flow is reflected in surface velocity profiles (Fig. 3b). At the vent, the velocity profile of a developing flow shows no plug flow and the decreasing velocity gradient next to the lateral boundary may indicate an increase in viscosity next to the flow margin. Away from the crater, however, the velocity profile becomes flatter at the center cor-

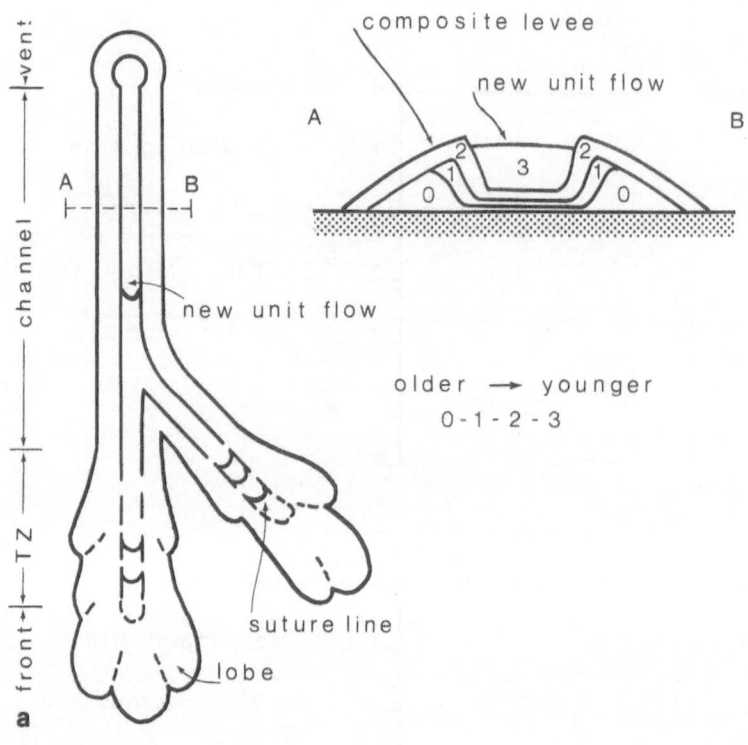

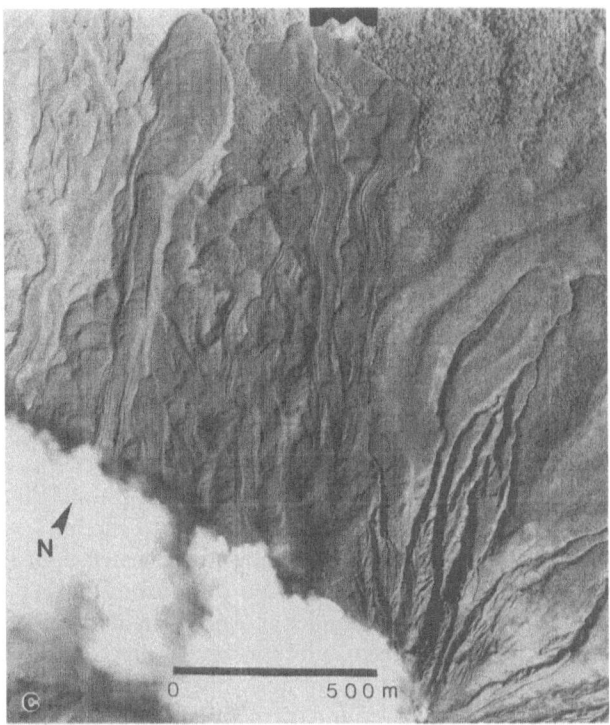

Fig. 4. a Map view and cross-section of composite flow. *Numbers* decrease with increasing age of lava. Note the composite nature of the levee, the transition zone (TZ), and the composite front with lobes and suture lines. *b* A unit flow front is flowing down a collapsed composite channel in March of 1983. This unit flow reached and fed the composite frontal zone, located approximately 500 m downstream, a few days after the photograph was taken. Note the layer of lava plastered onto the levee, to the *right* of the photograph, by the flow that passed before this one. *c* Aerial photograph of the NW flank of Arenal Volcano. The $A1_h$ lava field (to the *left*) has no vegetation, the $A2_h$ lava field (to the *right*) is covered by scarce vegetation. Note the composite levees, the transition zone, and the composite flow front with lobes and suture lines. Aerial photograph of the Instituto Geografico Nacional taken on March 5, 1980

responding to the growth of the cooler, high viscosity crust. The asymmetric velocity profile near the flow front during the collapsing phase indicates complex flow and thermal structures.

During the collapsing phase the velocity of the flow decreases exponentially to zero as the lava from the channel proper is drained into the front (Borgia et al. 1983). When no more lava can drain into the frontal zone (i.e., the channel proper is empty), the front stops, despite the fact that the front maintains its volume and is still fluid. Indeed, if a second lava flow reaches that front, the first will reactivate its downhill motion. Thus, the continued movement of the front is maintained by the pressure generated by the advancing lava in the channel proper. Because unit flows stop when the channel is empty, we observe

that the length of Arenal flows is completely controlled by the volume and duration of the lava supply. We do not see that the actual increase in viscosity
of the flow front with distance from the vent has a relevant role in determining
the final flow length.

2.3 The Composite Flow

When many individual unit flows traverse the same path at short time intervals
(a few days), they form a *composite flow*. A composite flow (Fig. 4a) consists
of a composite channel zone and a composite frontal zone. During the evolution of a composite flow, each lava unit flowing down the channel adds a
new layer of lava to the levee (Fig. 4a, b). Thus, composite levees show concentric layers, which are younger outward (from the levee nucleus).

The unit flows that come down the composite channel feed the composite
frontal zone. Unit flow fronts "sink" into the back of the composite frontal
zone, leaving parabolic suture lines on the surface of the composite flow (Fig.
4a, c). The lava units that form a composite frontal zone partially retain their
identity producing a complex morphologic and velocity structure in the flow.
In fact, a composite frontal zone generally consists of various lobes, which
may move discontinuously and independently from each other. Therefore,
there is no definite boundary between the composite channel zone and the
composite frontal zone, instead a transition zone exists (TZ in Fig. 4a).

During February-May 1982, 13 unit flows were observed to flow down the
same composite flow channel at time intervals ranging from 5 to 9 days. No
interruption of effusion was observed during the formation of the 13 flows,
nor during the 4-year (1980–1983) period of our fieldwork at Arenal. In fact,
from 1968 to 1984, only one major interruption of the effusion was reported
(Wadge 1983), when effusion shifted from crater A to crater C.

This continuous effusion allowed detailed observation of the formation of
composite flows from unit flows. The unit flows were separated from each other by a zone of empty channel between the collapsing zone of one unit flow
and the frontal zone of the subsequent unit flow. The zone of empty channel
lengthened as the flows moved downhill because the collapsing zone moves at
a higher velocity than the front (cf. Cigolini et al. 1984). It appears that the
collapsing zone of one flow and the front of the subsequent flow begin very
close together near the crater, increasing their separation with time. In fact, we
observed that the length of empty channel was less than 100 m on the upper
flanks and more than 500 m on the lower flanks when the channel proper of
the first unit flow was completely collapsed.

Since composite flows grow with time by the repeated emplacement of flow
units, which deposit new layers of lava, a composite flow will have a finite
growth rate in both areal and vertical extent. This growth rate is a function of
distance from the vent and time. It is observed that the top surface of the composite flow grows with time toward a constant slope (Fig. 5).

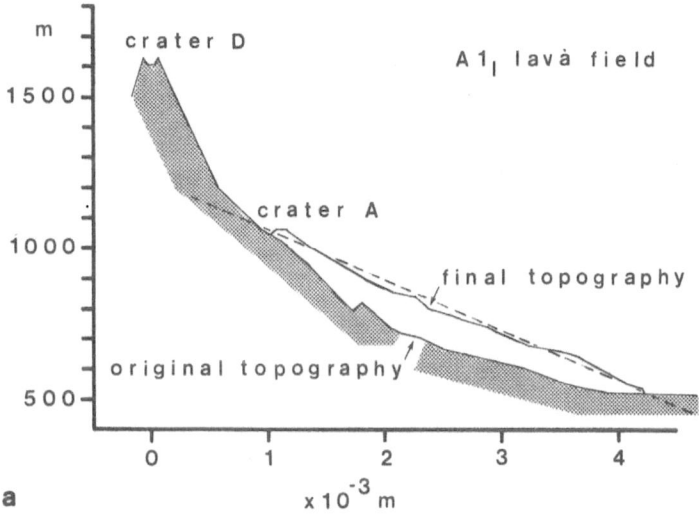

Fig. 5. a Original and final topography of the $A1_l$ lava field erupted during 1968–1973 from the lower crater (*A*). The last composite flow has almost reached a constant slope (*dashed line*). Original topography based on the Instituto Geografico Nacional 1:50000 Hoja "Fortuna" map of 1966; final topography is from the topographic map of Fig. 1. *b* Arenal Volcano looking south during the July 1984 explosive activity. The two lava fields ($A1_l$ and $A1_h$) emplaced during the current activity blanket the western flank. Note the relatively constant slope of the $A1_h$ lava field and the abrupt break in slope (indicated by the *arrow*) at the contact between the two lava fields

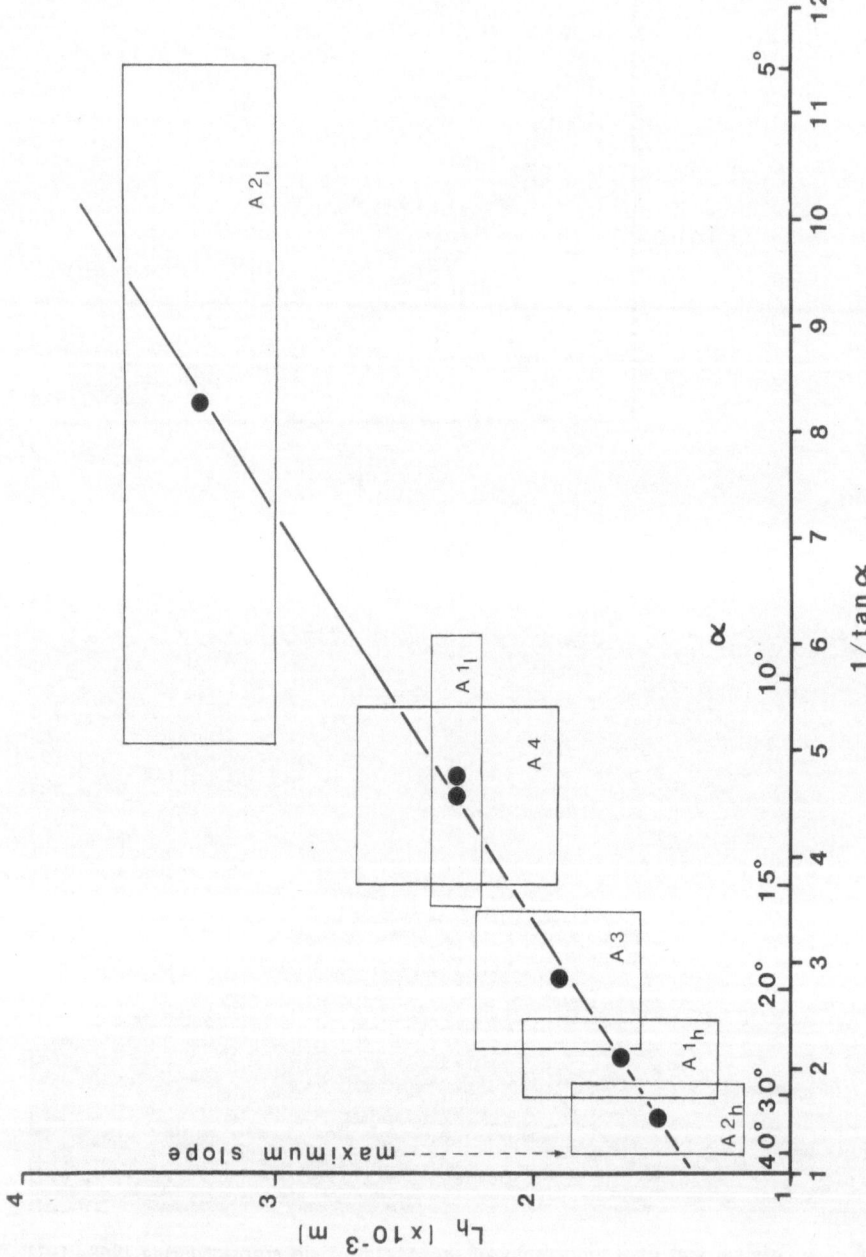

Fig. 6. Horizontal length of lava fields (L_h) versus inverse of slope. *Error boxes* are one sigma range. Length of lava fields is measured as the radial distance between crater and edge of lava field. Unusually long composite flows are excluded to compensate for shorter composite flows that do not reach the edge of the lava field and, thus, are not measurable. The regression line is $L_h = 261.4 \, (1/\tan\alpha) + 1113.8$, $r = 0.99$

The morphology of older (pre-1980) composite flows, including prehistoric lava flows, shows that they, too, formed by repeated emplacement of unit flows. An estimate for the current eruptive phase is 40–60 composite flows formed by 500–1000 unit flows.

2.4 The Lava Field

Composite flows that are erupted from the same crater during the same period of activity form a *lava field*. The morphology of a lava field is the result of the superposition of unit flow and composite flow features. A well-developed lava field tends to acquire a conoidal (concave upward) shape with the vertex at the effusion center. The topographic profile of a lava field tends to have a constant slope near the crater but is concave in the distal part. The constant slope on the upper flanks results from the last "layer" of composite flows erupted, whereas the concave part of the profile results from a general reduction of composite flow length with time. Hence, the older composite flows are exposed near the lower margin of the lava field under younger composite flows which have steeper slopes. Therefore, the isochrons for a lava field tend to become younger from the distal areas upward toward the crater.

At Arenal, six lava fields may be identified (Borgia et al. 1988). Each lava field formed during an eruption similar in character to the present one, and each has a volume that is a significant contribution to the total volume of the volcano ($\approx 10-15\%$). When the inverse of the average slope is plotted against average horizontal length of these lava fields, a linear correlation is found (Fig. 6). That is, lava fields (and the corresponding composite flows) are shorter on steeper slopes. This notion seems counterintuitive, yet has been verified by direct observation of the growth of the two current lava fields (cf. lava fields $A1_l$ with $A1_h$ in Fig. 6; also see Wadge 1983).

2.5 The Lava Armor

The summation of lava fields of a volcano constitutes the *lava armor*. Because each lava field represents a major constructive episode, the shape of Arenal is controlled by the shape of the lava fields. This fact is even more important due to the strong rains that erode tephras and breccias from the steep upper flanks of the volcano. The lava armor does not possess a perfect symmetry. The various lava fields have different average slopes, and their intersections tend to be parabolic and characterized by prominent breaks in slope (Fig. 1). This suggests that the lava fields formed from craters which were off-center relative to each other. Since lava fields are shorter on steeper slopes, a cross-section of the volcano will show lava fields erupted from craters at lower elevations exposed at the base of the cone and lava fields erupted at higher elevations exposed on the steep slopes near the summit (Fig. 7a; also cf. Borgia et al. 1988).

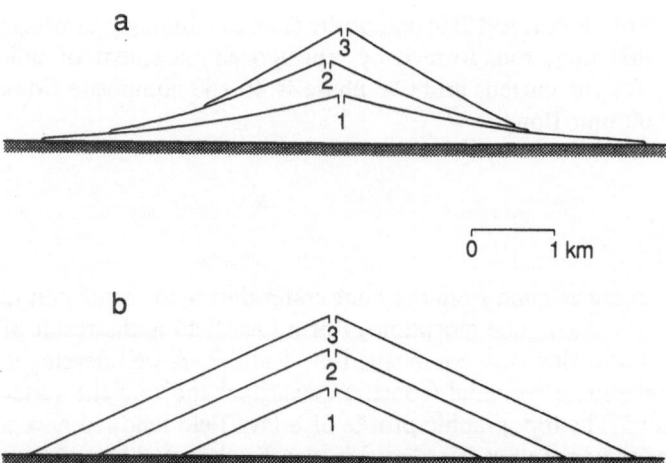

Fig. 7a, b. Schematic cross-sections of Arenal composite cone indicating the geometrical relationships between lava fields. *a* Model for Arenal. Note that shorter flows occur on steeper slopes, and that older flows crop out in the distal base of the cone. *b* Typical model envisioned for composite cones. Note that older lava constitutes the nucleus and is not exposed

This view is different than the typical cross-section envisioned for composite volcanoes (Fig. 7b).

In conclusion, the shape of Arenal Volcano may be approximated by the shape of the lava armor, which we may describe qualitatively with the summation equation:

$$\text{volcano} \approx \text{lava armor} = \sum_{i=1}^{I} \sum_{j=1}^{J_i} \sum_{k=1}^{K_{ij}} (\text{unit flow})_{ijk} , \qquad (1)$$

where the indices i, j, and k refer to lava field, composite flow, and unit flow, respectively, and I, J_i and K_{ij} are the total numbers of lava fields in the volcano, composite flows in (lava field)$_i$, and unit flows in (composite flow)$_{ij}$, respectively.

3 Analysis of Flow Dynamics and Volcano Growth

In the following sections we approach the description of flow dynamics and volcano growth from a strictly macroscopic point of view. We apply physically constrained first-order approximations to the detailed observations of the Arenal Volcano system to produce a simple model based on volume conservation. This model, following Eq. (1), integrates the dynamic behavior of a unit flow to describe the evolution of a composite flow; it then integrates over the composite flow in order to deduce the morphology of a lava field; and finally it integrates over the lava field to define the shape and structure of Arenal vol-

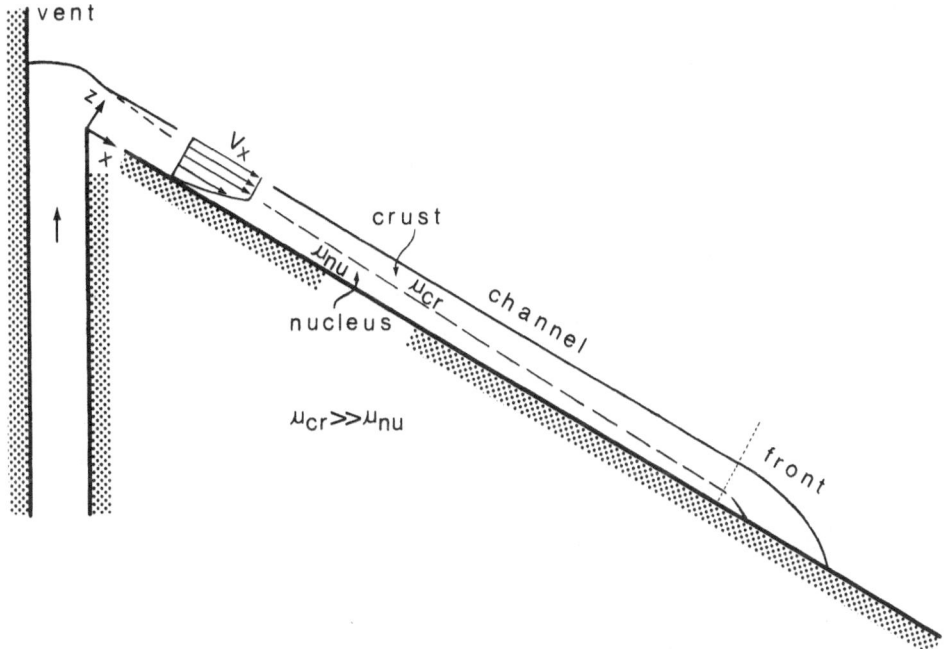

Fig. 8. Sketch of the unit flow model. See text for explanation

cano. Many of the real dynamic complexities of this volcanic system can thus be taken into account.

We assume a cartesian reference system, centered at the active crater, with the x-axis parallel to the lava flow, the y-axis horizontal and perpendicular to the flow, and the z-axis upward (Fig. 8).

3.1 Unit Flow Dynamics

3.1.1 Mass Balance

The motion of a unit flow of constant density satisfies conservation of volume of lava (i.e., conservation of mass with constant density) according to the expression (see Table 1 for list of symbols): flux of lava out of the vent (Q_v) equals flux of lava in the channel proper (Q_c) equals flux of lava at the front (Q_f):

$$Q_v = Q_c = Q_f . \tag{2}$$

Time changes in effusion rate at the vent can generate changes in flow depth and flux that will travel toward the front (Baloga 1987). In such a case, Eq. (2) would require the incorporation of time-dependent effects. Since we have never observed such surges at Arenal during the development of unit flows, we consider reasonable the steady state assumption given by Eq. (2).

Table 1. List of symbols

		Dimensions
A_c	Cross-sectional area of channel proper	m^2
A_f	Cross-sectional area of flow front	m^2
A_{fs}	A_f at separation ($t = t_s$)	m^2
A_l	Cross-sectional area of levees	m^2
A_v	Cross-sectional area of channel at vent	m^2
a	$(\varrho g \sin \alpha A_{fs})/(C_2 q f)$ Eq. (17 b)	s^{-1}
b	Intercept on L_h versus $1/\tan \alpha$ plot (Fig. 6)	m
C_1	Percentage of weight of lava in channel transmitted to flow front	
C_2	Proportionality constant between v_f and σ_e	$Pa\,s\,m^{-1}$
f	A_c/A_l	
g	Acceleration of gravity	$m\,s^{-2}$
h_c	Height of flowing lava in channel	m
h_{cr}	Thickness of the crust	m
h_f	Height of flow front	m
I	Total number of lava fields in the volcano	
i	Index for lava field (i = 1 to I); index for the arbitrary variable ζ	
J_i	Total number of composite flows in (lava field)$_i$	
j	Index for composite flow (j = 1 to J_i)	
K_{ij}	Total number of lava flows in (composite flow)$_{ij}$	
k	Index for lava flow (k = 1 to K_{ij})	
L	Maximum length of channel full of lava	m
L_c	Length of channel filled with lava	m
L_{com}	Distance a composite flow travels beyond maximum unit flow length (L_{max})	m
L_h	Horizontal component of composite flow length (L_{tot})	m
L_{max}	Maximum distance a unit flow travels ($= L+l$)	m
L_{tot}	Maximum distance a composite flow travels ($= L_{max}+L_{com}$)	m
l	Distance a unit flow travels during the collapsing phase ($= q f L$)	m
m	Slope on L_h versus $1/\tan \alpha$ plot (Fig. 6)	m
Q_c	Flux of lava in channel	$m^3\,s^{-1}$
Q_{cf}	Flux of lava from channel proper into frontal zone	$m^3\,s^{-1}$
Q_f	Flux of lava at flow front	$m^3\,s^{-1}$
Q_{fl}	Flux of lava from frontal zone to levees	$m^3\,s^{-1}$
Q_{fs}	Q_f at separation ($t = t_s$)	$m^3\,s^{-1}$
Q_V	Flux of lava in channel at the vent	$m^3\,s^{-1}$
q	Fraction of lava able to drain out of channel during collapsing stage	
R	Horizontal radial coordinate for lava armor	m
r	Correlation coefficient	
T_{su}	Temperature of surface of flow	K
T_{nu}	Temperature of lava flow nucleus	K
t	Time coordinate	s
t_0	Time when a unit flow begins to form	s
t_s	Time at which a unit flow separates from its source; duration of developing phase; period of formation of unit flows	s
t_∞	Time when a unit flow stops	s
u_{ζ_i}	Unit step function	
v_c	Average velocity of lava in channel	$m\,s^{-1}$
v_f	Velocity of front	$m\,s^{-1}$
v_{fs}	Velocity of flow front at separation	$m\,s^{-1}$
v_s	Average velocity of lava in the channel at flow surface	$m\,s^{-1}$
v_v	Average velocity of lava in the channel at vent	$m\,s^{-1}$

Table 1 (continued)

		Dimensions
v_x	Velocity component in the x-direction	$m\,s^{-1}$
w_c	Width of channel proper	m
w_f	Width of flow front	m
x	Position of flow front at $t = t_x$; x-axis, parallel to flow direction	m
y	Y-axis horizontal and perpendicular to flow direction	m
Z	Vertical coordinate for lava armor	m
z	z-Axis upward and perpendicular to flow direction	m
α	Topographic slope	degrees
Δh	Thickness of overhead of lava at vent	m
Δh_0	Maximum value of Δh, Δh at $t = t_0$	m
Φ	Functional dependence of σ_c on L_c	Pa
η	Dummy variable	s
μ_{cr}	Average viscosity of the crust	Pa s
μ_{su}	Viscosity of surface of flow	Pa s
μ_{nu}	Viscosity of lava flow nucleus	Pa s
ϱ	Mass density of lava	$kg\,m^{-3}$
ϱ_{su}	Density of flow surface	$kg\,m^{-3}$
ϱ_{nu}	Density of flow nucleus	$kg\,m^{-3}$
σ_c	Stress applied to the flow front due to weight of crust in the channel proper	Pa
σ_e	Entrance pressure ($= \sigma_v + \sigma_c$)	Pa
σ_v	Stress applied to the flow front due to overhead of lava at the vent	Pa
τ_0	Yield strength	Pa
ξ	Arbitary variable	
ξ_i	Value of the arbitary variable at i	

The flux of lava at the front is given by

$$Q_f = A_f v_f = (A_l + A_c) v_f = A_c v_c = A_v v_v , \qquad (3)$$

where A_f, A_l, A_c, and A_v are the cross-sectional areas of the front, the levees, the channel proper, and the channel at the vent, respectively, and $A_f = A_l + A_c$ (Fig. 3a). v_f is the velocity of the front, v_c is the average velocity of the lava in the channel, and v_v is the average velocity in the channel at the vent. Equations (2) and (3) may be used to record flux of lava during eruption of unit flows by monitoring the cross-sectional area and the velocity of the flow front.

A volume balance on the frontal zone (Borgia et al. 1983) gives

$$Q_{cf} = Q_{fl} = v_f A_l = (v_c - v_f) A_c , \qquad (4)$$

where Q_{cf} and Q_{fl} are the fluxes from the channel proper to the frontal zone and from the frontal zone to the levees, respectively. The term Q_{fl} also includes the debris which is formed at the front and which may not become part of the levees in a strict sense (for example, the bottom debris). Rearranging Eq. (4)

$$\frac{v_f}{v_c - v_f} = \frac{A_c}{A_l} = f , \qquad (5)$$

where f is a dimensionless parameter, a function of rheology, topography, and time. The parameter f is a measure of how much lava the front uses to advance relative to how much it uses to build up levees. This ratio has a primary role in determining the final dimensions of a flow. The velocity of the front is directly measurable, while the average velocity of the channel is given by

$$v_c = \frac{\int\limits_{-w_c/2}^{+w_c/2} \int\limits_0^{h_c} v_x \, dy \, dz}{\int\limits_{-w_c/2}^{+w_c/2} \int\limits_0^{h_c} dy \, dz} , \tag{6}$$

where w_c and h_c are the width and height of the flowing lava in the channel proper, respectively, and v_x is the x component of the velocity. v_c is difficult to estimate because the distribution of v_x in the y−z plane is generally unknown. If the surface velocity (v_s) is measurable, then, assuming plug flow in the upper cold crust of thickness h_{cr}, a parabolic velocity distribution in the nucleus, and neglecting lateral effects, we may estimate v_c by

$$v_c \approx (3 \, w_c)^{-1} \int\limits_{-w_c/2}^{+w_c/2} (2 + h_{cr}/h_c) v_s \, dy . \tag{7a}$$

Equation (7a) partially takes into account the influence of the shape of the channel by leting h_{cr} and h_c be functions of y. In the case where a Bingham rheology is a reasonable assumption and an estimate of the yield strength (τ_0) exists, then

$$h_{cr} = \frac{\tau_0}{\varrho \, g \sin \alpha} , \tag{7b}$$

where ϱ, g, and α are the density of the lava, the acceleration of gravity, and the angle of slope, respectively.

A_c and A_l may be estimated by measuring the topographic profile perpendicular to the direction of flow before, during, and after the flow has passed, then evaluating which portion of the flow was stationary (A_l) and which portion was not (A_c). Thus, by measuring velocities or areas, f may be computed from Eq. (5) on different topography and at different times after effusion. This data set may allow the prediction of f for future flows.

3.1.2 Role of Normal Stresses

Borgia et al. [1983, Eqs. (9) and (10)] postulate that the velocity of the front is proportional to the entrance pressure (σ_e)

$$v_f \propto \sigma_e . \tag{8}$$

The entrance pressure is the pressure that the lava flowing in the channel proper exerts on the flow front as it enters the frontal zone, i.e., is the pressure acting in the direction of the front at the channel proper − frontal zone boundary due to the presence of lava in the channel proper. Borgia et al. (1983) suggest that the frontal zone actually "feels" the amount of lava present in the channel proper. This hypothesis is not justified for isothermal, steady-state, low Reynolds numbers, and quasi-unidirectional flow. In this case, the pressure within the flow is approximately hydrostatic; thus, the stress distribution is everywhere local and no stresses may be transmitted along the flow. However, if the flow has a thermal structure the situation is more complex.

Assume the hypothetical case shown in Fig. 8. A lava sheet is flowing down a constant slope and has a marked thermal structure such that it may be described as a two-layered flow: a cooler, more viscous "crust" on top of a hotter, less viscous "nucleus". By virtue of its higher position in the flow, the crust experiences smaller shear stresses than the nucleus. Therefore, velocity gradients are smaller and velocities larger in the crust relative to the nucleus. This fact is further enhanced by the higher viscosity of the crust. Hence, the crust flows into the flow front at a faster rate than the nucleus (Baloga 1987). As a consequence, the front is mainly composed of high viscosity lava, and offers a greater resistance to flow than the lava in the channel behind. As the flow front advances, it maintains constant volume by continuously "discarding" to the side in a "fountain flow" geometry (Rose 1961) a volume of lava equal to that received from the channel proper. This discarded lava is left behind to form new levees.

In the limiting case of a very high viscosity contrast between crust and nucleus, the flow may be approximated by a solid upper layer "gliding" on an inviscid lower layer. For a nonaccelerating flow front, the rigid crust will apply a force to the front equal to the component of its weight in the flow direction. In the real case, however, the decrease in viscosity with depth is a continuous function and lateral boundaries are present, thus only a very small fraction of the weight of the crust may be transmitted to the flow front.

We suggest that the development of the crust and the related top debris drastically influences the detailed dynamics of a lava flow. The crust is not passively carried downhill but actively modifies the stress distribution in the flow, thereby controlling the rates of processes such as flow front velocity, levee formation, growth of surges, flow avalanching, surface folding, and formation of pressure ridges. We believe that the proportionality between front velocity and entrance pressure is physically founded and essential in understanding lava flow behavior.

The entrance pressure consists of two parts:

$$\sigma_e = \sigma_v + \sigma_c \ . \tag{9}$$

The first part, σ_v, is the contribution to σ_e due to the overhead of lava at the vent. In the hydrostatic approximation

$$\sigma_v = \varrho g \Delta h \tag{10}$$

Δh is the thickness of the overhead of lava. The second part of the entrance pressure, σ_c, is the contribution due to the weight of the crust in the channel proper. Borgia et al. (1983) propose for this term a relation of the form

$$\sigma_c = \Phi(L_c) \approx C_1 \varrho g \sin \alpha L_c , \tag{11}$$

where L_c is the length of channel filled with lava, and C_1 is a nondimensional constant describing what fraction of weight in the channel is actually transmitted to the front. Borgia et al. (1983) demonstrate with front velocity data that this approximation is valid during the final stage of the collapsing phase. It is not clear whether Eq. (11) retains its validity throughout the life of a flow. Perhaps more elaborate functions of L_c are appropriate. For example, a dependence on L_c^2 may be a better approximation for a crust which thickens downstream at a constant rate. Or, if convection is negligible, an error function solution to the heat transfer problem for the formation of the crust would suggest a dependence on $L_c^{3/2}$. However, since we do not have enough data to constrain the growth of the crust, we assume the linear dependence between σ_c and L_c given by Eq. (11) as a first-order approximation.

Equations (9), (10), and (11) show that when a flow begins to form ($t = t_0$), there is no channel so that the entrance pressure is given only by the term σ_v (Fig. 9a). As the channel length increases (developing phase, $t_0 \leq t \leq t_s$) both σ_v and σ_c contribute to σ_e. Once the flow separates from the crater (collapsing phase, $t_s \leq t \leq t_\infty$), then $\sigma_v = 0$ and $\sigma_e = \sigma_c$. Finally, when the channel is empty and the flow stops ($t = t_\infty$), $\sigma_e = \sigma_v = \sigma_c = 0$. Summarizing:

$$Q_v/A_v \propto v_f \propto \sigma_e = \sigma_v + \sigma_c ; \tag{12a}$$

$$\text{at } t = t_0 \qquad \sigma_c = 0, \ \sigma_e = \sigma_v ; \tag{12b}$$

$$\text{at } t_0 \leq t \leq t_s \qquad \sigma_e = \sigma_v + \sigma_c ; \tag{12c}$$

$$\text{at } t_s \leq t \leq t_\infty \qquad \sigma_v = 0, \ \sigma_\varepsilon = \sigma_c ; \tag{12d}$$

$$\text{at } t = t_\infty \qquad \sigma_e = \sigma_v = \sigma_c = 0 , \tag{12e}$$

where Eq. (12a) is obtained by substituting Eqs. (2), (3), and (8) into (9) for v_f.

Equation (12a) in the form ($Q_v/A_v \propto \sigma_v + \sigma_c$) may explain the recurrent formation of unit flows that we observed in 1980–1983. Since σ_c increases with time as L_c increases [cf. Eq. (11)], according to Eq. (12a) either (Q_v/A_v) increases correspondingly or σ_v decreases. Since Q_v cannot increase indefinitely, nor A_v decrease indefinitely, σ_v must decrease (Fig. 9a); that is, as the channel becomes longer and longer, less and less pressure from the overhead of lava at the crater is needed to drive the flow. Such a decrease of σ_v will produce a fall in the level of lava at the vent [a decrease of Δh in Eq. (10)] and might explain the concentric ridges of lava observed in the active vent. A critical channel length is reached when $\sigma_v = 0$. Since σ_v cannot be negative, A_v must now decrease. Thus, the flow thins out near the vent where the crust thickness and its effect are minimal and eventually separates from its source. This flow now enters the collapsing stage and a new lava flow begins to develop at the vent (Fig. 9b). Because v_c is greater than v_f (Cigolini et al. 1984), the collapsing zone rapidly moves away from the newly formed flow front.

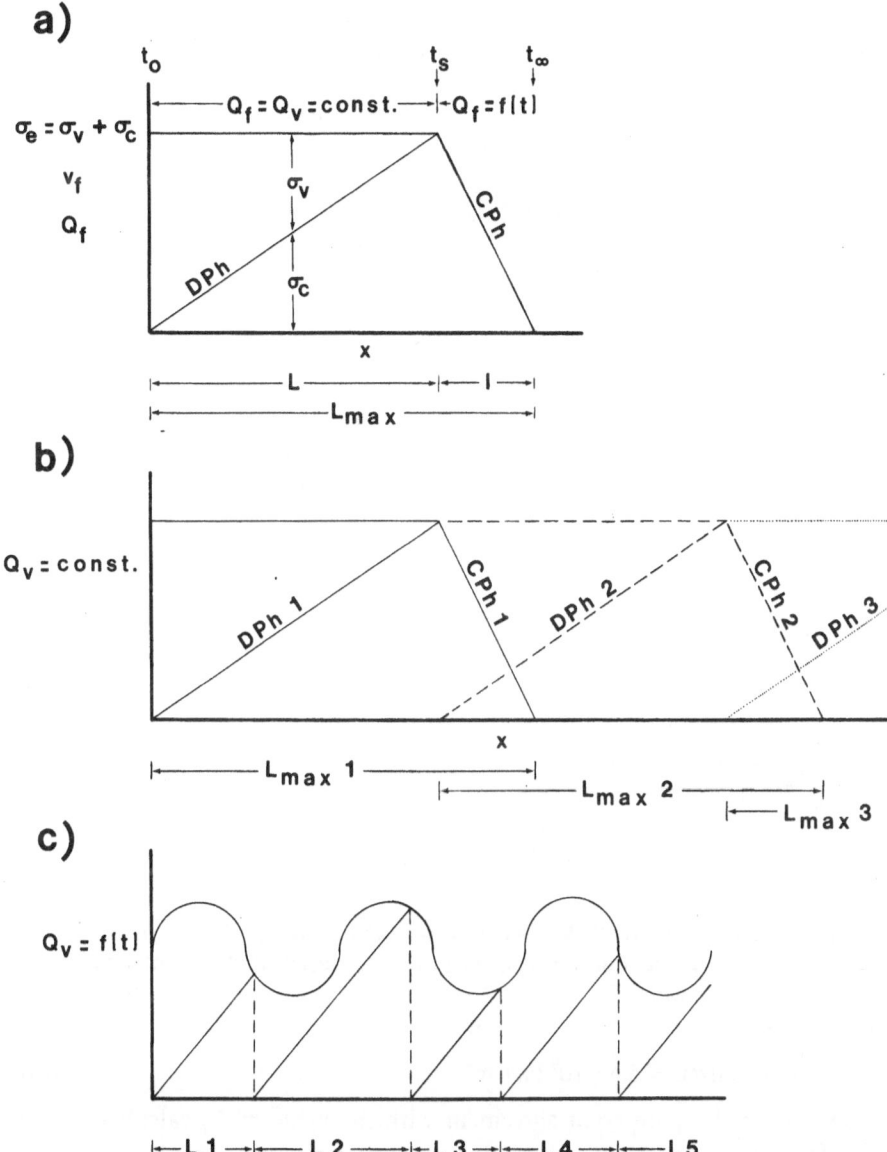

Fig. 9a–c. Qualitative plots of entrance pressure (σ_e) and velocity of the front (v_f) versus channel length (x) for constant Q_v (a and b) and variable (periodic) Q_v (c). Note that in this last case different flow lengths are obtained (L2 > L4 > L1 > L3). *Dashed lines* represent relative magnitudes of these parameters for a new flow formed after separation of the previous flow (*solid lines*). The relations between these quantities are summarized as follows:

$$v_f \propto \sigma_e = \sigma_v + \sigma_c \qquad \text{Eqs. (8) and (9)}$$
$$\sigma_v = \varrho g \Delta h \qquad \text{Eq. (10)}$$
$$\sigma_c = C_1 \varrho g \sin \alpha \, L_c \qquad \text{Eq. (11)}$$
$$Q_v/A_v \propto \sigma_v + \sigma_c \qquad \text{Eq. (12a)}$$

Solid, dashed, and *dotted lines,* as well as *increasing numbers,* refer to subsequent flows; *DPh* developing phase; *CPh* collapsing phase

The possibility that lava flows may lose contact with source conditions has been also suggested by Baloga and Pieri (1986) and Baloga (1987) based on theoretical considerations. We emphasize that the separation of unit flows from the vent is controlled by the weight of the crust onto the front. Thus, this separation may not occur on shallow slopes where the x-component of the gravity force (in the flow direction) is insignificant relative to the z-component.

If Eq. (11) is substituted into (12a) for σ_e with the conditions given by Eq. (12d) we find

$$v_f \propto C_1 \varrho g \sin\alpha L_c \tag{13a}$$

or

$$C_2 v_f = \varrho g \sin\alpha L_c , \tag{13b}$$

where C_2 is the proportionality constant and includes C_1. Solving Eq. (13b) for L_c at separation, we obtain the maximum attainable channel length full of lava ($L = L_c$ at $t = t_s$)

$$L = \frac{C_2 v_{fs}}{\varrho g \sin\alpha} , \tag{13c}$$

or from Eq. (3)

$$L = \frac{C_2 Q_{fs}}{\varrho g \sin\alpha A_{fs}} , \tag{13d}$$

where the subscript s indicates that v_f, Q_f, and A_f are the parameters at separation. From Eq. (13c) and Table 2 we calculate

$$C_2 = \frac{\varrho g \sin\alpha L}{v_{fs}} = \text{order of } 7 \times 10^9 \, \text{Pa s m}^{-1} . \tag{14a}$$

Equation (14a) may be written in terms of t_s if we observe that $t_s = (L/v_{fs})$. Thus, noting that t_s is well approximated by the average period at which unit flows were observed to flow down the same composite channel during January–June 1983 (Table 2), we calculate:

$$C_2 = \varrho g \sin\alpha t_s = 6.6 \times 10^9 \, \text{Pa s m}^{-1} . \tag{14b}$$

This value is in quite good agreement with the value of C_2 calculated using Eq. (14a).

In addition, for constant (Q_c/A_c), σ_v at t_0 must equal σ_c at t_s . Thus, combining Eq. (10) with Eq. (11) solving for C_1, and using the values in Table 2 for the flow of Feb.–May 1982 we find

$$C_1 = \frac{\Delta h_0}{\sin\alpha L} = \text{order of } 0.5\% , \tag{14c}$$

where Δh_0 is Δh at $t = t_0$.

The efficiency at which the crust may transmit normal stresses, due to its weight, down the channel to the front is less than 1%. Therefore, more than

Table 2. Data used in the calculations for the Feb. – Mar. 1982 unit flow

Parameter geometry	Value	Units	Obtained from	References[a]
$h_c \approx h_f$	6	m	Measurements	3
w_f	50	m	Measurements	3
w_c	29	m	Measurements	3
A_{fs}	471	m^2	$1/2\,\pi\,h_f\,w_f$	
A_c	174	m^2	$h_f \times w_c$	
α	20° to 30°		Measurements	1 and 3
L	400 to 800	m	Estimation	1
Δh_0	1 to 2	m	Estimation	1
L_{max}	1.0 to 1.2×10^3	m	Measurements	1
Flow parameters				
a	5.28×10^{-6}	s^{-1}	Calculations	3
f	0.75		Calculations	3
g	9.8	$m\,s^{-2}$	Standard	
Q_v	0.3	$m^3\,s^{-1}$	Calculations	4
$Q_{fs}/A_{fs} = v_{fs}$	0.6 to 1.4×10^{-3}	$m\,s^{-1}$	Estimated from measurement of v_f	1 and 5
q	0.54		Calculations	3
t_s	4.5 to 7.5×10^5	s	Average over 13 flows	1
ϱ	2.7×10^3	$kg\,m^{-3}$	Measurements	2
Linear regression of L_h versus $1/\tan \alpha$ (Fig. 6)				
m	261.4	m	Calculated from Fig. 6	1
b	1113.8	m	Calculated from Fig. 6	1
r	0.99		Calculated from Fig. 6	1
Relevant parameters for Arenal's lava flows				
T_{su}	920 to 1020	°C	Measurements	1 and 5
T_{nu}	1050 to 1100	°C	Estimation	5 and 6
Viscosity				
μ_{su}	1.4×10^{10}	Pa s	Calculations	5
μ_{nu}	1.6×10^7	Pa s	Calculations	5
Density				
ϱ_{su}	2.2×10^3	$kg\,m^{-3}$	Measurements	2 and 5
ϱ_{nu}	2.7×10^3	$kg\,m^{-3}$	Measurements	2 and 5
Yield strength				
τ_0	7.9×10^4	Pa	Calculations	5

[a] References: *1* this work; *2* Cigolini and Borgia 1980; *3* Borgia et al. 1983; *4* Wadge 1983; *5* Cigolini et al. 1984; *6* Reagan et al. 1987.

99% of the x-component of the weight of the crust is balanced by the viscous forces rising from shearing the crust and the nucleus.

Finally, dividing Eq. (11) by Eq. (13c) at separation [$t = t_s$, Eq. (12d)] and solving for σ_e with the values for C_1 and C_2 given by Eqs. (14c and a), respectively, we may write for the developing phase:

$$\sigma_e = C_1\,C_2\,v_f = 3 \times 10^4\,\text{Pa} \ . \tag{14d}$$

This is a maximum value for σ_e. After separation σ_e will decrease to zero as the lava from the channel is drained through the flow front to the levees.

3.1.3 Velocity and Position of the Flow Front

The velocity of the flow front is given by Eqs. (2) and (3) before the flow separates from the vent ($t_0 \leq t < t_s$), and is given by Eq. (13b) after the flow has separated ($t_s \leq t \leq t_\infty$). Thus,

$$v_f = \begin{cases} Q_f/A_f & \text{for} \quad t_0 \leq t < t_s \ ; \\ (1/C_2)\varrho g \sin\alpha \, L_c & \text{for} \quad t_s \leq t \leq t_\infty \ . \end{cases} \tag{15a}$$

Let us now define the unit step function u_{ζ_i} to be zero for every value of the arbitrary variable ζ smaller than ζ_i and one for every value of ζ equal to or larger than ζ_i. Then, we may write Eq. (15a) in the equivalent form

$$v_f = (u_{t_0} - u_{t_s})Q_f/A_f + u_{t_s}(1/C_2)\varrho g \sin\alpha \, L_c \ . \tag{15b}$$

Integrating Eq. (15b) we find

$$x = \int_{t_0}^{t_s} Q_f/A_f \, d\eta + \int_{t_s}^{t_\infty} (1/C_2)\varrho g \sin\alpha \, L_c \, d\eta \ , \tag{16}$$

where η is a dummy variable. Equation (15b) and (16) allow estimation of velocity and position of the flow front at any time after effusion has started. Borgia et al. (1983) evaluated the second integral in Eq. (16) assuming that only L_c is time-dependent. Field data show good agreement with their solution and indicate that variations in the slope around the average value produce only second-order changes in the predicted velocity and position of the front. With this approximation, Eq. (16) becomes

$$x = \int_{t_0}^{t_s} Q_f/A_f \, d\eta + u_{t_s} [Q_{fs}/(a A_{fs})](1 - e^{-at}) \ , \tag{17a}$$

where

$$a = \frac{\varrho g \sin\alpha}{C_2 q f} = \text{constant} \tag{17b}$$

and

$$l = q f L = Q_{fs}/(a A_{fs}) \ . \tag{17c}$$

This solution is obtained by using Eqs. (5), (6), (8), (12), and (13) of Borgia et al. (1983, p. 315–317) and our Eqs. (13d) and (16). In addition, l and q are, respectively, the distance the flow travels and the fraction of lava which is able to drain out of the channel during the collapsing stage. The value of the parameter a is obtained by least square interpolation of field measurements of flow front position versus time during the collapsing stage (Borgia et al. 1983). Equation (17b) allows the calculation of C_2 independently from Eq. (14a). Solving Eq. (17b) for C_2 and using the values given in Table 2 for the Febr.-

Mar. 1982 unit flow, we obtain $C_2 \approx 5.23 \times 10^9 \, kg \, m^{-2} \, s^{-1}$. This value is remarkably similar to the value of C_2 obtained before [Eq. (14a); $C_2 \approx 7 \times 10^9 \, kg \, m^{-2} \, s^{-1}$].

The qualitative meaning of Eq. (17) is shown in Fig. 9, where σ_e, Q_f, and v_f are plotted against the total channel length (x). In Fig. 9a and b, Q_f is assumed to be constant since changes in effusion rate seem to be negligible during the time of formation of a unit flow. Before the flow separates from the vent, the plot of Eq. (11) is a straight line passing through the origin. Because $\sigma_v = \sigma_e - \sigma_c$, σ_v eventually becomes zero once the line given by Eq. (11) intercepts $\sigma_e = \sigma_e(x)$. At this point the flow separates from the vent and its velocity will exponentially decrease to zero (Borgia et al. 1983). At the same time a new flow will start to grow, repeating the cycle. From Fig. 9c it is clear that flows of different lengths may form if the effusion rate varies, and that an increasing number of unit flows are generated during a period of decreasing effusion rate.

Equation (17a) may be further simplified if the rate at which Q and A vary is small, that is, if a quasi-steady state assumption becomes applicable. This assumption is certainly valid for Arenal after 1974 (Wadge 1983) and particularly during the period of our observations. Equation (17a) then becomes

$$x = \frac{Q_f}{a A_f} \{u_{t_0} - u_{t_s} a t + u_{t_s} (1 - e^{-at})\} \ . \tag{18}$$

This result may be used to approximate the position of the flow front at any time if the average values of a, A_f, and Q_f can be estimated from measurements on previous lava flows. The maximum distance a flow can reach is $L_{max} = L + l$. Therefore, using for L and l the expressions given by Eqs. (13) and (17c), respectively, we find

$$L_{max} = \frac{C_2 (1 + q f) Q_{fs}}{\varrho g \sin \alpha \, A_{fs}} = \left(1 + \frac{1}{q f}\right) \frac{Q_s}{a A_{fs}} \ . \tag{19}$$

Equation (19) allows the prediction of the final length of a unit flow if the average values of a, q, f, and v_{fs}/A_{fs} are known. Using the values given in Table 2, which were calculated for the unit flow of Feb.–Mar. 1982, we find $L_{max} \approx 650$ m. This value is smaller than the eventual length of that particular flow by a factor of less than two. However, considering the approximation of a linear dependence of σ_e on L_c, the fact that the parameters q and f are calculated from measurements of only one flow and that there is no assumption on lava rheology, we consider this to be a good correlation. In addition, note that in Eqs. (17), (18), and (19) the parameters q and f appear only as the product q f. Thus, we may use the same Eq. (19) to estimate the average value of this parametric product for a unit flow. Solving Eq. (19) for q f and using the values given in Table 2, we write

$$q f = \frac{v_{fs}}{a L_{max} - v_{fs}} \approx 0.21 \ . \tag{20}$$

From Eq. (19) we may obtain two other independent expressions for C_2:

$$C_2 = \frac{\varrho g \sin \alpha \, L_{max}}{(1+qf)v_{fs}} = 8.8 \times 10^9 \, \text{kg m}^{-2} \, \text{s}^{-1} \tag{21a}$$

and

$$C_2 = \frac{(a L_{max} - v_{fs})\varrho g \sin \alpha}{a v_{fs}} = 1.0 \times 10^{10} \, \text{kg m}^{-2} \, \text{s}^{-1} . \tag{21b}$$

These two values of C_2 are also very close to those calculated before. In addition, we observe that all the parameters in Eqs. (20) and (21) may be obtained by relatively simple field measurements of flow front position versus time. Thus, we suggest the use of Eqs. (20) and (21) to obtain average estimates of the parameters q, f, and C_2.

In Eq. (19) there is a direct proportionality between L_{max} and the parameters q and f. Thus, the flow length is larger if (1) a greater amount of lava is used to advance the flow relative to levee formation (large f), and (2) a greater amount of lava can be drained out of the channel during the collapsing stage (large q). For instance, flow in a narrow valley or lava tube will have larger f and q and will produce longer flows relative to open channel flow on a flat surface. A similar conclusion was reached by Pieri and Baloga (1986) based on insulation effects.

The direct proportionality between the maximum length of lava flows and the flow rate in Eq. (19) is a well-known relation (Walker 1972, 1973; Wadge 1978). Walker (1972) concluded that high effusion rates produce far-reaching lava flows while low effusion rates produce shorter composite flows made of many flow units.

Equation (19) also shows that the maximum unit flow length is inversely proportional to the cross-sectional area of the flow. Since A_f increases little along a lava flow and is approximately constant for lava flows erupted from the same crater, then L_{max} is directly proportional to the final volume of the unit flow. Malin (1980) notes that, for the Hawaiian volcanoes, this last relation offers a better correlation with flow length than does effusion rate.

Finally, Eq. (19) has another important corollary. It shows that, all else constant, the maximum length of lava flows is inversely proportional to the sine of the slope: as α increases, L_{max} decreases. This concept is not counterintuitive if we observe that on a steeper slope a greater fraction of the weight of the crust may be transmitted to the flow front. Thus, on a steeper slope a shorter channel is needed to trigger separation. The length of Icelandic basalt flows also seems to agree with this relationship. Very long flows were found to form on very gentle slopes, while flows barely reached the bottom of very steep slopes (Gregg 1956). It will be shown below that an equivalent relation for composite flows, and thus lava fields, has fundamental consequences for the morphological evolution of the volcano.

It seems that Eq. (19) offers an acceptable qualitative understanding of the relationships between some of the most common parameters used to describe lava flows. However, we believe that in different volcanoes different parameters may be predominant. Our model describes only the emplacement of unit flows which are volume limited (i.e., their rheology allows them to attain "source free" regimes) and are not thermally limited as are the flows discussed by Pieri and Baloga (1986) or Guest et al. (1987). Thus, we emphasize that our results may not, a priori, be applicable to other volcanoes.

3.2 Composite Flow Dynamics

Unit flows flowing down the same composite channel feed the composite frontal zone, which travels further downhill than L_{max}. Thus, the total length of a composite flow, L_{tot}, is given by

$$L_{tot} = L_{max} + L_{com} , \tag{22}$$

where L_{com} is the distance the composite flow travels beyond the maximum length of unit flows.

Each unit flow front mantles the composite channel with a new layer of quenched lava to build up the composite levees. The amount of lava used by the front to build composite levees relative to the amount used to advance the front is described by the parameters f and q discussed in the previous section. To a first-order approximation, Q_v, f, and q are constant, such that there will be a maximum length of composite flows beyond which unit flows cannot feed the composite frontal zone efficiently. In other words, all the lava available to a unit flow is used to accrete a new lava layer to the levees in such a way that once the unit flow reaches the composite front, no more lava is left to feed the frontal zone. In addition, the approximately regular shifts in the direction of effusion at the vent, as described above, tend to make the total volume of lava available to a composite flow constant. Thus, on the average, L_{com} is expected to be constant.

Using the approximation that a composite flow tends to acquire a constant surface slope over time, and noticing that $L_h = L_{tot} \cos \alpha$ where L_h is the horizontal component of the total length, then

$$L_h = \frac{C_2(1+qf)Q_{fs}}{\varrho g A_{fs}} (\tan \alpha)^{-1} + L_{com} \cos \alpha = m(\tan \alpha)^{-1} + b , \tag{23a}$$

where

$$m = \frac{C_2(1+qf)Q_{fs}}{\varrho g A_{fs}} \tag{23b}$$

and

$$b = L_{com} \cos \alpha . \tag{23c}$$

This linear equation relates all the various parameters which enter into the definition of the horizontal length of a composite flow. Note that except for L_{com}, all other parameters refer to unit lava flows. Substituting the values for

Table 3. Calculated parameters

Parameter	Value	Units	From	Calculated for
C_1	0.5%		Eq. (14c)	Developing phase
C_2	7×10^9	$kg\,m^{-2}\,s^{-1}$	Eq. (14a)	Developing phase
C_2	6.6×10^9	$kg\,m^{-2}\,s^{-1}$	Eq. (14b)	Developing phase
C_2	5.2×10^9	$kg\,m^{-2}\,s^{-1}$	Eq. (17b)	Collapsing phase
C_2	8.8×10^9	$kg\,m^{-2}\,s^{-1}$	Eq. (21a)	Whole unit flow
C_2	1.0×10^{10}	$kg\,m^{-2}\,s^{-1}$	Eq. (21b)	Whole unit flow
C_2	4.9×10^9	$kg\,m^{-2}\,s^{-1}$	Eq. (23b)	Whole volcano
C_2	$7 \pm 2 \times 10^9$	$kg\,m^{-2}\,s^{-1}$	Table 3	Average
qf	0.41		Table 2	Collapsing phase
qf	0.21		Eq.(20)	Whole unit flow
m	533	m	Eq. (23b)	Collapsing phase
m	255	m	Eqs. (23b, 20)	Whole unit flow
σ_e	4×10^4	Pa	Eq. (14d)	Developing phase

the parameters given in Table 2, we find m = 533 m if we use the values for q, f, and C_2 calculated for the collapsing phase of a unit flow [Eq. (17)]. On the other hand, if we use for q, f and C_2 the more representative values derived from the maximum length of a unit flow [Eqs. (20) and (21)], then m = 255 m. This second value compares very well with the value of 261.4 m calculated by linear regression on the L_h versus $1/\tan\alpha$ plot (Fig. 7). The similarity between these values is remarkable since in Eq. (23b) we have extrapolated data for only one unit flow to the whole history of the volcano. From the same linear regression it is also found that b = 1113.8 m. Thus, composite flows travel on the average about 10^3 m beyond the distance reached by unit flows.

The linearity of this regression (r = 0.99) suggests that the ratio Q_{fs}/A_{fs} in Eq. (23) is fairly constant for Arenal. This is consistent with the observation that during the early part of the current eruption effusion rate and flow dimensions from the lower crater (A in Fig. 1) were both about an order of magnitude larger than those from the higher crater (C in Fig. 1; cf. Cigolini et al. 1984). Finally, by using Eq. (23b) we may calculate a sixth estimate of C_2 averaged over the whole volcano. In this last case $C_2 = 4.9 \times 10^9\ kg\,m^{-2}\,s^{-1}$. The consistency found in the estimated values of C_2 (cf. Table 3) over a large range of volumes (from unit flow to volcano) is remarkable and suggests that C_2 is indeed a relevant, physically meaningful parameter that characterizes the dynamic evolution of andesitic lava flows.

3.3 Growth of the Lava Field

A lava field is composed of the superposition of composite flows that are erupted from the same crater. Lava fields are symmetric relative to a vertical axis centered at the crater. Thus, to describe lava fields we use cylindrical coordinates centered at the crater, with Z and R in the vertical and radial directions, respectively. Fundamental parameters of a lava field are its vertical growth rate and its horizontal extent. The vertical growth rate is a function of the emplacement of composite flows. Since composite flows radiate from the crater, the lava field will have a radial symmetry. In addition, the radial symmetry requires that the superposition of flows of a finite width and thickness will result in a higher vertical growth rate next to the crater relative to the distal part of the lava field. In other words, the number of lava flows which pass per unit area is much higher near the vent than in the distal part of the field. This process seems to overwhelm the increase in individual flow thickness away from the vent (Baloga and Pieri 1986; Dragoni 1987). Thus, with time the elevation of the crater will increase and so will the average slope of the lava field. For instance, from 1974–1984, the elevation of crater C increased by about 100–200 m and the slope from about 30° to 40°. This gives an average vertical growth rate at the crater of $10-20 \, \text{m} \, \text{yr}^{-1} = 3-6 \times 10^{-7} \, \text{m} \, \text{s}^{-1}$. The increase in slope produces shorter composite flows, and as a consequence, the radial extent of a lava field is determined by the first layer of composite flows emplaced.

The topography of a composite flow may be described by its average slope and its average length. These parameters are related by Eq. (23a), which may be solved for $\tan \alpha$ to give

$$\tan \alpha = \frac{m}{L_h - b} .$$

(24)

Since the topographic profile of a lava field is the integral of the topography of composite flows, we then integrate Eq. (24) to get

$$Z = \ln (R-b) \, m + C_3 ,$$

(25)

where C_3 is the integration constant. Equation (25) gives the concave part of the topographic profile. The upper, linear profile is given by lines tangent to Eq. (25). These lines will have a slope given by Eq. (24), and an intercept equal to

$$\ln (L_{min} - b)^m + \frac{m \, L_{min}}{L_{min} - b} + C_3 ,$$

(26)

where L_{min} is the average length of the last erupted layer of composite flows.

The topographic profile of a flow field may then be approximated by

$$Z = (u_0 - u_{L_h}) \left[\frac{m L_h}{L_h - h} \left(1 + \frac{R}{L_h} \right) + \ln (L_h - b)^m + C_3 \right]$$

$$+ u_{L_h} [\ln (R-b)^m + C_3] ,$$

(27a)

where u_0 and u_{L_h} are the unit step functions for $R = 0$ and $R = L_h$, respectively. Equation (25) is undefined for $R \leq b$. However, this singularity does not affect our solution, because the slope of the volcano is observed not to be above 38°–40°. Above this angle, tensional stresses induced by gravity over-

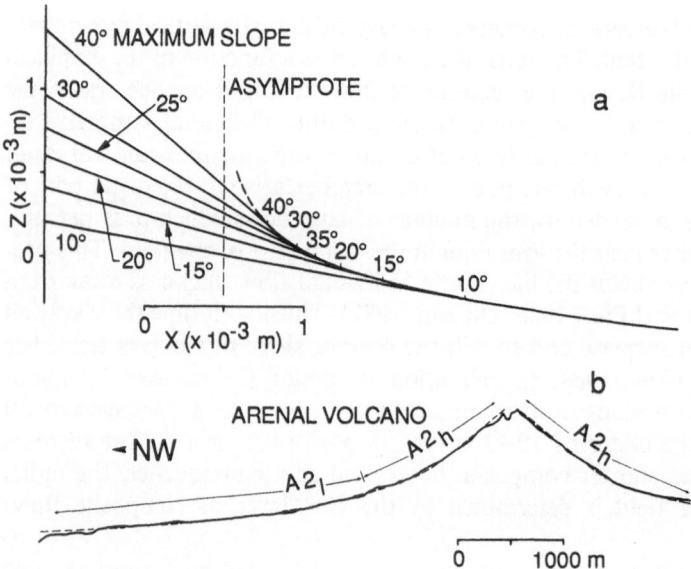

Fig. 10. a Curve generated by Eq. (27). *b* Graphical fit of Arenal's topography (*solid line*) to Eq. (28) (*dashed line*)

come the tensile strength of the lava, which breaks into boulders forming a talus slope (Borgia et al. 1983). This last condition can be specified by

$$\arctan\left(\frac{m}{L_h - b}\right) \leq 38°{-}40° \ . \tag{27b}$$

Figure 10a shows the curve generated by Eq. (27).

3.4 Growth of the Lava Armor

The lava armor consists of the whole of the lava fields. Therefore, the lava armor describes the shape of a volcano. As described above, the lava fields which constitute the lava armor of Arenal are not coaxial (i.e., each new crater is offset relative to the former). For this reason, the topography of the lava armor is not a perfectly symmetrical conoid. The intersection of the various lava fields are approximately parabolic in shape and exhibit clear discontinuities in the slope. Figure 1 shows the observed parabolic contacts between the lava fields. The topographic profile of a volcano then is the sum of a series of Eq. (26), each equation referring to one of the lava fields,

$$Z = \sum_{i=1}^{I} (u_0 - u_{L_{hi}})\left[\frac{m_i L_{h_i}}{L_{h_i} - h_i}\left(1 + \frac{R}{L_{h_i}}\right) + \ln(L_{h_i} - b_i)^{m_i} + C_{3_i}\right]$$
$$+ u_{L_{hi}}[\ln(R - b_i)^{m_i} + C_{3_i}] \tag{28a}$$

SW

NE

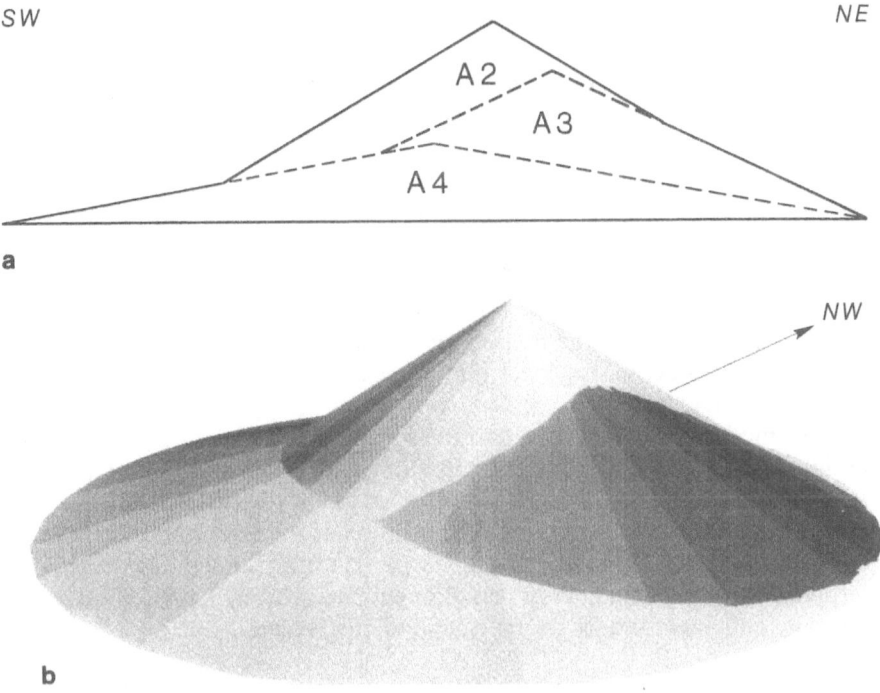

NW

a

b

Fig. 11 a, b. Computer-generated idealized model based on Eq. (28) for the construction of Arenal as superposed noncoaxial cones. The upper, intermediate, and lower cones model the *A2*, *A3*, and *A4* lava fields of Fig. 1. *a* Cross-section, *b* perspective view; notice the correspondence between the contacts of the lava fields in this sketch and the parabolic contacts of the eastern flank of Arenal in Fig. 1; also note that the A4 lava field (Fig. 1) does not extend to the southeast as far as the lower cone in this sketch, because the A4 lava field was "dammed" by the presence of Chato Volcano. Some of the discrepancies between the computer image and the actual geology of the volcano are a consequence that this computer model uses cones instead of conoids with a logarithmic lower profile

and

$$\arctan \left(\frac{m}{L_{h_i} - b_i} \right) \leq 38° - 40° \ . \tag{28b}$$

Figure 10b shows a graphical fitting of Eq. (28) to the NW-SE topographic profile of Arenal Volcano. The idealized model for Arenal's structure as superposed noncoaxial conoids is shown in Fig. 11.

The approximation of the topographic profile of a volcano given by Eq. (28) differs from that of Milne (1878, 1879) and Shteynberg and Solov'yev (1976). By summing the contributions of individual lava fields, Eq. (28) takes into account the fact that craters, and related lava fields, are not always coaxial. In addition, we assert that the structure and topography of Arenal are not determined by gravitiy-induced deformation of the cone, for which there is no geologic evidence, but by the accumulation of lava flows. Thus, the growth of the cone at Arenal is not geometrically uniform. This is at odds with the conclusion of a general morphometric study by Wood (1978) in which geometrically uniform cone growth is proposed to be a generalized relation for the growth of volcanoes. Walker (1973) suggested that the overall shape of volcanoes may be controlled by average effusion rate: flows are short and flanks steep for low effusion rates, while flows are long and flanks shieldlike for high effusion rates. Our analysis shows, however, that the angle of slope on which the lava flows are erupted is more important than effusion rate, at least for Arenal Volcano. Additional support for this model comes from experiments of structures built by an accumulation of wax flows. Shuver (1987, personal communication) has shown that the topographic profile of the resultant wax "volcano" is determined by the terminal location of individual flow fronts, and that flow length decreasing logarithmically with time results in a stratovolcanolike morphology.

Our model provides insight into the specific structure of the volcano. For example, from the asymmetric distribution and different slope of the lava fields we may estimate the position and elevation of the corresponding craters (Fig. 11). On this basis we propose the position of the three prehistoric craters E, F, and G in Fig. 1. Finally, we emphasize that only through the detailed interpretation of the dynamics of lava flow emplacement we have derived a geologically consistent model for the shape of the volcano.

4 Summary of Conclusions

Arenal Volcano is constructed by a hierarchical series of geologic units, i.e., unit flow, composite flow, lava field, and lava armor. The foundation of the building process are volume-limited unit flows which are emplaced at short time intervals to make up composite flows. Composite flows form lava fields, and lava fields in turn, constitute the lava armor. Tephra and lava breccias are selectively eroded from the steep slopes of the volcano by heavy rains, thus they contribute little to the actual shape of the cone.

This constructive process has primary consequences for the distribution of the age of lava on a composite cone. The lava of a unit flow increases from the crater to the flow front in the upper central portion of a flow, and from the flow front to the crater in the levees and at the flow bottom (Fig. 3). In a composite flow, the age of lava increases in a way similar to a unit flow with the additional complication that in the levees the lava becomes younger from the center of the levee outward (Fig. 4). The general trend for the age of the lava in a lava field is to be young toward the crater. Finally, for the lava armor, the age relation depends on the relative position of craters and related lava fields (Fig. 1). Therefore, lava of significantly different ages may be juxtaposed at all scales from the unit flow, to the composite flow, to the lava field, and to the lava armor. The time-sequential sampling of a composite volcano can thus be extremely complicated and requires careful attention because changes in composition may occur during the growth of a composite flow, a lava field, or the lava armor.

Detailed observations of the dynamic behavior of unit flows indicate that two dimensionless parameters determine the distribution of lava between an active component, that is flowing, and a passive stationary component (Borgia et al. 1983; Baloga and Crisp 1987). The first parameter, f, is a measure of how much lava the front uses to advance relative to how much lava the front uses to build up levees. The second parameter, q, is the fraction of lava that is able to drain out of the channel during the collapsing stage. Both parameters have a primary role in determining the final dimensions of a lava flow. These parameters may be calculated from observations of lava flowing onto different topography and at different times after effusion. This data set may allow the prediction of f and q, and as a consequence, the final flow length along possible future flow paths is also predictable.

The development of a thermal structure within the flow plays a critical role in the dynamic evolution of a unit flow. The cold, highly viscous crust at the surface of the flow is not just passively carried downhill but actively modifies the stress distribution in the flow. The hydrostatic approximation, which would suggest the stress distribution to be local, is no longer valid. The presence of boundaries, particularly the flow front, is felt all along the flow. Thus, the crust controls the rate of processes such as front velocity, levee formation, growth of surges, flow avalanching, surface folding, and formation of pressure ridges.

We hypothesize that the velocity of the front is proportional to the entrance pressure, σ_e, the pressure acting in the direction of the flow at the channel proper-frontal zone boundary. The entrance pressure has two components. The first, σ_v, is the pressure due to the overhead of lava at the vent; the second, σ_c, is generated by the weight of the crust in the channel onto the flow front. For a given volume flow rate of lava there is a critical channel length beyond which σ_c accelerates the flow and triggers the separation of the flow from its source near the vent. Thus, the unit flows are volume-limited. Based on this hypothesis we derive a relation for the velocity and position of the flow front at any time after effusion has started, assuming the time functions of f, q, and flow rate are known. We find that the length of a unit flow is directly proportional to f, q, and the flow rate and it is inversely proportional to the cross-sectional area of the channel and to the sine of the slope. These relations also hold for composite flows. Finally, by making the approximation that a composite flow grows to a constant slope we derive equations for the evolution of lava fields and the growth of the volcanic structure. These relations explain the asymmetric distribution, areal extent, and slope of the various lava fields at Arenal and allow us to infer the probable position of buried craters and contacts.

Our analysis shows the importance of detailed observations and measurements of lava flow processes (flow front advancement, levee construction, crust development, and crater activity) to understand the mechanism of flow emplacement and volcanic structure at Arenal. The derived model of the growth of Arenal Volcano is a direct consequence of the mode of emplacement of unit flows. Remarkably, this model is based on mass conservation and does not assume any particular rheology for the lava. With comparable observations on erosion, volcanic structure, eruptive history, vent distribution, and mechanism of lava emplacement a more elaborate model may be derived for larger, more complex volcanoes.

Acknowledgments. Support for this research has been provided by Smithsonian Institution, Princeton University, Centro Investigaciones Geofisicas and Escuela Centro Americana de Geologia of Universidad de Costa Rica, Departamento de Geologia of Instituto Costaricense de Electricidad, Latin American Program of the Associated Colleges of the Midwest, and CNR-NATO Fellowship program. We particularly thank Luis Diego Morales for his extensive help during fieldwork, David Laur for producing the computer images of Fig. 11, Steve Baloga, Michael Carr, Katharine Cashman, Jason Morgan, Frank Spera, and Robert Tilling for relevant suggestions, and Henry Moore and Ken Wohletz for reviewing the manuscript. We are grateful to Henry Moore for pointing out an error in Eq. (29) of Borgia et al. (1983). Finally, we thank Don Juan for his unrivaled Costa Rican hospitality.

References

Angevine CL, Turcotte DL, Ockendon JR (1984) Geometrical form of aseismic ridges and seamounts. J Geophys Res 89:11287–11292

Baloga SM (1987) Lava flows as kinematic waves. J Geophys Res 92:9271–9279

Baloga SM, Crisp J (1987) Leveed lava flows on Mars. Proceedings of IUGG XIX Conference IAVCEI

Baloga SM, Pieri D (1986) Time-dependent profiles of lava flows. J Geophys Res 91:9543–9552

Becker GF (1885) The geometric forms of volcanic cones and the elastic limit of lava. Am J Sci 30:293–382

Bennett FD, Raccichini S (1977) Las erupciones del Volcan Arenal, Costa Rica. Rev Geograph Am Centr Costa Rica 5–6:7–35

Borgia A, Linneman SR, Spencer D, Morales LD, Brenes JA (1983) Dynamics of lava flow fronts, Arenal Volcano, Costa Rica. J Volcanol Geotherm Res 19:303–329

Borgia A, Poore C, Carr MJ, Melson WG, Alvarado GE (1988) Structural, stratigraphic, and petrologic aspects of the Arenal-Chato volcanic system, Costa Rica: evolution of a young stratovolcanic complex. Bull Volcanol 50:86–105

Cigolini C, Borgia A (1980) Consideraciones sobre la viscosidad de la lava y la estructura da las coladas del Volcan Arenal. Rev Geograph Am Centr Costa Rica 11–12:131–140

Cigolini C, Borgia A, Casertano L (1984) Intra-crater activity, aa-block lava, viscosity and flow dynamics: Arenal Volcano, Costa Rica. J Volcanol Geotherm Res 20:155–176

Dragoni M (1987) A dynamic model of lava flows cooling by radiation. Proceedings IUGG XIX Confernece IAVCEI 2:416

Gregg DR (1956) Eruption of Ngauruhoe 1954–1955. NZ J Sci Technol 37:675–688

Guest JE, Kilburn CRJ, Pinkerton H, Duncan AM (1987) The evolution of lava flow fields: observations of the 1981 and 1983 eruptions of Mount Etna, Sicily. Bull Volcanol 49:527–540

Hallworth MA, Huppert HE, Sparks RSJ (1987) A laboratory simulation of basaltic lava flows. Modern Geol 11:93–107

Lacey A, Ockendon JR, Turcotte DL (1981) On the geometrical forms of volcanoes. Earth Planet Sci Lett 54:139–143

Malin MC (1980) Length of Hawaiian lava flows. Geology 8:306–308

Mavridis H, Hyrmak AN, Vlachopoulos J (1986) Deformation and orientation of fluid elements behind an advancing flow front. J Rheol 3:555–563

Melson WG, Saenz R (1968) The 1968 eruption of Volcano Arenal: preliminary summary of field and laboratory studies. Smithson Cent Short-Lived Phenomena Rep 7-1968

Melson WG, Saenz R (1973) Volume, energy and cyclicity of eruptions of Arenal Volcano, Costa Rica. Bull Volcanol 37:416−437

Milne JFGS (1878) On the form of volcanoes. Geol Mag 5:337−345

Milne JFGS (1879) Further notes upon the form of volcanoes. Geol Mag 6:506−514

Pieri DC, Baloga SM (1986) Eruption rate, area and length relationships for some Hawaiian lava flows. J Volcanol Geotherm Res 30:29−45

Pinkerton H, Wilson L (1987) Factors affecting the length of lava flows. Hawaii Symposium on How Volcanos Work p 203 (abstract)

Reagan MK, Gill JB, Malavassi E, Garcia MO (1987) Changes in magma composition at Arenal Volcano, Costa Rica, 1968−1985: real time monitoring of open-system differentiation. Bull Volcanol 49:415−434

Rose W (1961) Fluid-fluid interfaces in steady motion. Nature 191:242−243

Shteynberg GS, Solov'yev T (1976) The shape of volcanoes and the position of subordinate vents. Izv Earth Phys 5:83−84

Wadge G (1978) Effusion rate and the shape of aa lava flow-fields on Mount Etna. Geology 6:503−506

Wadge G (1983) The magma budget of Volcan Arenal, Costa Rica, from 1968 to 1980. J Volcanol Geotherm Res 19:281−302

Wadge G, Francis P (1982) A porous flow model for the geometrical form of volcanoes − critical comments. Earth Planet Sci Lett 57:453−455

Walker GPL (1972) Compound and simple lava flows and flood basalts. Bull Volcanol 35:579−590

Walker GPL (1973) Length of lava flows. Philos Trans R Soc Lond 274:107−118

Williams H, McBirney AR (1979) Volcanology. Freeman, Cooper, San Francisco, 397 p

Wood CA (1978) Morphometric evolution of composite volcanoes. Geophys Res Lett 5:437−439

Wood CA (1982) On the geometrical form of volcanoes − comment. Earth Planet Sci Lett 52:451−452

Subject Index

Page numbers in *italics* refer to citations in figures or tables.